Tool Use in Animals

Cognition and Ecology

The last decade has witnessed remarkable discoveries and advances in our understanding of the tool-using behavior of animals. Wild populations of capuchin monkeys have been observed to crack open nuts with stone tools, similar to the skills of chimpanzees and humans. Corvids have been observed to use and make tools that rival in complexity the behaviors exhibited by the great apes. Excavations of the nut-cracking sites of chimpanzees have been dated to around 4000–5000 years ago. *Tool Use in Animals* collates these and many more contributions by leading scholars in psychology, biology and anthropology, along with supplementary online materials (available at www.cambridge.org/9781107011199), into a comprehensive assessment of the cognitive abilities and environmental forces shaping these behaviors in taxa as distantly related as primates and corvids.

Crickette M. Sanz is an assistant professor in anthropology at Washington University, St. Louis, where she teaches courses on primate behavior and human evolution. She is one of the principal investigators of the Goualougo Triangle Ape Project, which focuses on studying and conserving sympatric central chimpanzee and western lowland gorilla populations.

Josep Call is a comparative psychologist specializing in primate cognition at the Max Planck Institute for Evolutionary Anthropology. He is the co-founder and director of the Wolfgang Kohler Primate Research Center. His work focuses on the study of the problem-solving abilities of primates and other animals.

Christophe Boesch is the director of the Max Planck Institute of Evolutionary Anthropology's Primatology department. His work covers many areas of chimpanzee biology, which he has used to further understanding of the evolution of cognitive and cultural abilities in humans. He is also the founder and president of the Wild Chimpanzee Foundation.

Tool Use in Animals

Cognition and Ecology

Edited by

CRICKETTE M. SANZ
Washington University, St. Louis, USA

JOSEP CALL
Max Planck Institute for Evolutionary Anthropology, Leipzig, Germany

CHRISTOPHE BOESCH
Max Planck Institute for Evolutionary Anthropology, Leipzig, Germany

CAMBRIDGE
UNIVERSITY PRESS

University Printing House, Cambridge CB2 8BS, United Kingdom

Cambridge University Press is part of the University of Cambridge.

It furthers the University's mission by disseminating knowledge in the pursuit of education, learning and research at the highest international levels of excellence.

www.cambridge.org
Information on this title: www.cambridge.org/9781107011199

First published 2013

A catalogue record for this publication is available from the British Library

Library of Congress Cataloguing in Publication data
Tool use in animals : cognition and ecology / edited by Crickette Sanz, Washington University, St Louis, USA; Josep Call, Max Planck Institute for Evolutionary Anthropology, Leipzig, Germany; Christophe Boesch, Max Planck Institute for Evolutionary Anthropology, Leipzig, Germany.
pages cm
Includes bibliographical references and index.
ISBN 978-1-107-01119-9
1. Tool use in animals. 2. Primates – Behavior. I. Sanz, Crickette Marie, 1975– editor of compilation. II. Call, Josep, editor of compilation. III. Boesch, Christophe, editor of compilation.
QL785.T77 2013
569′.8–dc23

2012034006

ISBN 978-1-107-01119-9 Hardback
ISBN 978-1-107-65743-4 Paperback

Additional resources for this publication at www.cambridge.org/9781107011199

Contents

Contributors

Lucinda Backwell
Bernard Price Institute for Palaeontological Research
University of the Witwatersrand
Johannesburg
South Africa

Christophe Boesch
Department of Primatology
Max Planck Institute for Evolutionary Anthropology
Leipzig
Germany

Richard W. Byrne
Centre for Social Learning and Cognitive Evolution
Scottish Primate Research Group
School of Psychology and Neuroscience
University of St. Andrews
St. Andrews
UK

Josep Call
Department of Developmental and Comparative Psychology
Max Planck Institute for Evolutionary Anthropology
Leipzig
Germany

Matthew V. Caruana
Bernard Price Institute for Palaeontological Research
University of the Witwatersrand
Johannesburg
South Africa

Susana Carvalho
Leverhulme Centre for Human Evolutionary Studies
University of Cambridge

Cambridge
UK

Francesco d'Errico
UMR-CNRS PACEA
Université Bordeaux
Talence
France

Nathan J. Emery
School of Biological and Chemical Sciences
Queen Mary University of London
London
UK

Dorothy Fragaszy
Psychology Department
University of Georgia
Athens, GA
USA

Russell D. Gray
Department of Psychology
University of Auckland
Auckland
New Zealand

Gavin R. Hunt
Department of Psychology
University of Auckland
Auckland
New Zealand

Tetsuro Matsuzawa
Primate Research Institute
Kyoto University
Kyoto
Japan

William C. McGrew
Department of Archaeology and Anthropology
University of Cambridge
Cambridge
UK

Shannon P. McPherron
Department of Human Evolution
Max Planck Institute for Evolutionary Anthropology
Leipzig
Germany

Ellen J. M. Meulman
Anthropological Institute and Museum
Universität Zürich
Zürich
Switzerland

David B. Morgan
Lester E. Fisher Center for the Study and Conservation of Apes
Lincoln Park Zoo
Chicago, IL
USA

April M. Ruiz
Department of Psychology
Yale University
New Haven, CT
USA

Laurie R. Santos
Department of Psychology
Yale University
New Haven, CT
USA

Crickette M. Sanz
Department of Anthropology
Washington University
St. Louis, MO
USA

Alex H. Taylor
Department of Psychology
University of Auckland
Auckland
New Zealand

Sabine Tebbich
Department of Cognitive Biology
University of Vienna
Vienna
Austria

Imgard Teschke
Max Planck Institute for Ornithology
Seewiesen
Germany

Carel P. van Schaik
Anthropological Institute and Museum
Universität Zürich
Zürich
Switzerland

Elisabetta Visalberghi
Instituto di Scienze e Tecnologie della Cognizione
Rome
Italy

Part I

Cognition of tool use

1 Three ingredients for becoming a creative tool user

Josep Call
Max Planck Institute for Evolutionary Anthropology

The bird approaches the transparent vertically oriented tube and looks down its opening with apparent interest. Then it looks at the tube from the side and walks around the tube to look down into the opening with one scrutinizing eye once more. There is a worm located at the bottom of the tube, beyond the bird's reach. After a few seconds, the bird steps away from the tube, picks up a stick with its beak and inserts it down the tube's opening. Once inside, it grabs the tool again and applies downward pressure on it so that the tool dislodges the platform that is keeping the worm inside the bottom of the tube. The worm drops free from the bottom of the tube to be picked up by the bird, which quickly flies away.

Observations like this pose a double challenge to researchers in the field of comparative cognition. The first challenge is to explain why some species can come up with innovative solutions while others facing the same situation do not do so. For instance, pigeons presented with the same task as crows and left to their own devices may be incapable of producing the same solution, even after hours of exposure to the same problem. One possible explanation for this outcome is that this crow species, unlike the pigeons, may have a strong predisposition to using tools since this has offered it an adaptive advantage. In fact, the crow is in all likelihood a New Caledonian crow (*Corvus moneduloides*), well known for their propensity and dexterity at making and using tools to extract embedded food from substrates. The fascinating thing is that the above description is not about a New Caledonian crow, but a distantly related cousin, the rook (*Corvus frugilegus*). Rooks, unlike New Caledonian crows, do not usually use tools in this way, but they can do so in the laboratory, as Bird and Emery (2009) discovered. This revelation poses a second challenge to comparative researchers: How are rooks solving this problem? What cognitive mechanisms are responsible for the observed behavior and what experiences are necessary for this clever solution to emerge? Since all species are endowed with associative learning mechanisms, a key question to be explained is where interspecific differences come from. Nowhere is this challenge so acute as in the area of tool use in animals.

The main goal of this chapter is to discuss three ingredients that are sufficient, perhaps even necessary, to become a creative (i.e., flexible) tool user, defined as using multiple tools (not necessarily in combination) to solve multiple problems, particularly when tool-using solutions can be classified as innovations (see Reader & Laland, 2002). Possessing those ingredients may allow certain species that seemingly do not possess a propensity to use

Tool Use in Animals: Cognition and Ecology, eds. Crickette M. Sanz, Josep Call and Christophe Boesch. Published by Cambridge University Press.

tools in their natural habitats to become proficient tool users under certain circumstances. But before we get to the three ingredients I will briefly explore the relation between tool use and intelligence, and contrast two pathways by which even non-tool-using creatures can become proficient tool users. Then I will turn my attention to whether tool use can be considered a cognitive specialization, and will devote much of the rest of the chapter to exploring the cognitive mechanisms underlying flexible tool use. Surprisingly, there has been relatively little progress in the last 50 years in elucidating the processes that may underlie problem solving in general (and tool use in particular) in non-human animals. For those to whom my assertion may sound exaggerated, let me clarify that I am not referring to learning but to reasoning. I hope it will become clear in the rest of the chapter that these are two different processes that can play a role in problem solving. The comparative literature could have benefited from insights from cognitive psychology in this area, but such transfer of ideas has not taken place, at least not as much as it occurred many years ago. Therefore, one second overarching goal of this chapter is to bring closer data and concepts developed in cognitive psychology that can find some application in comparative psychology.

Is tool use an indicator of intelligence?

Tool use, defined as using an object to alter the position or form of another object or individual, has traditionally been regarded as an indicator of intelligence and complex cognition (e.g., Köhler, 1925; Thorpe, 1963; Parker & Gibson, 1979). In fact, before it was discovered that wild chimpanzees manufacture and use tools with regularity, some scholars considered tool using and tool making as a human Rubicon (e.g., Oakley, 1976) – something that separated humans from non-human animals. Even today, species that use tools seem to enjoy a special status, and new discoveries on tool-using behavior quickly grab the attention not only of academics but also of the mass media and the general public. From the point of view of animal intelligence, this fascination for tool use is perhaps a bit surprising given that other skills such as spatial navigation, which may involve equally impressive cognitive sophistication, do not enjoy such a prominent status.

One contributing factor to the special status of tool use is its narrow distribution in the animal kingdom. Although there are many species that use tools occasionally, and many more that with appropriate training can use tools, spontaneous and customary tool use is relatively rare among animals. Rarity, however, is not a synonym of intelligence, as there are many rare traits (e.g., electrocommunication) that are not assumed to entail advanced intelligence. Another contributing factor is that humans also use tools; in fact, it is safe to say that we are a species obsessed with tools and artifacts. By analogy, we may be prone to attributing some of our human cognitive qualities to those species that display a behavior that is strongly tied to our own species identity. Moreover, tool use has played such a pivotal role in theories of human evolution that seeing another species use tools automatically grabs our attention as it may provide clues about the evolution of technology.

Although the two previous arguments may have some appeal and explanatory power, they do not provide a strong case for tool use as an indication of intelligence. Other authors, however, have provided arguments grounded on the psychological processes involved in tool use to support the idea that using tools is cognitively demanding (e.g., Piaget, 1952; Parker & Gibson, 1979). At a very basic level, using an object to affect a second object involves more elements, and consequently requires more coordination, than acting on the second object directly. It is easy to underestimate how cognitively demanding this kind of coordination can be. However, some studies have shown that simply changing the number of tools available and their position can substantially affect subjects' performance on trap tasks (e.g., Girndt *et al.*, 2008; Seed *et al.*, 2009). However, this still does not seem a good reason to single out tool use, given that other behaviors like spatial navigation also involve coordinating between the subject, multiple external entities and their relations.

Other authors have focused their attention on the suddenness with which some solutions appear in order to discuss their cognitive significance. Similarly, solutions that represent a significant departure from previous attempts would also fall into this category. There is indeed something enthralling about observing an individual facing a reward that is located outside of reach and after several unsuccessful attempts to get it, turn around and pick up a tool and use it to retrieve the reward. Even if we had witnessed the same animal finding another solution by changing their spatial orientation or position, the tool-use example would still convey a stronger sense of cognitive sophistication. However, it is important to emphasize that not all cases of tool use necessarily imply the cognitive sophistication described above. In fact, tool use is a very broad functional category that includes very different examples whose cognitive substrate may differ substantially between and within species.

Some animals use a single tool for a single purpose in a particular context. Modifying slightly the problem shows that those cases of tool use are best described as inflexible specializations. For instance, archer fish (*Toxotidae*) can use water to down insects located above the water level. However, archer fish, as far as we know, do not use this skill in any other context. In contrast, other animals are capable of using multiple tools for multiple purposes in multiple contexts. Unlike the case of tool specialists previously alluded to, here alterations of the problem space invariably produce changes in the behavior that help the individual to adapt to the new problem space. In some cases the problem space is even moved outside of their natural ecological niche and yet subjects still can solve the problem. For instance, orangutans (*Pongo abelii*) can use water to raise the level of a peanut floating at the bottom of a tube so that they can grab it (Mendes *et al.*, 2007). The rooks mentioned at the beginning of the chapter using tools to get food that is located outside of their reach (Bird & Emery, 2009) would also fall into this category. Thus, whereas some examples of tool use are best characterized as behavioral specializations, other examples appear to be behavioral innovations – a term that Reader and Laland (2001: 788) defined as "the ability to respond to novel circumstances or stresses with new behaviour patterns."

It is this type of tool use that we have characterized as behavioral innovations that may be more properly labeled as intelligent because it possesses two key features: adaptability and creativity. Individuals are not only able to use old solutions to solve novel problems but also can generate new solutions for old problems when the original solutions no longer work, or even produce new solutions for novel problems. In some cases such solutions do not simply entail using a tool, but also manufacturing the tool or using a sequence of several tools to achieve a particular goal (e.g., Mulcahy *et al.*, 2005; Wimpenny *et al.*, 2009; Taylor *et al.*, 2010). In some cases, individuals can anticipate problems with the tools and they can select novel tools before having obtained feedback on how effective they are, just based on their features (e.g., Marín Manrique *et al.*, 2010, in press; Marín Manrique & Call, 2011). Finally, some species such as the great apes can use a tool for multiple purposes, and one purpose can be served by multiple tools.

Traditionally, primates – the great apes and capuchin monkeys (*Cebus apella*) in particular – were the best examples of creative tool users. In the last decade, however, birds – more specifically corvids – have emerged as serious contenders for the title of most creative non-human tool user on earth. New Caledonian crows in particular, which use tools to extract embedded food, manufacture hook-shaped tools with various materials (Hunt, 1996; Weir *et al.*, 2002). They can also use tools in sequence, such as using a short tool to get a longer one that can be used to get the food (Wimpenny *et al.*, 2009; Taylor *et al.*, 2010). Recently they have also been reported to use tools not just to obtain food but also to investigate their environment (Wimpenny *et al.*, 2011). We devote the remainder of this chapter to exploring the cognition underpinning flexible tool use.

Two routes for becoming a tool user

Despite its restricted appearance in the animal kingdom, with proper training numerous species that do not usually use tools can become proficient tool users. Some can even learn to use tools in sequence and distinguish the features of good and bad tools. How they manage this achievement is a different matter. Following Maier and Schneirla (1935) we can distinguish two main ways in which a species can solve a problem: learning and reasoning. Learning entails combining contiguous experiences. For instance, an individual who initially showed no preference for pulling from two parallel strings develops a preference for pulling the string that is closest to the food to which it is tied. Here the subject develops a preference for selecting those responses that are reinforced and discarding those responses that are not reinforced or negatively reinforced. Thus, reinforcement is the glue that binds stimuli and responses together. Although this is a very powerful mechanism for acquiring new responses, it also has a serious limitation. There has to be spatio-temporal contiguity between stimuli, responses and reinforcement for these elements to effectively bind together.

The other way to become a tool user is reasoning, which Maier and Schneirla (1935) broadly defined as combining separate experiences. For instance, a subject is given the opportunity to explore one part of a maze on one day and on a separate day she is allowed to explore the rest of the maze. Later on, the subject is tested on whether she can find the most efficient route to go from point A to point B, which entails navigating the entire maze, not just a part of it. Here we can distinguish between acquiring information in the absence of reinforcement but in the context of exploration and combining the information to find a solution. Obviously, information that has also been acquired by reinforcement can later be combined with information acquired through exploration. This means that learning and reasoning are not necessarily mutually exclusive. In fact, both learning and reasoning can contribute to problem solving, of which tool use is a particular example. Katona's (1940; see also Wertheimer, 1959) distinction between reproductive and productive thinking is relevant here. Reproductive thinking entails applying familiar procedures to solve problems that have been encountered before, or slight variations on those problems. In contrast, productive thinking entails inventing new procedures for solving a problem, either familiar or unfamiliar.

Learning and reasoning have things in common and things that distinguish them. Both mechanisms have in common that experience plays a crucial role for solving problems. Individuals engaged in trial-and-error learning do not engage in every possible attempt. Often their attempts are canalized and they are likely to first try things that have worked in the past in similar situations. Similarly, reasoning does not work in a vacuum of experience. Indeed, experience is necessary for reasoning to occur. According to Wertheimer (1959) the crucial question in problem solving is not whether past experience plays a role or not, but what kind of experience is implicated in generating solutions, blind connections or structural grasp. Thus, the distinction between both mechanisms (or between reproductive and productive thinking) is not on whether one is based on experience and the other one is not, but on what type of information is acquired and how it is managed to produce a solution to the task. This distinction is important, but it has often been neglected by comparative psychologists.

One crucial distinction between learning and reasoning lies in the amount of separation that may exist between two experiences and yet be able to combine them to produce a novel solution to a problem. At the most basic level, separation is strictly determined by spatio-temporal parameters. The closer events occur in time and space, the more likely they are to become associated. For instance, for instrumental learning to occur, responses and reinforcement have to occur in close temporal proximity, whereas this is not a necessary condition for reasoning. At a more abstract level, separation between experiences may be determined by their "symbolic" distance. Learning can cope reasonably well with deviations from stimuli dimensions, such as color or sound frequency. For instance, pigeons trained to peck on a blue disc show some stimulus generalization to blue-green discs. However, in some cases the new stimuli do not share perceptual features with familiar ones, but share instead functional features. For instance, although a hole on a platform and a barrier on a platform both impede the displacement of an object from point A to point B on the surface of the platform, they have different perceptual features. It is less clear that stimulus generalization, originally developed to account for coping with variations within stimuli dimensions, is also responsible for categorizing the hole and the barrier as equivalent from a functional point of view. Wertheimer (1959) argued that focusing on physical features could enable the transfer to highly similar features, whereas focusing on structural features could enable transfer to a greater variety of problems that differ in terms of physical features but share structural features. In both cases connections are established by similarity, but the nature of this similarity is either physical or structural. In the case of the barrier and the hole mentioned before, seemingly disparate stimuli (a barrier and a hole) are related because both impede the progression of the food on the table. In this sense, their symbolic distance (or functional distance) is smaller than the perceptual distance that exists between a hole and a barrier. Note that to be able to reduce the distance, and therefore facilitate the combination of separate experiences, the mind has developed a new type of representation that enables the system to classify both a gap and a wall as obstacles toward a goal. Being able to conceive objects and events at different representational levels is therefore one important way to facilitate connectivity, thus greatly enhancing the ability of a system to put together disparate pieces of information.

Are learning and reasoning the same process?

Over the years there have been several attempts to reduce these two processes into one (e.g., Maltzman, 1955; Epstein *et al.*, 1984; Weisberg, 1986). However, by conflating both mechanisms there is the danger of losing sight of one of the most important aspects of reasoning: relating disparate pieces of information. After all, if all there is to problem solving is trial and error and learning via reinforcement, automatic chaining and/or resurgence (see Epstein, 1987), one still has to explain why rooks can solve the tool-using task and pigeons presumably cannot. Some authors have argued that a pigeon (*Columba livia*) can solve problems just like chimpanzees (*Pan troglodytes*) by using the same learning mechanisms indicated above. For instance, Epstein and colleagues showed that pigeons were able to solve problems with multiple steps comparable to those in which chimpanzees stacked boxes under a banana hanging from the ceiling, beyond their reach. In the final solution the performance of the pigeon is truly impressive. Faced with a target that is located too high and out of reach, the pigeon diligently pushes the box under the target, hops onto it and triumphantly pecks at the target.

Despite the apparent similarity between the pigeon and the chimpanzee performance, the cognitive processes involved in the solution could be quite different. Simply put, automatic chaining is not equivalent to reasoning as is intended to be used here (see also Ellen & Pate, 1986). One crucial distinction is that chimpanzees, unlike pigeons, were not shaped to respond to each task separately before they were presented with both tasks in combination. As Epstein and colleagues convincingly showed, a pigeon that had been shaped to respond to two tasks separately could combine them to produce a solution based on chaining one response after the other in the absence of trial and error. However, one crucial aspect that was missing from the pigeon's behavior was spontaneity. It is true that the pigeons had never faced the target located at a higher position, and therefore the pigeon showed flexibility when it was able to adapt and solve this new challenge. But the key question is whether pigeons would have spontaneously produced those chains of responses without the benefit of having been reinforced for each task separately. The answer to this question seems to be negative, since pigeons failed to solve the task when one of the components was not trained (Epstein, 1987).

Chimpanzees spontaneously stack boxes in the context of play, which provides them with crucial experience that later they will be able to use for problem solving (Birch, 1945). The role of experience in problem solving has often been misunderstood and used as evidence that learning rather than insightful problem solving was responsible for the observed solutions. However, gestalt psychologists were quite clear in the crucial role that experience played in problem solving, both in terms of facilitating and hindering potential solutions (e.g., Duncker, 1945; Köhler, 1967). Accordingly, solutions cannot be reduced to automatic chaining simply because experience is an integral part of problem solving. Chimpanzees do require experience to produce insightful solutions, but note that such experiences are not identical to those later required to solve a task. Additionally, those experiences that will later be used to solve a task were often acquired in the context of exploratory play. Acquiring and using this information is quite different from training the pigeon to move the box under the target. All the experiences that the pigeons received were associated with food in the same context. Moreover, a pigeon placed in an experimental box experiences reduced attentional competition for stimuli. In fact, the experimenter selects for the pigeon both what stimuli to pay attention to and what responses to produce. This is obviously a much simpler problem than the one faced by chimpanzees because the number of possible combinations is greatly reduced. The real challenge would be to know whether pigeons placed in the same situation as chimpanzees and left to their own devices would produce a solution to the problem on their own.

The previous discussion, however, should not be taken as evidence that chimpanzees can engage in reasoning whereas other animals such as pigeons or rats cannot. On the contrary, there is evidence from other studies suggesting that rats use reasoning in spatial navigation tasks. In fact, Thorpe (1963) noted that insightful problem solving (and perhaps reasoning in general) may be widespread among animals and not restricted to a few species like chimpanzees that have been traditionally considered "smart" ones.

Is tool use a cognitive specialization?

To answer this question we have to go back to a distinction we made earlier in the chapter. For some species, tool use is best understood as a specialization analogous to other behavioral specializations, such as a fear grin in macaques, or morphological specializations such as the elongated beaks of hummingbirds (*Trochilidae*). In this sense, the use of stones in Egyptian vultures (*Neophron percnopterus*) to crack open ostrich's eggs, and the use of stones in digger wasps (*Sphex* spp.) to block the entrance of their burrows would qualify as tool-use specializations. Specializations are extremely efficient at accomplishing a particular function, typically hardwired, perhaps even based on a modular architecture, but they are also relatively inflexible.

Labeling cases of tool use as behavioral specializations should be revised if it can be shown that those species could spontaneously (without human intervention) apply their tool-using skills outside of their canonical ecological contexts. A first step toward increased flexibility and away from specialization consists of using tools for more than one function, especially when those new uses can be classified as innovations. For instance, if Egyptian vultures were to use the stone to cover the food from the prying eyes of competitors, or drop the stone against a predator/competitor to drive them away. In such cases, the label "specialization" that we have provided should be revised, especially if multiple non-canonical uses are observed.

Those cases in which individuals use multiple tools for multiple functions represent an even stronger challenge to the notion of specialization. Note that the key here is not displaying multiple cases of one tool for one function, but cases in which multiple tools are used for multiple functions, that is, when a tool and its function appear dissociated. Particularly interesting are double dissociations in which a particular tool is used for multiple functions (e.g., a stone to crack open nuts, to throw during agonistic displays) and a function is served by multiple tools (e.g., nut cracking with a stone or log). Such means–ends dissociation is one of the key indicators of intentional and flexible behavior. For those species displaying this kind of flexible tool use, it is unlikely that tool use is based on a narrow cognitive specialization as it was just described for Egyptian vultures and digger wasps. Instead, it is likely that tool use in these species is based on the same sensorimotor and conceptual abilities that can be recruited to solve a variety of problems, not just tool-using problems.

Seemingly distinct tasks such as spatial detours and tool use are considered equivalent from the point of view of problem solving (e.g., Köhler, 1925). Although some may argue that this putative similarity is purely functional, there is evidence showing that spatial and stimulus discrimination problems may recruit some of the same cognitive resources in terms of inhibitory and motor control (Walker *et al.*, 2006). Another case in point is the relation between gesture production and tool use. Focusing on apes, many of the features that enable flexible tool use can also be found in the way they use gestures. Not only do gestures display a means–ends dissociation like the one described for flexible tool use (Call & Tomasello, 2007), some neurophysiological evidence indicates that the same motor control areas are recruited both to use tools and to produce gestures (Roby-Brami *et al.*, 2012). Moreover, although much of the work on problem solving in rodents was mainly based on spatial tasks, the conclusions were similar to those arrived at by those scholars investigating tool-using primates.

Contrasting species that routinely use tools with their close relatives that do not use them can also be very informative. Although bonobos (*Pan paniscus*) and rooks do not normally use tools to obtain food in their natural habitats, they can do so if the situation (in the laboratory) requires it. In fact, they can be as proficient as chimpanzees and New Caledonian crows, respectively, which are habitual tool users in the field. Tebbich (Chapter 7) also found that woodpecker finches (*Cactospiza pallidus*), which use tools quite proficiently, do not appear to differ in cognitive abilities from small tree finches (*Cactospiza parvulus*), a species that does not use tools. Although future research may uncover differences between species, currently tool use in this taxon may be best conceived as a manifestation of the existing cognition rather than a specialization that evolved to use tools. Reader and Laland (2002) reached a similar conclusion when they argued that tool use and social learning were best understood as manifestations of general intelligence rather than special processes. Reader and Laland (2002) also showed a correlation between innovation, tool use and social learning in primates. This finding is interesting because it relates acquiring information (via social learning) with using information in innovative ways, of which tool use is just one manifestation. It would be equally interesting to know whether those species also show a strong exploratory tendency and play behavior.

The previous discussion, however, should not be taken as evidence that all species process information in the same way. Differences between species do exist (but see Macphail, 1987), but they have not evolved as a consequence of tool use. At the risk of oversimplifying the issue, this means that our rook–pigeon dichotomy is more virtual than real. If a rook could see the world through a pigeon's eyes, it would behave like a pigeon. This idea is certainly attractive from the point of view of simplicity: a common processing architecture for all species paired with a different input system. However, and without denying the importance of selecting certain types of information, one cannot ignore the fact that information still undergoes massive processing after input has occurred. Even if two individuals paid attention to the same stimuli, the way they were later processed may produce very different representations. Moreover, even if two individuals created equivalent representations, they may also differ both in what representations they chose to use and in their ability to relate them with other existing representations. Making a clear-cut distinction between input and processing becomes even more problematic when one considers that input can also largely be determined by the pre-existing representations that lead organisms to perceive stimuli in certain ways.

Psychological ingredients to be(come) a creative tool user

Spontaneity, innovation and flexibility are some of the key labels that we have used to characterize the behavior of species that solve problems in creative ways. Such labels apply not just to tool-using problems but also to other kinds of physical and social problems. In this last section of the chapter we re-focus our attention on tool use in particular, and list three key ingredients that would make a species an adaptable and creative tool user.

Information hoarding

The first ingredient is information seeking and hoarding. Many species acquire responses based on positive or negative reinforcement. Acquisition of responses by secondary reinforcement (i.e., features associated with features that are associated with reinforcement) would also belong to this category. Thus, learning based on reinforcement is an important source of useful knowledge. However, information can also be acquired in the absence of reinforcement. In fact, for some species this non-reinforced pathway may be its major source of information. Therefore, a predisposition for exploration and latent learning are important sources of information even in the absence of reinforcement. In fact, information acquired in non-goal-directed contexts is often crucial to later solve goal-directed problems. For instance, Birch (1945) reported that chimpanzees that were given the opportunity to manipulate objects in the context of play were more skillful at later solving tasks that involved those objects than chimpanzees that did not have that experience. Similarly, there is a literature on latent learning showing that rats that are given the opportunity to explore a maze (in the absence of reinforcement) prior to being tested on that maze learn faster than rats that are naive to the maze (see Thorpe, 1963). Moreover, the use of the information is quite flexible as they can solve problems that involve connecting different experiences to produce the most efficient solution.

Some authors (e.g., Thorpe, 1963) have established a strong connection between latent learning and insightful problem solving since latent learning may be the basis for deploying adequate information. Information acquisition can occur by observing the consequences of our own or others' behavior. Even actions that fail to produce the desired outcome are often a good source of information to reach adequate solutions because they can reveal relations that we had not mentally entertained, but once they are perceived, they can be comprehended and used. Learning about the consequences of others' behavior is an equally good source of information since others' behavior can also reveal relations between the elements of a problem. In fact, such appreciation of information is the basis for emulation learning.

One advantage of information acquired in the absence of reinforcement is that it may be more amenable to being used in more flexible ways when the situation demands it. In other words, knowledge that is associated with the delivery of a reward may become encapsulated and therefore less accessible than knowledge based on exploration and play. That play and exploration may perform a crucial role in the acquisition and storage of information that can be used for later problems is not a new idea (e.g., Bruner, 1972). However, knowledge acquired by exploration in the absence of reinforcement may also have its limitations since it may be difficult to recruit for solving problems compared to knowledge that has been associated with reinforcement, which works well with familiar problems and may even generalize to similar problems.

The information stored, however, can be of different types. Two organisms witnessing the same event, even if they belong to the same species, may encode it in different ways. Upon witnessing an experimenter hiding a toy under one of three distinct containers, one-year-old human infants, just like non-human great apes, focus on the spatial coordinates of the hidden toy when they are allowed to retrieve it. In contrast, three-year-old children focus on the features of the container and they search for the toy under the container where the object was hidden, regardless of its current spatial position (Haun *et al.*, 2006). Of particular interest in the case of tool use is information about object–object relations, or so-called structural representations which deal with relations between objects (Seed & Call, 2009). These representations specify functional properties of objects, typically in relation to other objects (e.g., solidity, connectivity). These structural representations contain different information from so-called perceptual representations, which refer to the configuration or appearance of objects, not their functional properties. Thus, one can represent one object next to, on top of, or affecting the position of, another object. In all cases, the perceptual information is the same but the representation formed differs dramatically.

There is some evidence showing that animals differ on how much they use these two different types of information. Presented with the string task, pigeons solve the problem by encoding the configural information, whereas apes encode not just the perceptual appearance – their analysis is also about object–object interactions (Seed *et al.*, in press). Apes interpret the food as the cause of the noise made by a baited cup when it is shaken, whereas dogs do not seem to perceive this relation (Call, 2004; Bräuer *et al.*, 2006). The crucial distinction here is between perceiving the reward and the cue as mere co-occurrence or as causally related (Call, 2006; Seed *et al.*, in press).

Information recombination

The second ingredient is information recombination or restructuration. Indeed, the combination of disparate pieces of information or the restructuration of information plays a crucial role in problem solving. In fact, information restructuring has been traditionally viewed as a crucial element in insightful problem solving. Some authors have criticized the use of the term "insight" because it lacks explanatory power. In other words, labeling a phenomenon does little to explain it. These critics are correct in pointing this out, but it is also true that other authors have gone beyond labeling and offered models to explain the cognitive process that culminates in insight (e.g., Davidson, 1995; Seifert *et al.*, 1995). For instance, Davidson (1986, 1995) has proposed the three-process theory of insightful thinking. She considers three processes whose participation is necessary for insight to occur: selective encoding; selective combination; and selective comparison. Selective encoding consists of seeing in a stimulus features that were not obvious before. Selective combination consists of putting together elements of a problem in a way that was not previously apparent. Finally, selective comparison consists of the discovery of new relations between new and old pieces of information.

Seifert *et al.* (1995) have proposed a different model that consists of a three-phase process culminating in insight. In the preparation phase individuals confront the problem by gathering information about it and attempting to solve it. Lack of success generates memory traces of failure and the ceasing of further solution attempts. Once the individual shifts attention away from the problem, the incubation phase begins, in which individuals are exposed to additional information, typically in the form of external stimuli. These authors, unlike others, do not consider that the temporal evolution of memory traces or random knowledge reorganization play a substantive role during the incubation phase. Finally, the illumination phase begins when the new information contacts the memory traces of failure, which leads to an evaluation of the new information in relation to the problem. The accommodation and assimilation of the new information may generate an improved representation of the problem that leads to the correct solution. Then and only then, the flash of insight appears, producing a physiological arousal that facilitates the acquisition of the solution and its eventual encoding into long-term memory. Unfortunately, none of these models, or any others that have been proposed, have yet found their way into the comparative literature.

The problem is further compounded by the fact that information recombination or restructuring can take different forms (Mayer, 1995). Some authors see it as perceiving stimuli in a different way (i.e., reorganizing visual information; Köhler, 1925), putting isolated pieces of information together or filling gaps (i.e., completing a schema; Selz, 1913), overcoming mental blockages produced by past experience (i.e., functional fixedness; Duncker, 1945), establishing an analogy at the level of structural organization and applying it to a new problem (Wertheimer, 1945) or reformulating the problem, either in terms of its goals (e.g., radiation problem; Duncker, 1945) or the function of the objects that can enable a solution to the problem (e.g., two-string problem; Maier, 1931).

Each of these proposals has some support in the literature. For instance, perceiving ambiguous figures in a different way is an example of focus switching. Mistakes that reveal a new spatial configuration of the elements of the problem are the catalyst to produce the solution. It is easy to confuse this last case with trial and error, but the crucial difference is that the subject improves based on the information newly acquired rather than on the reinforcement that it receives for producing a certain response. Moreover, the solution appears fully formed, one of the features indicated by Köhler (1925) and, at least in humans, it is accompanied by the realization that the problem has been solved even before executing the response (Davidson, 1995). Note that this "solution before execution" is also observed in other tasks such as inferential reasoning, in which subjects can acquire the correct response even before that response has been reinforced. Regardless of the type of restructuration process, all have in common that organisms create new representations by combining separate pieces of information, splitting existing ones or viewing familiar information in a new light.

One implication of the different usages of restructuration is that different aspects of the problem may determine the likelihood of a restructuration leading to a solution. First, it may depend on the "distance" (either in terms of spatio-temporal parameters or abstractness) between two pieces of information that need to be combined. In general, the larger the spatio-temporal distance between two experiences, the less likely their recombination is to occur. Moreover, species may differ on how well they can cope with such distance. It is conceivable that encoding information in certain formats (e.g., structural representations) ameliorates the detrimental effects of spatio-temporal distance because disparate pieces of information can be related with generalization mechanisms beyond stimulus generalization.

Additionally, structural representations, by virtue of their abstractness, may be in some cases immune to blockages produced by configural information at the perceptual level. Second, it may depend on how familiar individuals are with a certain usage of an object. A phenomenon known as functional fixedness (Duncker, 1945) occurs when the canonical use of an object interferes with a problem solution that entails using that object in a non-canonical way. A recent study has suggested that functional fixedness may have been responsible for the difficulty displayed by some chimpanzees when trying to solve the floating peanut task (Hanus *et al.*, in press). This means that functional information (i.e., what objects are for), just like perceptual information, can also interfere in problem solving.

The previous discussion highlights a key aspect of problem solving in general, and insightful problem solving in particular. Insight is not a process. Insight is best understood as the end-product of a process (Dominowski & Dallob, 1995; Seifert *et al.*, 1995) – the final step in a chain of events that begins with the acquisition and transformation of information and culminates with the recombination and solution to the problem. Note that Thorpe (1963) already distinguished between insightful problem solving and insight. Thus, when individuals solve a task by insight, it would be more appropriate to say that they learn by information recombination (based on one of the models proposed) that culminates in insight. Often the process that is proposed to explain insight, in particular the sudden appearance of a solution after a period without overt attempts, is that of mental trial and error. Although the idea of mental trial and error is appealing for its apparent simplicity – a sort of trial and error with the motor output suppressed and capable of simulating possible outcomes – it may not be useful to conflate it with the processes described before that culminate in insight. One important distinction is between trying multiple options mentally, perhaps even in a systematic manner, and insight. It is not clear that insight, at least the way a number of cognitive psychologists conceive it, requires the systematic examination of multiple possibilities. The literature on humans shows that those are two distinct phenomena. One can examine a problem analytically by considering the various alternatives before executing a plan for action, or the solution may appear suddenly and involuntarily rather than by analytical reasoning. In both cases information is being mentally manipulated, but the processes governing such manipulations may be quite different. Future studies are needed to determine what processes are used by various species.

Object manipulation

The third ingredient to become a creative tool user is a propensity to manipulate objects. Although one may be tempted to focus on certain anatomical features to explain differences among species, such features may be less important than may have been suspected. After all, capuchin monkeys and squirrel monkeys have similar hands and yet they differ vastly in their manipulative propensity. Capuchin monkeys readily manipulate objects, whereas squirrel monkeys do not. Similarly, the beak of New Caledonian crows is not that different from other similar-sized crows such as carrion crows, and yet both species show a different predisposition to using tools. This is not to say that anatomy is irrelevant. On the contrary, parrots, by virtue of the anatomy of their beaks, have more difficulty manipulating tools in the same way as members of the corvid family. However, the differences in object manipulation may be mainly attributed to psychological rather than anatomical differences between species. This idea also fits well with recent analyses on vocal production in primates and other animals. Traditionally, anatomical constraints (position of the larynx) had been used to explain the lack of articulate speech in a number of non-human species. However, it appears now that the lack of articulated speech is mostly a consequence of the lack of neural wiring of certain motor areas to control vocal track rather than the anatomy per se (e.g., Hammerschmidt & Fischer, 2008). Similarly, differences in manipulative ability may have to do with motivational rather than anatomical determinants.

Both primates and birds provide convergent evidence of a link between a propensity to manipulate objects during exploration and tool use (Parker & Gibson, 1979; Torigoe, 1985). Primates differ in their manipulative propensity, with capuchin monkeys and the great apes showing not only more manipulation of objects than other primates, but also more complex forms of manipulation, involving multiple elements and relations (Torigoe, 1985). Such manipulation takes place during exploration of those objects. Interestingly, these are also the primate species that use tools more proficiently to get objects that cannot be attained directly. Birds may offer a convergent picture with that of primates. Both parrots and corvids manipulate objects readily in the context of play and exploration, and they are also the species that have shown remarkable tool-using abilities (e.g., Heinrich, 1999; Pepperberg, 1999; Burghardt, 2006). However, unlike primates, no large-scale study has been attempted to relate object manipulation (both in terms of frequency and complexity) with tool use. Such data would be extremely useful in tracing the evolutionary roots of tool use and its cognitive substrate.

A metric of problem solving

My goal in this chapter has not been to promote one mechanism for problem solving over another, reasoning over learning, productive over reproductive thinking, but to argue that overlooking reasoning in favor of learning is a loss too great to incur. The invocation of learning (and generalization) to explain every case of problem solving (both in the physical and social realm) in the animal kingdom can be both misleading and unproductive. For one thing, it obscures the important fact that there are important differences in the way different species learn things, not just in terms of speed but also in the information they focus on. I have already noted that both learning and reasoning can produce flexible responses, and therefore both inform us about different aspects of problem solving. Rather than spending too much time and effort on terminological disquisitions, I propose that our primary goal should be to quantitatively measure behavioral performance in order to assess how individuals process information, that is, what information they acquire, how they store it and how they use it. Whether one wants to label the resulting process one way or another appears to me a less important endeavor.

Assessing how information is processed in problem-solving situations would involve measuring the type of information that subjects require to solve problems and whether they can combine separate experiences. The former can be assessed by presenting different groups of subjects with different types of information as it is commonly done in social learning experiments. In turn, the distance between experiences can be assessed by presenting two pieces of information required to solve the task separately in time and/or space. Or, if one desires to investigate symbolic distance, one can train subjects to solve one task and assess transfer to another functionally equivalent task. For instance, once subjects learn to solve a task with one type of barrier, one can see whether they generalize that to another type of barrier that shares no perceptual features with the original problem. Ideally, one can have different types of barriers that differ in order to see whether they are treated in similar ways.

Getting quantitative information on these measures may contribute to bypass the unproductive debate about whether individuals are using learning or reasoning to solve a task. Thus, species (or individuals) could be characterized not by whether they solve a problem by learning or reasoning, but by quantitative measures of how much information (or what type) they need to solve a task and how separate different pieces of information have to be to still be able to produce a viable solution. These measures can highlight the differences between species and also inform us about how various species combine information to solve novel problems. An individual that could combine disparate pieces of information that were acquired in a different context without trying every possible option during the problem-solving phase would show more flexibility than an individual that was incapable of this. Moreover, an individual that can "complete" a problem with the least pieces of information or more abstract ones would also show more flexibility.

Conclusion

Flexible tool use is an indicator of intelligence if it is understood as the ability to adapt to new situations with innovative solutions. Individuals can produce novel solutions by learning what responses produce the desired consequences and/or combining separate experiences to generate appropriate solutions. It is best to keep learning and reasoning separate, since at the very least, they may be able to be distinguished quantitatively in terms of the source of the knowledge, the type of knowledge and the distance between experiences that can be used to solve novel problems. We have postulated three psychological ingredients to become a flexible tool user. First, acquiring and storing information even in the absence of reinforcement. Here, exploration and latent learning play a crucial role in building up a database of knowledge. Second, relating pieces of information that were acquired independently in terms of time and space. Of particular importance for facilitating the connectivity between disparate pieces of information is its encoding. Structural (i.e., functional) representations, in particular, may play a crucial role in the development of flexible tool use. Third, possessing a marked propensity to manipulate objects in the context of exploration, as this may facilitate the discovery of functional properties of objects and object–object relations. The first two ingredients are more general, and can be applied not just to tool use but also to other tasks (e.g., spatial detours). In contrast, the third ingredient is specific for tool use. The three ingredients represent a mixture of cognitive, motivational and sensorimotor aspects that are hypothesized to play important roles in the emergence of flexible tool use. It is important to distinguish between the various tool-using skills. Tool use in some species is better understood as a narrow specialization to solve a particular problem, whereas other species use tools in a more open-ended fashion. For the latter, tool use may not represent a cognitive specialization but a consequence of their sensorimotor skills and their conceptual knowledge, which is also recruited to solve other tasks. Other species, such as rats, show some of the ingredients that we have proposed to foster creative tool use, such as exploration, information hoarding and recombination of information, but do not possess the manipulative propensities of some primates and some birds. However, creative tool use is a result of possessing multiple ingredients, which may be the reason why creative tool use in not so widespread in the animal kingdom.

References

Birch, H. G. (1945). The relation of previous experience to insightful problem-solving. *Journal of Comparative Psychology*, **38**, 367–383.

Bird, C. D. & Emery, N. J. (2009). Insightful problem solving and creative tool modification by captive nontool using rooks. *Proceedings of the National Academy of Sciences USA*, **106**, 10370–10375.

Bräuer, J., Kaminski, J., Riedel, J., Call, J. & Tomasello, M. (2006). Making inferences about the location of hidden food: social dog – causal ape. *Journal of Comparative Psychology*, **120**, 38–47.

Bruner, J. S. (1972). Nature and uses of immaturity. *American Psychologist*, **27**, 687–708.

Burghardt, G. M. (2006). *The Genesis of Animal Play: Testing the Limits*. Cambridge, MA: MIT Press.

Call, J. (2004). Inferences about the location of food in the great apes (*Pan paniscus*, *Pan troglodytes*, *Gorilla gorilla*, *Pongo pygmaeus*). *Journal of Comparative Psychology*, **118**, 232–241.

Call, J. (2006). Descartes' two errors: reasoning and reflection from a comparative perspective. In S. Hurley & M. Nudds (eds.) *Rational Animals* (pp. 219–234). Oxford: Oxford University Press.

Call, J. & Tomasello, M. (2007). Comparing the gestural repertoire of apes. In J. Call & M. Tomasello (eds.) *The Gestural Communication of Apes and Monkeys* (pp. 197–220). New York: LEA.

Davidson, J. E. (1986). The role of insight in giftedness. In R. J. Sternberg & J. E. Davidson (eds.) *Conceptions of Giftedness* (pp. 201–223). Cambridge: Cambridge University Press.

Davidson, J. E. (1995). The suddenness of insight. In R. J. Sternberg & J. E. Davidson (eds.) *The Nature of Insight* (pp. 125–156). Cambridge, MA: MIT Press.

Dominowski, R. L. & Dallob, P. (1995). Insight and problem solving. In R. J. Sternberg & J. E. Davidson (eds.) *The Nature of Insight* (pp. 33–62). Cambridge, MA: MIT Press.

Duncker, K. (1945). *On Problem Solving*. Washington, DC: American Psychological Association.

Ellen, P. & Pate, J. L. (1986). Is insight merely response chaining? A reply to Epstein. *Psychological Record*, **36**, 155–160.

Epstein, R. (1987). The spontaneous interconnection of four repertoires of behavior in a pigeon (*Columba livia*). *Journal of Comparative Psychology*, **101**, 197–201.

Epstein, R., Kirshnit, C., Lanza, R. P. & Rubin, L. (1984). "Insight" in the pigeon: antecedents and determinants of an intelligent performance. *Nature*, **308**, 61–62.

Girndt, A., Meier, T. & Call, J. (2008). Task constraints mask great apes' ability to solve the trap-table task. *Journal of Experimental Psychology: Animal Behavior Processes*, **34**, 54–62.

Hammerschmidt, K. & Fischer, J. (2008). Constraints in primate vocal production. In K. Oller & U. Griebel (eds.) *The Evolution of Communicative Creativity: From Fixed Signals to Contextual Flexibility* (pp. 93–120). Cambridge, MA: MIT Press.

Hanus, D., Mendes, N., Tennie, C. & Call, J. (in press). Comparing the performances of apes (*Gorilla gorilla*, *Pan troglodytes*, *Pongo pygmaeus*) and humans (*Homo sapiens*) in the floating peanut task. *PLoS ONE*.

Haun, D. B. M., Call, J., Janzen, G. & Levinson, S. C. (2006). Evolutionary psychology of spatial representations in the Hominidae. *Current Biology*, **16**, 1736–1740.

Heinrich, B. (1999). *Mind of the Raven*. New York: Harper Collins Publishers.

Hunt, G. R. (1996). Manufacture and use of hook-tools by New Caledonian crows. *Nature*, **379**, 249–251.

Katona, G. (1940). *Organizing and Memorizing: Studies in the Psychology of Learning and Teaching*. New York: Columbia University Press.

Köhler, W. (1925). *The Mentality of Apes*. New York: Liverright.

Köhler, W. (1969). *The Task of Gestalt Psychology*. Princeton, NJ: Princeton University Press.

Macphail, E. M. (1987). The comparative psychology of intelligence. *Behavioral and Brain Sciences*, **10**, 645–695.

Maier, N. R. F. (1931). Reasoning in humans II: the solution of a problem and its appearance in consciousness. *Journal of Comparative Psychology*, **12**, 181–194.

Maier, N. R. F. & Schneirla, T. C. (1935). *Principles of Animal Psychology*. New York: Dover.

Maltzman, I. (1955). Thinking: from a behaviorist point of view. *Psychological Review*, **62**, 275–286.

Marín Manrique, H. & Call, J. (2011). Spontaneous use of tools as straws in great apes. *Animal Cognition*, **14**, 213–226.

Marín Manrique, H., Gross, A. N. & Call, J. (2010). Great apes select tools based on their rigidity. *Journal of Experimental Psychology: Animal Behavior Processes*, **36**, 409–422.

Marín Manrique, H., Sabbatini, G., Call, J. & Visalberghi, E. (in press). Tool choice on the basis of rigidity in capuchin monkeys. *Animal Cognition*.

Mayer, R. E. (1995). The search for insight: grappling with Gestalt psychology's unanswered questions. In R. J. Sternberg & J. E. Davidson (eds.) *The Nature of Insight* (pp. 3–32). Cambridge, MA: MIT Press.

Mendes, N., Hanus, D. & Call, J. (2007). Raising the level: orangutans use water as a tool. *Biology Letters*, **3**, 453–455.

Mulcahy, N. J., Call, J. & Dunbar, R. I. M. (2005). Gorillas and orangutans encode relevant problem features in a tool-using task. *Journal of Comparative Psychology*, **119**, 23–32.

Oakley, K. P. (1976). *Man the Tool-maker*. 6th edn. Chicago, IL: University of Chicago Press.

Parker, S. T. & Gibson, K. R. (1979). A developmental model for the evolution of language and intelligence in early hominids. *Behavioral and Brain Sciences*, **2**, 367–408.

Pepperberg, I. M. (1999). *The Alex Studies: Cognitive and Communicative Abilities of Grey Parrots*. Cambridge, MA: Harvard University Press.

Piaget, J. (ed.) (1952). *The Origins of Intelligence in Children*. New York: Norton.

Reader, S. M. & Laland, K. N. (2001). Primate innovation: sex, age and social rank differences. *International Journal of Primatology*, **22**, 787–805.

Reader, S. M. & Laland, K. N. (2002). Social intelligence, innovation, and enhanced brain size in primates. *Proceedings of the National Academy of Sciences USA*, **99**, 4436–4441.

Roby-Brami, A., Hermsdörfer, J., Roy, A. C. & Jacobs, S. (2012). A neuropsychological perspective on the link between language and praxis in modern humans. *Philosophical Transactions of the Royal Society of London B*, **367**, 118–128.

Seed, A. & Call, J. (2009). Causal knowledge for events and objects in animals. In S. Watanabe, A. P. Blaisdell, L. Huber & A. Young (eds.) *Rational Animals, Irrational Humans* (pp. 173–188). Tokyo: Keio University.

Seed, A. M., Call, J., Emery, N. J. & Clayton, N. S. (2009). Chimpanzees solve the trap problem when the confound of tool-use is removed. *Journal of Experimental Psychology: Animal Behavior Processes*, **35**, 23–34.

Seed, A., Hanus, D. & Call, J. (in press). Causal knowledge in corvids, primates and children: more than meets the eye? In T. McCormack (ed.) *Tool Use and Causal Cognition*. Oxford: Oxford University Press.

Seifert, C. M., Meyer, D. E., Davidson, N., Patalano, A. L. & Yaniv, I. (1995). Demystification of cognitive insight: opportunistic assimilation and the prepared-mind perspective. In R. J. Sternberg & J. E. Davidson (eds.) *The Nature of Insight* (pp. 65–124). Cambridge, MA: MIT Press.

Selz, O. (1913). *Uber die Gesetze des geordneten Denkverlaufs*. Stuttgart: W. Spemann.

Taylor, A. H., Elliffe, D., Hunt, G. R. & Gray, R. D. (2010). Complex cognition and behavioural innovation in New Caledonian crows. *Proceedings of the Royal Society of London B*. doi: 10.1098/rspb.2010.0285.

Thorpe, W. H. (1963). *Learning and Instinct in Animals*. London: Methuen and Co.

Torigoe, T. (1985). Comparison of object manipulation among 74 species of non-human primates. *Primates*, **26**, 182–194.

Walker, S. C., Mikheenko, Y. P., Argyle, L. D., Robbins, T. W. & Roberts, A. C. (2006). Selective prefrontal serotonin depletion impairs acquisition of a detour-reaching task. *European Journal of Neuroscience*, **23**, 3119–3123.

Weir, A. A. S., Chappell, J. & Kacelnik, A. (2002). Shaping of hooks in New Caledonian crows. *Science*, **297**, 981.

Weisberg, R. W. (1986). *Creativity: Genius and Other Myths*. New York: Freeman.

Wertheimer, M. (1945). *Productive Thinking*. New York: Harper.

Wertheimer, M. (1959). *Productive Thinking*. Chicago, IL: University of Chicago Press.

Wimpenny, J. H., Weir, A. A. S., Clayton, L., Rutz, R. & Kacelnik, A. (2009). Cognitive processes associated with sequential tool use in New Caledonian crows. *PLoS ONE*, **4**, e6471.

Wimpenny, J. H., Weir, A. A. S. & Kacelnik, A. (2011). New Caledonia crows use tools for non-foraging activities. *Animal Cognition*, **14**, 459–464.

2 Ecology and cognition of tool use in chimpanzees

Christophe Boesch
Max Planck Institute for Evolutionary Anthropology

Introduction

Humans, as the most technological species, tend to assume that tool use is a sign of higher intelligence and that, over the course of our evolution, tools conferred a decisive advantage in the struggle to adapt to different environments (Mithen, 1996; Wynn, 2002; Wolpert, 2003; Dietrich *et al.*, 2008). As such, animal species that use tools are considered more intelligent, while those that do not are judged as being less intelligent. This amounts to an anthropocentric judgment whereby humans adopt a human criterion to judge the adaptive skills of other species (Barrett *et al.*, 2007; Goodrich & Allen, 2007). However, both *phylogeny* and *ecology* must be taken into account before one makes judgments about when and where we might expect tools to be used (Bluff *et al.*, 2007; Hansell & Ruxton, 2008).

Tool use as an adaptation

Physical adaptations

If one remembers that, in most cases, tools are an extension of one's body that allow an individual to solve tasks that cannot be solved with the body alone (Goodall, 1970; Beck, 1980; Boesch & Boesch, 1990), we must acknowledge that some primate species possess more efficient physical specializations than humans. For example, baboons have very hard, sharp teeth, which allow them to break open hard-shelled fruits that humans would be unable to open without the help of a tool (Kummer, 1968). Similarly, orangutans and gorillas, which are clearly physically stronger than humans, have been seen accessing food resources using sheer force in situations where humans would need to rely on tools (Schaik & Knott, 1996; Cipolletta *et al.*, 2007). In addition to sheer force, it has been argued that hands help in tool use and this would then explain some of the distribution of tool use in the animal kingdom, although we should not forget that birds hold tools with their beaks and some otters use tools as well. Therefore, independent of the cognitive capacities required to use tools, tool use by animals should not be expected to occur in all situations where humans might use them. Our natural tendency to anthropomorphize

Tool Use in Animals: Cognition and Ecology, eds. Crickette M. Sanz, Josep Call and Christophe Boesch. Published by Cambridge University Press.

hinders us from reaching a better understanding of the evolution of tool use, and it is imperative that we look directly to animals for answers about when tools might be beneficial.

Challenges imposed by ecological niche

Furthermore, each individual animal lives in a specific ecological niche, where they encounter daily challenges that must be solved, and some of these challenges might be more efficiently solved with the help of tools. Moreover, such challenges will vary according to the species and the specific population studied. For example, a capuchin monkey living in the dry open areas of northeast Brazil only faces a limited number of challenges similar to those encountered by a capuchin monkey in wetter regions of Brazil (Moura & Lee, 2004; Visalberghi *et al.*, 2007; Canale *et al.*, 2009; Souto *et al.*, 2011; Spagnoletti *et al.*, 2011). Moreover, both face totally different conditions than capuchin individuals living in captivity. In addition, each species possesses different physical adaptations, which might directly affect the possibility of using tools. The result might be that, within each population, an individual's development of causal understanding – the understanding of the relationship between a set of factors (causes) and a phenomenon (the effect) – will be promoted by being exposed to situations where such understanding confers a competitive advantage. In turn, this will increase the exposure of juveniles to specific object–object interactions that promote the development of folk physics and select adaptive population-specific cognitive traits responsible for their specific tool-use abilities (Fragaszy & Visalberghi, 2004; Boesch, 2007; Bluff *et al.*, 2010).

So the question should really be, what is the best way to solve a challenge faced by an individual within his ecological niche, rather than, why doesn't a particular species/individual use tools? Such an ecological approach toward tool use avoids the anthropocentric assumption that it is better to use tools in most situations and allows us to look at the ecological and morphological factors that affect tool use. It is important to note that, in the real world, there is a constant interaction between brain, body and world (Barrett *et al.*, 2007; Boesch, 2007, 2010), and that we need to put ourselves into the situation of the animals as much as possible if we want to understand the solutions they adopt.

Macaques, a widely distributed genus, have not been observed to be particularly adept tool users; however, whenever facing the appropriate conditions, such as those of the Piak Nam Yai Island in the Andaman Sea, they have been observed to become skilled tool users and develop customary forms of tool use to extract food items in the intertidal zones of the coastal habitats and select tools based on the food being processed (Gumert *et al.*, 2009). "Appropriate conditions" are going to be species-specific, making any claims about limitation in tool-using abilities hazardous, before we have gained an in-depth knowledge of the species-wide population variations.

Some have argued that mental reasoning and understanding of the technical aspects of tool use can only be demonstrated in "new situations on the first trial" (Tomasello & Call, 1997; Bluff *et al.*, 2007; Penn & Povinelli, 2007). However, such a criterion cannot apply to natural situations, and thus, by definition, would seem to thwart any claims about reasoning and understanding from natural observations. If such a biased approach to the study of animal behavior could be adhered to for the few animal species that can be maintained in ethically and ecologically valid captive conditions, it would prevent any understanding of higher cognition for all species in which captive conditions cannot be implemented with ecologically valid environment. Besides this, below we will see that some tool techniques result from very specialized ecological challenges that cannot be fully mimicked in captive conditions. Nevertheless, some birds have proven to be the ideal subjects and careful experimentation has allowed for much progress in detailing some of the cognitive requirements of tool use (see reviews by Emery & Clayton, 2004; Hunt *et al.*, 2006; Bluff *et al.*, 2007; Seed *et al.*, 2009a).

In the present review, I propose to concentrate on natural observations of tool use in capuchins and chimpanzees in an attempt to understand causal understanding in primates. This will allow me to illustrate the central role ecology plays in tool use, especially for complex tool use aimed at retrieving food embedded in natural substrate. For many complex real-world situations, experiments are not possible and only natural observations will guide our thinking about animal reasoning and produce hypotheses about the way animals might understand and alter their environments.

Tool use and anthropomorphism

Anthropomorphism, the attribution of human feelings and intention to other species, is an inherent tendency of each human observer (Keeley, 2004; Goodrich & Allen, 2007; Wynne, 2007). This is potentially dangerous as such an attribution can be purely a consequence of our own way of thinking and have nothing to do with the way an animal is thinking. At the same time, without an intimate understanding of the real challenges an animal faces and the way it can arrive at solutions, there is no way we can make progress in understanding the ways animals react in their environment. By definition, we can only achieve that in an anthropomorphic way as we are humans. The nut-cracking behavior in Taï forest chimpanzees or the underground termite fishing in Goualougo forest chimpanzees has to be solved within the technical constraints of those forests and only human experiments can give us a feeling of what the real challenges are. So, basically, it is our anthropomorphic understanding of the technical components of a tool-oriented task that forms our basis for evaluating the appropriate solutions and estimating the cognitive challenges.

Having said that, no observer thinks animals have to solve the task in a human way, and we do consider the specificity of the animal's way of addressing the technical challenges, but this remains a challenge for us as we are humans and cannot really make judgments about the perceptions, understanding, previous experiences or alternatives that another species brings with them when solving a task. This problem is especially acute in captive experimental situations where all dimensions of the settings and the solutions are an enforced human design, often with very limited ecological validity (Allen, 2002; Boesch, 2007, 2010), and it becomes particularly difficult to untangle what animals have brought in apart from the human impositions.

The importance of ecology, cognition and culture

The "eco-cultural model" was originally proposed to explain the differences in cognitive development observed between humans belonging to different cultures (Segall *et al.*, 1999; Carpendale & Lewis, 2004), and I proposed to expand it to include all large-brain animal species capable of learning (Boesch, 2007). This model stresses that behavior and cognition are not purely driven by genetics, but result from the year-long ontogenetic interactions of the individual with his social group and his ecological niche. The eco-cultural model predicts that tool-use frequency should vary according to the ecological challenges faced by different species. The more ecological niches are encountered by a species, the more diverse technological solutions should be observed in that species, when all other factors would remain constant. In other words, with the same level of learning ability and the same physical ability, more tool uses should be observed in the species with larger diversity of ecological niche (Boesch, 2007). Such a model would concur with the "technical intelligence hypothesis," which was proposed to relate the technical and cognitive demands of using tools and thereby explain the origin of the ape/monkey grade-shift in intelligence (Byrne, 1997, 2004). The classic example of the influence of ecology and culture is nut cracking with hammers, which has been observed in chimpanzees and capuchin monkeys (Boesch *et al.*, 1994; Visalberghi *et al.*, 2007; Spagnoletti *et al.*, 2011); to be able to crack the nuts, those nuts must be present and large proportions of chimpanzee and capuchin monkey populations do not live in forests where the nuts are available, and therefore were not seen to use hammers to crack them. On the other hand, many populations of chimpanzees live in forest where the nuts are present, but they do not crack those (Boesch *et al.*, 1994; Boesch, 2003).

Tool use as complex behavior

A tool is defined as an external object detached from its substrate use to attain a goal (Goodall, 1970; Beck, 1980). In a sense, a tool is perceived as an extension of the body that is used to achieve a goal that cannot be directly achieved with the use of only hands or teeth. To be able to achieve such a task, an individual needs to have a certain level of causal understanding about the effect of one's own body on objects as well as on the body's actions on external objects (Greenfield, 1991; Visalberghi & Tomasello, 1998; Boesch & Boesch-Achermann, 2000; Fragaszy & Visalberghi, 2004; Visalberghi & Fragaszy, 2006; Fragaszy, 2007; Hansell & Ruxton, 2008; Wimpenny *et al.*, 2009; Bluff *et al.*, 2010). As the latter functions only through the force exerted on the tool to achieve a certain goal, it was proposed that this requires some more complex forms of causal understanding. While new examples of tool use have been uncovered from wild animals, causal-understanding abilities were suggested to be very limited in primates but at the center of the evolution of human intelligence (Mithen, 1996; Wolpert, 2003). While the causal understanding in each species must indeed be critically considered, it is only with a detailed and precise consideration of all field evidence that we will be able to comprehend what level of causal understanding and technical expertise animals have acquired.

Flexible tool use, the use of different types of tools in different contexts and to reach different types of rewards, has been observed more frequently in chimpanzees and humans than in any other animal species (see below; Boesch & Boesch-Achermann, 2000). This distribution of flexible tool use in primates suggest that an elaborate understanding of causal relationships between different external objects favors the development of tool use to new and different contexts, and this might be an important prerequisite for the evolution of flexible tool use (Boesch & Boesch-Achermann, 2000).

More categorical definitions of tool use have become important due to the growing evidence that tool-use skills could be found in many different animal species, including birds and the increasing evidence of flexible tool use, both in captive and wild animals (Goodall, 1970; Beck, 1980; Fragaszy, 2007; Hansell & Ruxton, 2008; Wimpenny *et al.*, 2009). In an attempt to single out some of the cognitive demands of tool use, some authors have proposed a "tree model" of tool use using a sequential hierarchical approach concentrating on the number of spatial relations produced by the actor as well as the order in which they occur (Greenfield, 1991; Matsuzawa, 2001). Developing on this and focusing on capuchin monkey tool use, a "spatial relation model" was proposed that identifies four specific aspects of tool use, with a special emphasis on the spatial relations necessary to master how to place and manipulate the tool correctly in connection with the food reward (Visalberghi & Fragaszy, 2006; Fragaszy, 2007). Such spatial relations are qualified depending on if they are performed concurrently or sequentially, and in a dynamic fashion or not (see Table 2.1). Recently, a new taxonomy of tools has been proposed based on the number of tools used or combined, stressing the importance of metatools that were suggested to included all secondary tools as well as combined tools (Wimpenny *et al.*, 2009). This new taxonomic approach tends to underestimate the complexity of the decisions made when using tools, as revealed by a chaîne opératoire approach (Bar-Yousef *et al.*, 1992; Carvalho *et al.*, 2008; Bar-Yousef & van Peer, 2009; Boesch *et al.*, 2009). Furthermore, since I am mainly interested in the ecological challenges solved with tools, I want to categorize them according to the technical challenges addressed. Therefore, I will discuss tools within a framework that attempts to distinguish four levels of increasing complexity in the decision process and come back to the spatial relation model approach when discussing the effect of cognition on tool use.

Simple tool use

Simple tool use includes all instances where a single tool is used to perform all the actions necessary to obtain the reward. Examples of simple tool use are: an Egyptian vulture breaking an egg with a stone held in its beak; a Californian otter pounding an oyster on a stone placed on its belly; a chimpanzee fishing termites with a twig; and a human cutting a piece of meat with a knife (see review by Goodall, 1970; Beck, 1980). The actions performed with the tool can be numerous, but they are all performed with that single tool, and therefore the causal understanding remains rather straightforward and the delay to obtain the reward is limited as the connection between the body and the reward is direct via the tool.

Table 2.1 Cognitive aspects of tool use based on my definitions of the different types of tool use (adapted and expanded from Visalberghi & Fragaszy, 2006; Fragaszy, 2007).

Type of tool use	Numbers of relationships	Temporal actions	Static/dynamic	Permissive/ specific	Direct/indirect
Simple tool use					
a-Probe a hole	1	N/A	Static	Permissive	Direct
b-Break through resin	1	N/A	Static	Permissive	Direct
c-Fish for termite	1	N/A	Static	Specific	Direct
d-Dip for ants	1	N/A	Static	Permissive	Direct
e-Push food in trap-tube	2	Concurrent	Dynamic	Specific	Direct
Combined tool use					
a-Pound nut/surface	1	Sequential	Static/dynamic	Permissive	Direct
b-Pound held nut/surface	2	Concurrent	Dynamic	Permissive	Direct
Sequential tool use					
a-Tree honey	≥3–5	Sequential	Dynamic	Specific	Indirect/indirect/direct
b-Underground termite	2	Sequential	Static	Specific	Indirect/direct
c-Underground honey	2–3	Sequential	Dynamic	Specific	Indirect/indirect/direct
d-Secondary tool	2–3	Sequential	Static	Specific	Indirect/direct
Composite tool use	≥3	Concurrent	Static	Specific	Indirect/direct

Tool-use examples are as follows:
Simple tool use: (a) The use of a stick to inspect inside a hole, as seen in many animal species; (b) The use of a heavy stick to break open the resin blocking the nest entrance of a bee hive, as seen in different chimpanzee populations; (c) The classic example of the Gombe chimpanzees using small twigs, bark or herbs to fish for termites by inserting it in holes of the termite mound; (d) The use of a stick or wands placed in a driver nest entrance to let soldiers bite and climb on it; (e) The classic tube test in which an animal has to push food out of a tube with a stick while at the same time avoiding making the food fall into a trap.
Combined tool use: (a) The classic nut cracking in chimpanzees, in which a nut is placed on a hard surface and then pounded with a hammer; (b) The same sequence but in a tree, where the tool user must stabilize the nut with its free hand at the same time, in order to prevent it from falling down.
Sequential tool use: (a) The use of a set of up to five different tools to open, access and extract honey from bee hives located in trees; (b) The use of two types of tools to locate and fish termites out of underground mounds; (c) The use of up to three different tools to locate, access and extract honey from underground bee hives; (d) The use of a tool to produce a tool that will be used, for example, to cut meat.
Composite tool use: The production of a spear with a sharp-edge stone attached to a straight branch with binding material.

Body → Tool → Reward

Hence, such simple tool use has been labeled as a first-order problem by Visalberghi and Fragaszy (2006), as they imply a single relationship between the body and a tool producing a single dynamic spatial relation (Table 2.1). Simple tool use is a widely observed ability in the animal kingdom, but one striking difference is that, in most species, a single or very few different types of simple tool use have been seen (Beck, 1980; Schaik & Knott, 2001). On the other hand, the flexible use of simple tools is limited to only a couple of species, which suggests that flexibility in tool use is a demanding aspect (Beck, 1980; Byrne, 1995). As simple as they may be, such tool use still requires an understanding of the relationship between tool rigidity and function, on the one hand, and an understanding between tool orientation and position of food reward as well as the relationship between trajectory and substrate on which tools move, on the other (Santos *et al.*, 2006).

As in dense and rich habitats, where opportunities to access embedded resources with tools are high, flexible tool use could rapidly result in an adaptive advantage, and it is intriguing that such flexibility is so rarely seen in the wild (Beck, 1980; Boesch & Boesch-Achermann, 2000; Visalberghi *et al.*, 2007; Hansell & Ruxton, 2008; Spagnoletti *et al.*, 2011). However, each simple tool use requires the understanding of and control over a specific causal and spatial relationship. It is therefore conceivable that the mastery of multiple relationships such as this rapidly becomes quite demanding, both in terms of cognitive understanding and flexibility. All chimpanzee and human populations have been seen to use a variety of simple tools so regularly that we can identify where individuals originate from on the basis of their specific tool repertoire. As would be expected from the predictions of the eco-cultural model, these two species face the largest number of different ecological niches within the primate family (Wolfheim, 1983) and, therefore, face the most diverse sets of ecological challenges.

Combined tool use

In some specific ecological contexts, such simple tool use might not be efficient to solve a given technical challenge and some innovations can then become necessary. Although a single tool can successfully extract ants or termites from their nests, it will not successfully crack hard-shelled fruits, like nuts. In that case, the nut must be placed on a hard surface and pounded with a hammer in order to be successful (Boesch & Boesch, 1984; Visalberghi *et al.*, 2007). An infant who has forgotten to place the nut on a hard surface will quickly notice that the nut will become encased in the soil and will never break open, no matter how hard one pounds (Boesch & Boesch-Achermann, 2000). Therefore, in this situation, the nut must not only be placed in relation to the hammer but, at the same time, must also be placed on a hard anvil in order for the pounding to be successful. Chimpanzees and capuchin monkeys are able to flexibly do this, as they select the hardness of both the anvil and the hammer in accordance with the hardness of the nut they intend to crack (Boesch & Boesch, 1984; Visalberghi *et al.*, 2009; Spagnoletti *et al.*, 2011). This combination of two types of relationships requires a precise understanding of causal relationships and this results in further delay in achieving success as the tool user needs to bring the two objects together before being able to act (Boesch & Boesch, 1984; Inoue-Nakamura & Matsuzawa, 1997).

Body → Hammer + Anvil → Reward

Combined tool use, which has also been termed a "second-order problem" as individuals are required to manage two spatial relationships between objects concurrently (Visalberghi & Fragaszy, 2006), has also been observed in primates, although clearly less frequently than simple tool use (e.g., nut cracking [see above] or honey extraction [see below]). Wild capuchin monkeys have been seen to efficiently master both the spatial relationships between positioning the nut on the anvil as well as the pounding of the nut with a hard hammer.

Sequential tool use

In addition to combined tools that are used at the same time, animals have been observed to use tools one after the other before reaching the reward. Two types of sequential tools can be distinguished. True sequential tools that are used one after the other in a sequence before reaching the goal, and secondary tools where one tool is used to make a tool, that can then be used to reach the goal (see Watanabe, 1972 in Sumita, 1985).

Such sequential tool use is observed when important food resources are not always directly accessible with a single tool in nature, as they are either too well protected from predators or are too distantly located to be reached directly with only one tool. Two major classes of resources are concerned here. The first, honey, is a very attractive food resource for many animal species, and therefore bees construct well-protected hives as they rely on them for their survival and reproduction. Many hives are located in small openings high up in trees and accessing them is a real challenge, even for honey badgers, which are armed with specialized morphological tools such as powerful claws and teeth. Some hives are impressively large; for example, some stingless bees in the forests of Gabon produce hives that are up to 100 cm deep, making it mandatory to use tools to reach the honey for animals that are physically too large to enter the nest. The second class of resources includes underground foods such as tubers, termites, ants and honey. For all of these, the resource must first be located, then accessed and, finally, the food must be extracted. Often many tools are necessary for consumption to be successful.

The technical challenges of accessing such resources may discourage some animal species, who will then simply neglect them, while others may acquire morphological adaptations in order to specialize for them. This is the case with honey badgers in African forests, which specialize on honey, and with pangolins, which specialize on termites. However, in other habitats, food availability might make such resources highly prized, and sequential tool use will be the logical consequence.

In the past, the cognitive challenges of attaining such underground resources were proposed to be too demanding for non-human primates and to therefore be a uniquely human specialty (Hatley & Kappelman, 1980; Laden & Wrangham, 2005). Recent observations have shown that the exploitation of such protected food resources is well within the ability of chimpanzees and that they do so regularly, using complex tool sets (Sanz *et al.*, 2004; Hernandez-Aguilar *et al.*, 2007; Boesch *et al.*, 2009; Sanz & Morgan, 2009). Tool set use, which is the use of different tools one after the other in order to reach a reward, is in fact a sequential tool use as the order in which each tool is used is not independent from the others.

Sequential tool use adds a layer of complexity, as the likelihood of randomly finding the correct sequence to use the tools decreases proportionally with the number of different tools used and, at the same time, requires a longer delay in obtaining the reward. That each tool is of a different type and fulfills a different function implies that individuals must keep track of different causal relationships in a sequential order, of which only the last step allows access to the reward. Continuing with the example of honey extraction in chimpanzees, tool 1, called the pounder, is used to forcefully hit the wax covering the entrance to the bee hive until it is broken, which is sub-goal 1. Once this is achieved, tool 2, the enlarger, is used to open the honey chambers within the nest, which is sub-goal 2. Finally, a collector, tool 3, is used to extract the honey, which is the reward (Boesch *et al.*, 2009; Sanz & Morgan, 2009; see Figure 2.1). As such nests are very large, an individual might have to reuse an enlarger (tool 2) to access a new honey chamber (sub-goal 2) to be able to extract more honey with a collector (tool 3). Therefore, tool users must differentiate between the spatial relationships of "tool 1," with its specific action to object

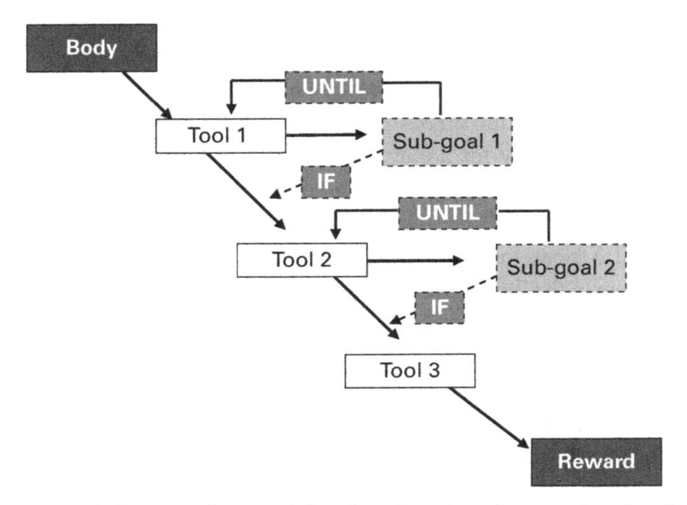

Figure 2.1 General diagram of sequential tool use based on the complex situations seen in chimpanzees. "Tool 1" will be reused *until* "sub-goal 1" is reached. Only *if* "sub-goal 1" is reached will "tool 2" be used, and then *until* "sub-goal 2" is reached. This can go on for up to five different tools until the reward is reached (Boesch *et al.*, 2009; Sanz & Morgan, 2009). If only one tool is used, this diagram reduces itself to the simple tool use presented above.

1 – intact bee hive entrance – from the spatial relationship of "tool 2," with its other specific action to object 2 – chamber walls within the broken bee hive entrance – from the spatial relationship of "tool 3" – collecting honey from deep inside the hive. Tool size, weight and hardness is selected and modified by tool users in accordance with such specific actions and functions (see Boesch *et al.*, 2009; Sanz & Morgan, 2009). Thus, the memorizing of different steps within a specific sequence and the hierarchical differentiation between them is what makes sequential tool use challenging.

Furthermore, the same tool is often used repeatedly until the sub-goal is reached, at which point the individual will shift to using a second type of tool (see Boesch *et al.*, 2009; Sanz & Morgan, 2009; Figure 2.1). In some sequential tool use, this requires quite extensive use of the tool and the time delay before the use of the second tool can be quite considerable. For example, when using a pounder (tool 1) to break open the bee hive, a Goualougo chimpanzee could perform the action over 280 times before attaining sub-goal 1 and switching to tool 2 (Sanz & Morgan, 2009). Furthermore, quite often the sequential tool use needed to reach a bee hive is challenging and chimpanzees are regularly unsuccessful, as bee hives are often well protected. Underground hives, in particular, are very hard to locate (Boesch *et al.*, 2009; Sanz & Morgan, 2009). Thus, sequential tool use imposes not only an important time delay before getting at the reward, but requires that an individual apply a "conditional recurrent hierarchical sequence" as each tool is inserted within a sequence and each has to achieve a certain function that is dependent upon the functions performed by previous tools and that will be performed by subsequent ones.

The puzzling observation that Central African chimpanzees employ sequential tool use regularly but West African chimpanzees have been observed to do so only very rarely begs the question of why some populations seem to use sequential tools much more often than others. Taï chimpanzees have been seen to use sequential tools only in the context of nut cracking, when they use sticks to extract the pieces of kernel out of nut shells after breaking them open with hammers, while in 30 years of observations, we saw them using simple tools only to extract honey from honeybee hives (Boesch & Boesch-Achermann, 2000). At the same time, chimpanzees in Central Africa have been seen to employ sequential tool use extensively for honey extraction from both honeybee and stingless bee hives. Therefore, it does not seem to be a cognitive limitation, but more likely an ecological one, in the sense that the density of bee hives could be lower in Taï forest compared to Central African forests. In addition, honeybee hives seem to be constructed in larger holes in trees in the Taï forest, permitting its exploitation directly with the hand, while this has not been seen in Central African populations, where honeybees may face higher predation pressure and select tree holes that are narrower and therefore more difficult to reach. Similarly, differences in the presence of stingless bees between West and Central African forests, as well as differences in the amount of honey found in those nests, could contribute to the fact that Central African chimpanzee populations seem to raid them more often and with more tools. It is still premature to make any firm conclusions, but ecology might be largely responsible for the differences we see in sequential tool use between chimpanzee populations.

Secondary tools that are produced with the help of another tool, as is so typical in stone knapping, rests on a similar technical process, where tool 1 is used to reach sub-goal

1, which produces tool 2, which is the tool that will reach the reward (McPherron, 2000; Ambrose, 2001; Sharon, 2009). Up to now, using one tool to make another tool has not been observed in chimpanzees, but the sequential dimension is very similar, and planning to transform either the tool or the object has many striking similarities. Making secondary tools is still, by nature, the imposition of form on the raw material resulting from detailed planning and a balance with the technical constraints of the raw material and the function of the tool (e.g., Wynn, 1993; McPherron, 2000). The essential difference, however, lies in the additional complexity of concurrently mastering the external causal relationships between two objects external to the actor, by which the first external object transforms the second one, while previous tool transformations seen in other animal species were always directly done by the actor. Before concluding about the inability of chimpanzees or other animal species to master secondary tools, we must first see them face an ecological challenge that would require them to shape one tool with another tool, such as cutting the meat of large dead prey, or perforating a hole in a hard-to-reach solid surface that would require a pointed end on a large branch (see also a captive bonobo successfully making a secondary tool: Toth & Schick, 1993).

Composite tool use

Composite tools are tools that are made of at least two different material elements that are kept together so as to function as one tool (Ambrose, 2001). This type of tool use, which has been proposed to be unique to humans as it has not yet been seen in non-human animals (Goodall, 1970; Beck, 1980; Boesch & Boesch, 1990), appeared in the course of our evolution when we entered the Middle Stone Age (Ambrose, 2001, 2010). Analysis of such composite tools suggests that they were mainly used to produce spears, knives and scrapers involving at least three tool components, such as a handle or shaft, a stone insert and a binding material (Ambrose, 2001). They seem to have developed as an adaptation to the hunting of large prey species for which long-distance weapons and increased cutting technology would have conferred an adaptive advantage to the tool users. Chimpanzees have not yet been seen to hunt large mammals, and therefore have not yet been seen to face the challenges that would make composite tools beneficial. It is clear that the technical challenges of making such tools are important and novel, and require an essential planning component and hierarchical sets of actions. However, the sequential hierarchical actions discovered recently in Central Africa suggest that the differences might be smaller than was previously thought and more observations of wild chimpanzee tool use should help to clarify the differences in such a domain between human Middle Stone Age technology and chimpanzees (see Mercader *et al.*, 2007).

A special ecological challenge: non-visible food resources

To illustrate the cognitive demands that ecological challenges may represent, I discuss here the special situation of non-visible food resources before moving to the general question of tool use as a cognitive challenge (see below).

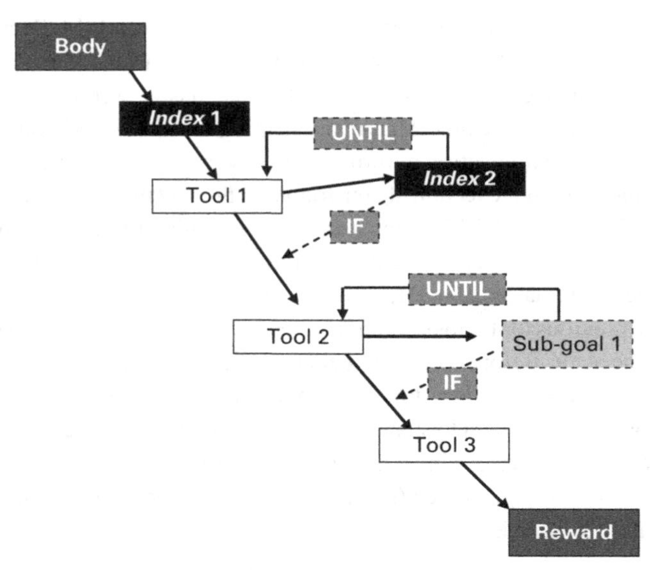

Figure 2.2 Schematic diagram of sequential tool use for non-visible resource, as used by chimpanzees in Goualougo Triangle, Congo, or in Loango National Park, Gabon. First, an individual searches for an indirect index of the presence of the resource, like a termite mound elevation, and once found, "tool 1" is used to locate another indirect index, index 2, which indicates the precise underground location of the termite chamber. "Tool 2" is then used to fish for termites by Goualougo chimpanzees. In Loango, chimpanzees must first find the opening of the nest marked by a small wax tube, index 1, and then tool 1 is used to perforate the ground until the nest is located, normally by smell, which is index 2. Then tool 2 can be used to dig a tunnel to reach the honey in the nest, which is sub-goal 1. Thereafter, and depending on the depth of the nest, tool 3 can be used to help extract the honey.

Animals may face situations that are difficult even for human researchers to understand, and for that I would like to turn my attention to the extraordinary case of sequential tool use aimed at underground non-visible food resources. When chimpanzees in the Goualougo forest intend to feed on underground termites, they first need to locate them. To do so, they use sticks to puncture the ground until they find the termite mound chamber (Sanz *et al.*, 2004).

Similarly, when Loango chimpanzees feed on honey from underground stingless bee hives, they must first locate the invisible honey chamber below the ground (Boesch *et al.*, 2009). The particular challenge of such an endeavor is that the reward – termite soldiers or honey – is not directly visible, so its location must first be inferred with the help of indices that suggest the presence of the reward – such as a small elevation increase on the ground surface or a tiny opening of a wax tube. Nevertheless, such indices constitute only a very rough indication of the reward's location ("index 1" in Figure 2.2). This is very different from ant-dipping or classic termite fishing, where the location of the nest is directly visible and determining whether the ants or termites are present can be done with a simple scratch on the surface of the nest entrance (Goodall, 1968; Boesch & Boesch, 1990).

Index 1 is only a rough indicator of the location of the food resource, because the prey species – termites and bees – do their best to hide the resource from the many

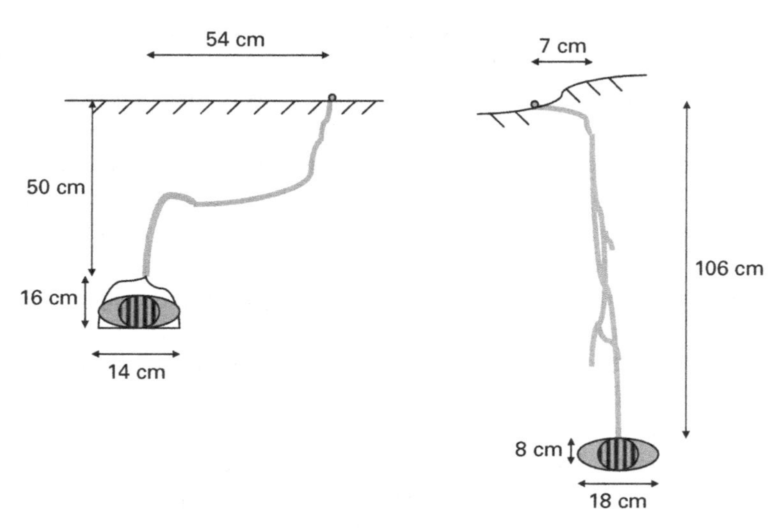

Figure 2.3 A vertical schematic illustration of the structure of two Melipone underground bee nests of the species predated by chimpanzees in Loango National Park, Gabon. We extracted those nests that were intact and active ourselves, as otherwise there was no way we could understand the technical challenges the chimpanzees face when trying to access honey in underground nests. The honey chamber within the nest is located underground, anywhere from 20 cm to over 100 cm from the surface, and, as illustrated, can be located laterally quite far away from the tube exit at the surface. The very thin wax tube made by the bees can follow either a relatively straight path or a very convoluted path to the chamber, which nothing on the surface would allow an individual to predict.

predators present in the forest. In the case of underground bee honey, the nest can be up to 100 cm under the surface of the ground, and sometimes up to 70 cm laterally from the only visible sign of a nest, the 3 mm wide wax tube marking the entrance of the nest (see two examples of underground bee nests in Figure 2.3). After finding "index 1," chimpanzees will select a tool to puncture/perforate the ground, in order to better locate the resource. By doing this, they are looking for a second indirect sign of the reward location, "index 2." From personal trials and by looking at the actions of the other chimpanzees, index 2 can remain indirect in the sense that it can be a change in the physical property of the ground that a chimpanzee has to then relate to the food (Figure 2.2). For example, a chimpanzee must relate the perceived change in the resistance of the soil as it forcefully pushes the stick through the ground trying to detect the honey chamber or termite mound. Sometimes a chimpanzee will then smell the end of the stick, as if confirming the localization of the resource. Index 2 will then be used to direct the subsequent actions of the individual (Figure 2.2), and additional tools might be needed before an individual gets to eat the food. Without such indices, the actions would be performed purely at random and with very low success rates. Indices are a cognitive challenge as they are not the reward but *stand for the reward* and guide the chimpanzee. Furthermore, they are not visual cues, but instead involve other sensory modes, as lower soil resistance implies the presence of a termite mound or honey chamber.

After finding index 2, chimpanzees often dig tunnels that are barely large enough to let a human hand enter, indicating that they know exactly where they are aiming at. They cannot do that by simply following the bee tube, as this is much too fragile to resist the tool-assisted digging process. Personal testing at locating the nest proved to be extremely difficult as the soil can be so hard that I was unable to perforate the ground beyond 30–40 cm. In fact, local pygmy trackers told me that the elderly in their village advised them not to try to search for the honey of this species of bee as it "is too deep." When successful, it took me up to 40 minutes just to locate the honey chamber (Figure 2.3). At the time of writing, Luisa Rabanal, the student working on this technique, and myself are still unclear about how the chimpanzees are able to locate the nest so precisely and how often they are successful in reaching the honey. Luckily, the chimpanzees are habituated to the presence of camera traps and we are presently in the process of filming chimpanzees as they try undisturbed to solve this technical challenge.

Thus, an elaborate understanding of unseen nest structure, combined with a clear appreciation that tools permit the location of unseen resources and a precise three-dimensional sense of geometry to reach the honey chamber from the correct angle is demonstrated by the chimpanzees when extracting underground honey. In addition, due to the important variation in nest structure and the impossibility of predicting it from visible cues, chimpanzees have to demonstrate a high sense of flexibility in implementing the sequence shown in Figure 2.2, as angles, depth, direction and size of the nest, as well as hardness and type of soil vary for each nest, requiring a permanent adaptation to the local situation in the selection and the number of tools, as well as the modification to do to them (Figure 2.3) (Boesch *et al.*, 2009). Finally, the tool users' techniques also vary, as they may use one or both hands or use both their hands and their feet to perforate the ground with the tool.

Tool use as a cognitive challenge

Different aspects of tool use can represent a cognitive challenge and different authors have tried to address them and place them within a coherent framework so that different types of tool use can be compared. Tool selection has been considered as an important cognitive challenge as it requires a flexible adaptation to the technical constraints (e.g., Boesch & Boesch, 1990; Bluff *et al.*, 2007). Tool selection has been studied with some captive monkeys and apes in different artificial experiments and results have shown that, after training, some limited understanding of the functional property of tools could be demonstrated (Povinelli, 2000; Santos *et al.*, 2003, 2006; Herrmann *et al.*, 2008). Intriguingly, all species seem to give more importance to proximity – when a tool touches an object – rather than to connectivity – when a tool is attached or supports the object – leading to a lot of surprising and mysterious mistakes. In the real world, confusing proximity with connectivity would not be very adaptive, and neither nut cracking nor ant dipping would be successful and might even be deadly when it comes to jumping between branches high up in trees.

In order to get a grip on all these different aspects of tool use and, therefore, make comparisons between the different tool-use possibilities, I suggest adapting and expanding the relation spatial model in order to describe the cognitive challenges of tool use. This model was initially proposed to address the tool-using skills of capuchin monkeys (Visalberghi & Fragaszy, 2006; Fragaszy, 2007), and studies found that, in its more complex forms, capuchins were able to concurrently master two different spatial relationships when using tools with a dynamic dimension (see Table 2.1 under combined tool use (b)). I have expanded on this by adding one more dimension than was originally proposed and have used my four levels of different tool-use types (see above). In this model, each tool use is classified with respect to four orthogonal properties; the number of spatial relationships that need to be mastered for the tool use to be accurate (1) if such spatial relationships have to be managed successively or concurrently; (2) if the relationships are static in time or needed to be dynamically adapted during the course of actions; (3) if the tools needed to be used in a specific manner or could be used in a more permissive manner (for example, whether or not both sides of a stick could be used); and (4) if direct body contact is involved or not.

The concurrent mastering of two spatial relationships has recently been suggested to be crucially important in explaining some of the difficulties observed when captive chimpanzees attempt to solve the tube-trap problem (Visalberghi & Limongelli, 1994; Povinelli, 2000; Mulcahy & Call, 2006; Seed *et al.*, 2009b), as it requires an additional load on the subject's attention system, cross-modal matching if different sensorial modes are used, and increased response variability. The hierarchic sequential mastering of more than one spatial relationship might similarly represent a cognitive challenge, as cross-modality, attention and response variability with two spatial relationships is equally required. Furthermore, since actions are not concurrent, the more tools an individual uses, the more demands there will be on remembering them and placing them in the appropriate sequential order. Tool use has been suggested to be the outcome of coordinated multiple cognitive processes, each of which can respond differently to experience and to immediate circumstances (e.g., Spencer *et al.*, 2001; Fragaszy *et al.*, 2009).

Of special relevance here are the works comparing the performance of capuchin monkeys versus chimpanzees, our two best primate tool users. Comparative approaches using exactly the same procedure showed that capuchin monkeys fail a causal reasoning task that chimpanzees could solve (Visalberghi & Limongelli, 1994), and children and chimpanzees viewing similar films fare comparably well in another comparative causal reasoning test (O'Connell & Dunbar, 2005). More directly related to tool-use reasoning, capuchin monkeys and chimpanzees were able to maintain focus on moving a cursor toward a goal and monitoring the effect of each choice, but only chimpanzees could master the notion of detour to reach a goal and plan the continuity of the path to that goal, which made them more able to redirect their behavior when faced with multiple possibilities (Fragaszy *et al.*, 2009). In addition, the performance of chimpanzees, but not capuchin monkeys, increased with experience. Both are relevant to our understanding of the differences in the tool-use skills of capuchin monkeys and chimpanzees, as monitoring the effect to reach a goal is important for simple tool-use efficiency, while planning a path to a goal with distinct detours and incorporating different tools is at the base of performing sequential tool use, and reaches a higher level of complexity than does simple tool use. Capuchin monkeys have been observed to use simple and combined, but not yet sequential, tools.

Tool use in the wild imposes an additional need for flexibility on tool users, as in all tool-use contexts, variations are found in the shape, structure, location and hardness of the food, as well as in the size, hardness, weight and color of the tools. This requires additional flexibility in the individuals using the tools, as this will increase with the number of different tool-use contexts. This might contribute to the observations that tool use in birds is limited to clearly defined contexts, normally that of extracting insects and grubs from holes in branches, and the time delay to solution is relatively short. By contrast, chimpanzees use tools in a wide variety of contexts, such as during social play, to attain food, for defense and hygienic purposes and as a means of communication (Goodall, 1970; Boesch & Boesch, 1990). The more tools that are made from different materials and used in different contexts, the more the individuals will have to make decisions specific to each situation. Nut cracking, for example, is not only about hitting a nut with a hard hammer, but the hardness, the weight and the color of each hammer is different, while the placement of the nut on the anvil, and therefore its stability and accessibility (the angle to hit it), will vary as well. In reality this makes each situation different from the previous ones and represents a new challenge each time.

As can be seen in Table 2.1, the cognitive demands of sequential tool use have reached a challenging level of complexity, and it is not surprising that such an ability was previously thought to be restricted to humans; this also used to be thought of the ability to extract underground food resources (Hatley & Kappelman, 1980; Laden & Wrangham, 2005). Interestingly, following reviews of cognitive achievements such as those demonstrated from comparative experiments with captive monkeys and apes, the "unobservability hypothesis" states that only humans, but none of the non-human primates, are able to reason and understand causality about unobservable entities (Povinelli, 2000; Penn & Povinelli, 2007; Penn *et al.*, 2008; Vonk & Subiaul, 2009). If true, this would match nicely with the proposed dichotomy in the ability to extract unseen resources with sequential tool techniques. However, if this hypothesis might apply to some captive individuals that have experienced only very deprived ecological conditions throughout their lives, it seems totally at odds with the new discoveries of underground resource exploitation by chimpanzees in Central Africa, as well as some Tanzanian chimpanzees who have been observed to extract underground tubers (Sanz *et al.*, 2004; Hernandez-Aguilar, 2007; Boesch *et al.*, 2009; Sanz & Morgan, 2009). This is not the only hypothesis constructed from captive experimental works that is hard to reconcile with new evidence of sequential tool use in wild chimpanzees; for example, some have suggested that captive chimpanzees show strong limitations in causal reasoning compared to humans (Premack, 2007), while others have argued that great apes are unable to flexibly generalize from one task to another equivalent one (Martin-Ordas *et al.*, 2008). In particular, the flexibility of wild chimpanzees in applying sequential tool solutions to such different but equivalent tasks as the extraction of kernels from nuts, honey from both tree and underground bee hives, soldiers from underground termite mounds and soldiers from deep driver ant nests seems impossible to reconcile with such cognitive limitations.

An attempt to experimentally identify the cognitive requirements for sequential tool use has been tried with Caledonian crows (Taylor *et al.*, 2007; Wimpenny *et al.*, 2009). However, due to the limitations of captivity, the procedures developed did not faithfully mimic natural conditions, but instead forced the subjects to adopt a sequential approach not by nature of the challenge, but simply by making them physically unable to directly reach "tool 2," which would have allowed them to directly get at the food reward. The experimental setup forced the birds to select first tool 1 to extract tool 2 from a tube before being able to use tool 2 to reach the food. Therefore, the experimental procedure not only imposed an artificial sequential sequence, but also imposed the order in which the tools could be used, so that the sequence was imposed by the procedure rather than being a decision made by the subject when facing a demanding ecological challenge. Such experiments did not really test the ability of crows to spontaneously understand sequential tool use. In the real world, it is the subject itself that has to decide about the sequence and order of use of the different tools. Second, in the real world, the sequence is often more complex due to the fact that it is not length alone that is important, but rather a combination of length, width and weight. When underground termite fishing, the Goualougo chimpanzees insert a thin fishing probe into a tunnel made with a wider, stout puncturing tool (Sanz & Morgan, 2009). Similarly, when gathering tree honey, a thinner and longer collector or fluid dip is used only after a short, heavy, thick pounder has broken open the bee hive (Boesch *et al.*, 2009; Sanz & Morgan, 2009). The striking feature of natural sequential tool use is that chimpanzees perform them with a clear structure in mind as they follow a precise sequence, despite encountering numerous and important difficulties in attaining sub-goals; different types of branches can be used as pounders in Central African forests, and when the first pounder does not succeed in opening the nest, the chimpanzees continue selecting only branches that could function as pounders until the opening is made (often keeping unsuccessful pounders for later reuse; Sanz & Morgan, 2009). Similarly, chimpanzees in Mahale National Park, Tanzania, have been observed to regularly arrive at ant fishing sites carrying up to three manufactured fishing probes with them, as the probes wear out rapidly when used, which the chimpanzees were anticipating (Nishida & Hiraiwa, 1982; personal observation). From video recordings, Goualougo chimpanzees arrived sometimes at termite nests carrying two readily prepared types of sticks (Sanz *et al.*, 2004), demonstrating that they correctly anticipated the future need of sequential tools.

Viewing tool use as a problem-solving task and considering that the problems are set by the ecological niche of each population predicts a less rigid relationship between tool-use skills and phylogeny, as illustrated by the extraordinary observations of New Caledonian crows manufacturing and using complex tools (Hunt, 1996; Kenward *et al.*, 2005; Hunt *et al.*, 2006). Similarly, we should expect that some ecological challenges in nature might require similar causal understanding without requiring a tool and, therefore, the observations that some tool-use abilities have already evolved in non-tool-using primates is not totally unexpected (Santos *et al.*, 2003, 2006). Furthermore, the evolution of a complex causal understanding of external objects in chimpanzees has been proposed to be selected by hunting, as this requires thinking, anticipating and reacting to the movements of the prey species (Boesch & Boesch-Achermann, 2000). In particular, the transfer of hunting skills to tool-use situations that are perceptually quite different but conceptually quite similar would favor the development of the type of flexibility that might be important for acquisition of flexible tool use as well as the sophisticated sequential tool use seen in chimpanzees.

Tool use as a cultural expression

According to the "eco-cultural model" (Segall *et al.*, 1999; Carpendale & Lewis, 2004; Boesch, 2007), we should expect not only the ecology to be fundamental in explaining the tool use observed within a species, but also the cultural environment to shape tool use. The best-studied example of such a cultural tool use is the nut-cracking behavior observed in chimpanzees (Sugiyama & Koman, 1979; Boesch & Boesch, 1984, 1990; Boesch *et al.*, 1994; Boesch, 2003). The chimpanzees of the Taï forest crack five species of nuts and the chimpanzees of Bossou, Guinea, crack an additional species. All of these species of nuts have a large distribution range within the tropical forests of West and Central Africa, going from Sierra Leone–Liberia to Gabon– Congo. *Coula edulis* is one of the most abundant canopy trees in these forests and also has one of the least hard nuts of those that the chimpanzees crack. Furthermore, observations in the Taï forest have shown that adolescent individuals have enough strength to crack them open directly with their teeth (e.g., Tina, an adolescent female with handicapped hands was seen to crack them successful for many months; Boesch & Boesch-Achermann, 2000). Despite this, no evidence of nut cracking has been found in *Coula*-rich forests east of the Sassandra River in Côte d'Ivoire, where chimpanzees live (see Figure 2.4), or in many forested regions of Gabon, Tanzania, Uganda and Congo. Only in the Ebo forest in southern Cameroon were a handful of observations of nut cracking reported (Morgan & Abwe, 2006). These observations made nut cracking one of the most cited cultural tool-using behaviors (Boesch, 2003).

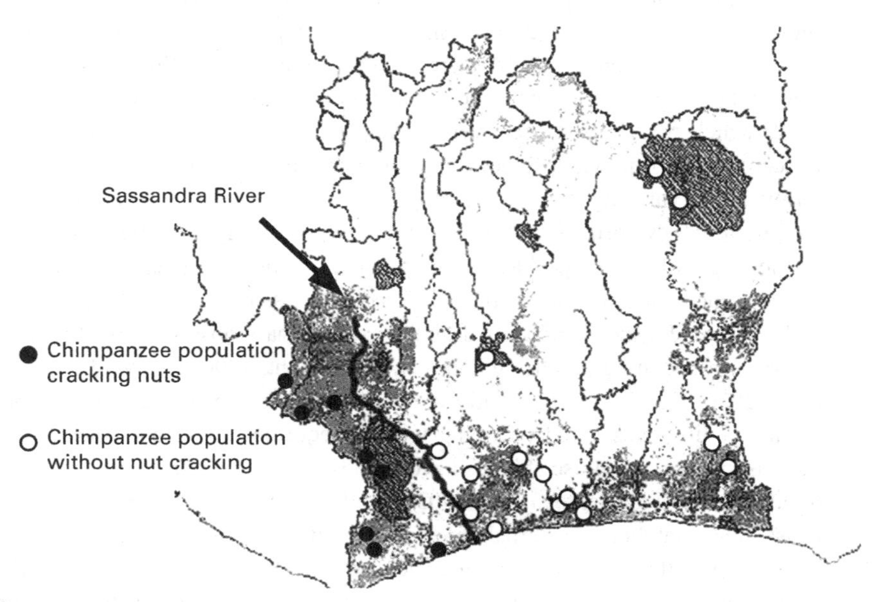

Figure 2.4 Nut cracking has been found in wild chimpanzees of Côte d'Ivoire only on the western side of the Sassandra River, despite the fact that *Coula edulis* and *Panda oleosa* nuts are found in all forests in the south of the country. Twenty-two locations in the country with the confirmed presence of chimpanzees have been visited for signs of nut-cracking sites in 1994–1995 (Boesch *et al.*, 1994).

Ant dipping in chimpanzees has also been shown to be significantly affected by cultural preferences when this behavior is compared between different populations. Two main ant-dipping techniques have been identified. First, the chimpanzees of Gombe National Park in Tanzania have been described to use relatively long wands to dip for the soldiers or driver ants directly from the nest entrance and then use their two hands to bring the ants into their mouths (Goodall, 1968, 1970; McGrew, 1974). This technique is very efficient as it allows individuals to capture about 700 ants per minute (Boesch & Boesch, 1990). Second, the chimpanzees of Taï National Park in Côte d'Ivoire have been described to use much shorter sticks to dip the soldiers out of the nest entrance and then put the end of the stick directly into their mouth to remove the ants with their teeth (Boesch & Boesch, 1990). This technique is proven to be much less efficient, as only about 125 ants are captured per minute, but this method is nevertheless systematically used by all group members of four chimpanzee communities observed in Taï forest.

A precise comparison of the techniques used by chimpanzees in the Taï forest and in Bossou, Guinea, has confirmed that some aspects of the way the chimpanzees in Bossou dip for ants are influenced by the species of ants they dipped for, as well as the location they dipped for them (Möbius *et al.*, 2008). However, most of the differences observed between the Bossou and Taï populations were not explained by ecological differences (Möbius *et al.*, 2008). A larger review of chimpanzee populations has supported that conclusion in the sense that we see a mixture of ecological effects as well as important cultural effects in the presence of the ant-dipping behavior (Schöning *et al.*, 2008). Therefore, in agreement with the prediction of the eco-cultural model, tool use in chimpanzees is affected by both the ecology experienced by group members as well as by the specific cultural environment they face (Boesch, 2007, 2010).

Cumulative cultural evolution in tool use

Complex cultural tool use as observed in nut cracking and honey extraction strongly suggest a cumulative cultural process. Let me take here the example of nut cracking, and include the observations of tool technology from different chimpanzee populations. We can see that flailing insects, conspecifics or snakes with branches or saplings has been reported from all chimpanzee populations, as has clubbing playmates or social competitors with wooden sticks (named as "universal " in Figure 2.5). One innovation was added to this universal in Gombe and Taï chimpanzees when they used tree trunks or roots to pound hard fruit against; this is similar to clubbing, but in this case, the club was a different sort of object and served a different purpose, namely to access food (Figure 2.5). A second innovation was seen in West Africa, where hammers were used to pound on food that had been placed on a hard surface. From this last innovation, two possible further novelties, both examples of sequential tool use, were incorporated: the Taï chimpanzees added a stick in order to extract nut remains from cracked shells; and the Bossou chimpanzees used mobile anvils on which they placed nuts. A final innovation that was observed a few times in the

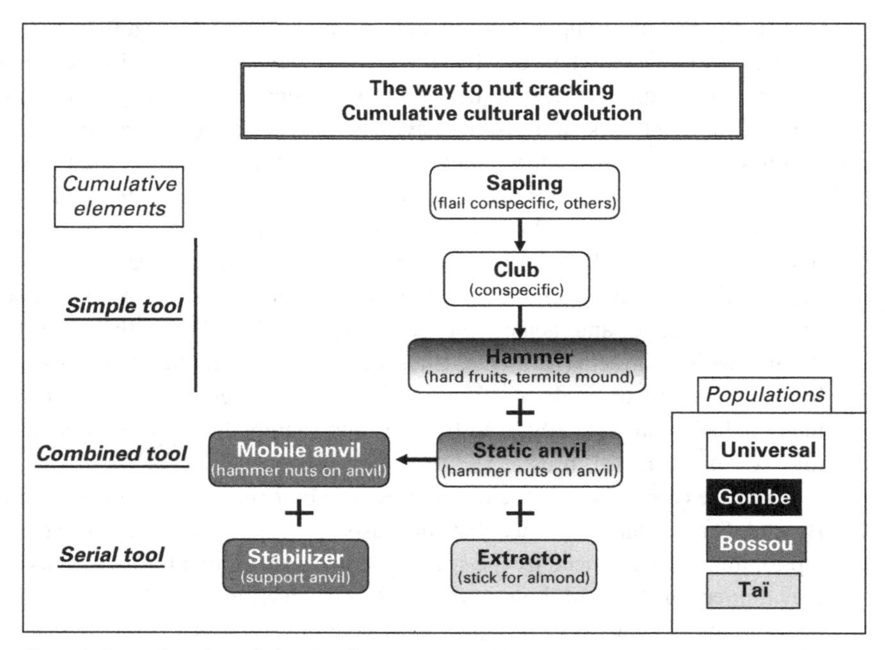

Figure 2.5 Cumulative cultural evolution leading to nut-cracking technique, whereby a behavior seen in all chimpanzee populations is elaborated through successive additions of new, different elements. First, a simple tool is integrated into the pounding movements, as seen in Gombe, Bossou and Taï chimpanzees. Then, the introduction of a second tool led to the invention of a combined tool use as seen in Taï and Bossou chimpanzees. For each step of the accumulation process seen in a given population, I have listed the different resources gained or involved.

Bossou chimpanzees was the placement of a stone under the mobile anvil to stabilize it before hitting the nut with the hammer (Figure 2.5). For each step in the cumulative process, the tool technique was efficient to improve the access to a given food resource and is still used effectively in at least one chimpanzee population, but it was later elaborated for a new food source. Thus, nut cracking in chimpanzees resulted from a cumulative cultural process developing from a universal, simple chimpanzee technique.

The cumulative innovations were most likely not due to one single individual using them, for the different techniques are all still in use in different populations. As the Taï chimpanzees are separated from these other groups by about 3000 km and several major rivers, it would be unreasonable to assume one inventor for all these innovations. Thus, this might be a very good example of what has been labeled a "collective cast of mind" in the human literature, whereby the invention of one group member that has disseminated to the whole group could be improved by another invention from another individual, and that collective product would then disseminate again to the whole group before being improved once again (Boesch, 2012). This accumulation of innovations, which improves the product at each step, is what has been labeled a "cumulative cultural evolution."

Conclusion

The observations of sequential tool use that have recently been observed in Central African chimpanzees could become the flagship example of how much can be missed about the behavioral and cognitive abilities of a species when important ecological niches of an animal species' range are not studied. After over four decades of field studies on wild chimpanzees, new studies following chimpanzees in the Congo basin regions have uncovered such sophisticated tool-use skills. Following the "eco-cultural model," we must expect that the greater number of different ecological niches a species experiences, the greater number of different solutions they will develop. Thus, we must expect more new discoveries about chimpanzees, as they are one of the primate species with the largest distribution ranges in Africa, ranging from dry savannah regions, like in Senegal and Mali, to the deepest rainforests in Côte d'Ivoire, Congo, Gabon and the DRC.

The tool-use framework discussed here has, in conjunction with some careful captive experiments, allowed us to pinpoint some important technical skills that require cognitive skills that, at present, seem to explain some species differences. The number of spatial relationships that must be controlled during tool use seems to place important cognitive demands on the individual performing it and could explain why, at present, no sequential tool use has been observed in wild capuchin monkeys. Similarly, the flexibility demands of natural ecological challenges in terms of synchronicity of actions, the dynamic dimension of the actions itself, as well as the specificity of the actions requiring more flexibility are all important demands on the individuals that need to be solved for tool use to be successful. The discoveries of sequential tool use and cumulative culture evolution in wild chimpanzees have shed new light on their tool-use abilities. Further new discoveries on chimpanzee skills are overdue and will help us to understand about their abilities and the influence of ecology on such technical behavior.

Viewing tool use as a way to solve natural feeding challenges has proven to have good explanatory power for the distribution of tool-using abilities in the animal kingdom, as it does not relate directly to phylogenetic proximity. Furthermore, the food-storing abilities of certain birds have already pointed to the need to adopt an ecological approach to cognition (Emery & Clayton, 2004), and not stick strictly to a phylogenetic viewpoint based on the anthropocentric belief that, because humans are the best tool users and the most intelligent species, those species most closely related to us must be more elaborate in these dimensions than others. At present, the ecological approach favored by the technical intelligence hypothesis or the eco-cultural model seems much better supported. This also supports the view that we need to be more modest about claiming species differences in such abilities, as long as we have not gained a representative knowledge of ecological cognitive responses of each of those species (see, for example, Boesch, 2010; Heinrich *et al.*, 2010). Such an approach also has the potential future advantage of allowing us to identify those factors that were important to explain the emergence of human technical skills and cognition. The ecological approach suggested here would certainly predict a domain-general view of the evolution of cognition, as is also demonstrated by the presence of tool-use skills in non-tool-using monkeys and birds (Santos *et al.*, 2003, 2006; Huber & Gajdon, 2006).

Tool use as a problem-solving task must be studied and thought about in relation to the ecological challenges a species faces, and the total disconnection of tool-use abilities from the natural challenges make some captive experimental studies difficult to interpret and use to understand real-world animal abilities. I realize that, for birds, the coupling of captive and wild studies is easier because of the increased possibility of producing ecologically valid settings, and that this is more difficult for large mammal species like the great apes. In addition, it may be impossible in captive conditions to mimic the complex technical challenges encountered in the wild. In that light, discussing causal cognition by focusing solely on work done with laboratory animals will, in essence, totally underestimate what animals are capable of (Tomasello *et al.*, 2005; Penn & Povinelli, 2007; Penn *et al.*, 2008), and, therefore, face the risk of being restricted to understand the detrimental effect of captivity on animal cognition (Bard *et al.*, 2005; Leavens *et al.*, 2005; Boesch, 2007, 2010; Ijzendoorn *et al.*, 2009; Lyn *et al.*, 2010). Some species differences could potentially be understood if the experimental designs were exactly the same (e.g., Santos *et al.*, 2006; Fragaszy *et al.*, 2009), but this would only be the case if we assume that the species being compared will react similarly to captive conditions, an assumption that has yet to be proven. Too often, the methods used in the experiments diverge in some ways and this leads to very contradictory results (see, for example, Leavens *et al.*, 2005; Boesch, 2007; Seed *et al.*, 2009b; Lyn *et al.*, 2010) and never-ending discussions about which methods reveal the essence of the animal skills. An enlightening example is the 21-year-old famous trap-tube test developed to study causality understanding in different animal species, by presenting them with a food bait that needed to be pushed out with a tool from a transparent tube while avoiding a trap in which the food would fall (Visalberghi & Trinca, 1989; Visalberghi & Limongelli, 1994). This tube test was very successful and used in many studies with many species, but with hugely contradictory results (e.g., Povinelli, 2000; Martin-Ordas & Call, 2009; Seed *et al.*, 2009b), so that readers reach total confusion as each minimal change in the experimental procedures can lead to major changes in the performance of the animals. In careful experiments done with humans, it could be shown that gratuitous and unsupported assumptions about how humans' folk physics knowledge works was at the base of the imprecision in the experiments (where too many variables were intervening at the same time and could be perceived with different importance), as well as of the errors in the interpretations (as we should expect different responses when following different rules) (Silva *et al.*, 2005; Silva & Silva, 2006).

New observations and discoveries of tool use in macaques, capuchin monkeys, orangutans and chimpanzees will continue to emerge as we study an increasing number of different populations of those species, and they will make us alter our view of species differences in cognition as well as for tool use. This is a very exciting perspective and we hope that human destruction of nature will not prevent this from happening!

References

Allen, C. (2002). A skeptic's progress. *Biology and Philosophy*, **17**, 695–702.

Ambrose, S. (2001). Paleolithic technology and human evolution. *Science*, **291**, 1748–1753.

Ambrose, S. (2010). Coevolution of composite-tool technology, constructive memory and language: implications for the evolution of modern human behavior. *Current Anthropology*, **51**(1), S135–S149.

Bar-Yousef, O. & van Peer P. (2009). The chaîne opératoire approach in middle Paleolithic archeology. *Current Anthropology*, **50**(1), 103–117.

Bar-Yousef, O., Vandermeersch, B., Arensburg, B., *et al.* (1992). The excavations in Kebara Cave, Mt. Carmel. *Current Anthropology*, **33**, 497–534.

Bard, K., Myowa-Yamakoshi, M., Tomonaga, M., *et al.* (2005). Group differences in the mutual gaze of chimpanzees (*Pan troglodytes*). *Developmental Psychology*, **41**(4), 615–624.

Barrett, L., Henzi, P. & Rendall, D. (2007). Social brains, simple minds: does social complexity really require cognitive complexity. *Philosophical Transcriptions of the Royal Society of London B*, **362**, 561–575.

Beck, B. B. (1980). *Animal Tool Behavior*. New York: Garland Press.

Bluff, L., Weir, A., Rutz, C., Wimpenny, J. & Kalcelnik, A. (2007). Tool-related cognition in New Caledonian crows. *Comparative Cognition and Behavior Reviews*, **2**, 1–25.

Bluff, L., Troscianko, J., Weir, A., Kalcelnik, A. & Rutz, C. (2010). Tool use by wild New Caledonian crows *Corvus moneduloides* at natural foraging sites. *Proceedings of the Royal Society of London B*, **277**, 1377–1385.

Boesch, C. (2003). Is culture a golden barrier between human and chimpanzee? *Evolutionary Anthropology*, **12**, 26–32.

Boesch, C. (2007). What makes us human (*Homo sapiens*)? The challenge of cognitive cross-species comparison. *Journal of Comparative Psychology*, **121**(3), 227–240.

Boesch, C. (2010). Away from ethnocentrism and anthropocentrism: towards a scientific understanding of "what makes us human." *Behavioral and Brain Sciences*, **33**(2/3), 26–27.

Boesch, C. (2012). From material to symbolic cultures: culture in primates. In J. Valsiner (ed.) *The Oxford Handbook of Culture and Psychology*. Oxford: Oxford University Press.

Boesch, C. & Boesch, H. (1984). Mental map in wild chimpanzees: an analysis of hammer transports for nut cracking. *Primates*, **25**, 160–170.

Boesch, C. & Boesch, H. (1990). Tool use and tool making in wild chimpanzees. *Folia Primatologica*, **54**, 86–99.

Boesch, C. & Boesch-Achermann, H. (2000). *The Chimpanzees of the Taï Forest*. Oxford: Oxford University Press.

Boesch, C., Marchesi, P., Marchesi, N., Fruth, B. & Joulian, F. (1994). Is nut cracking in wild chimpanzees a cultural behaviour? *Journal of Human Evolution*, **26**, 325–338.

Boesch, C., Head, J. & Robbins, M. (2009). Complex toolsets for honey extraction among chimpanzees in Loango National Park, Gabon. *Journal of Human Evolution*, **56**, 560–569.

Byrne, R. (1995). *The Thinking Ape*. Oxford: Oxford University Press.

Byrne, R. (1997). The technical intelligence hypothesis: an additional evolutionary stimulus to intelligence? In A. Whiten & W. Byrne (eds.) *Machiavellian Intelligence II: Extensions and Evaluations* (pp. 289–311). Cambridge: Cambridge University Press.

Byrne, R. (2004). The manual skills and cognition that lie behind hominid tool use. In A. Russon & D. Begun (eds.) *The Evolution of Thought: Evolutionary Origin of Great Ape Intelligence* (pp. 31–44). Cambridge: Cambridge University Press.

Canale, G., Guidorizzi, C., Kierulff, M. and Gatto, C. (2009). First record of tool use by wild populations of the yellow-breasted capuchin monkey (*Cebus xanthosternos*) and new records for the bearded capuchin (*Cebus libidinosus*). *American Journal of Primatology*, **71**, 366–372.

Carpendale, J. & Lewis, C. (2004). Constructing an understanding of mind: the development of children's social understanding within social interaction. *Behavioral and Brain Sciences*, **27**, 79–151.

Carvalho, S., Cunha, E., Sousa, C. & Matsuzawa, T. (2008). Chaines opératoires and resource-exploitation strategies in chimpanzee (*Pan troglodytes*) nut cracking. *Journal of Human Evolution*, **55**, 148–163.

Cipolletta, C., Spagnolietti, N., Todd, A., *et al.* (2007). Termite feeding behavior of western gorillas (*Gorilla gorilla gorilla*) at Bai Hokou, Central African Republic. *International Journal of Primatology*, **28**, 457–476.

Dietrich, S., Toth, N., Schick, K. & Chaminade, T. (2008). Neural correlates of early Stone Age toolmaking: technology, language and cognition in human evolution. *Philosophical Transactions of the Royal Society of London B*, **363**, 1939–1949.

Emery, N. & Clayton, N. (2004). Mentality of crows: convergent evolution of intelligence in corvids and apes. *Science*, **306**, 1903–1907.

Fragaszy, D. (2007). Relational spatial reasoning and tool use in capuchin monkeys. *A Primatologia no Brasil*, **10**, 521–546.

Fragaszy, D. & Visalberghi, E. (2004). Socially biased learning in monkeys. *Learning and Behavior*, **32**(1), 24–35.

Fragaszy, D., Kennedy, E., Murnane, A., *et al.* (2009). Navigating two-dimensional mazes: chimpanzees (*Pan troglodytes*) and capuchins (*Cebus apella* sp.) profit from experience differently. *Animal Cognition*, **12**, 491–504.

Goodall, J. (1968). Behaviour of free-living chimpanzees of the Gombe Stream area. *Animal Behaviour Monograph*, **1**, 163–311.

Goodall, J. (1970). Tool-using in primates and other vertebrates. In D. S. Lehrmann, R. A. Hinde & E. Shaw (eds.) *Advances in the Study of Behavior*, Vol. **3** (pp. 195–249). New York: Academic Press.

Goodrich, G. & Allen, C. (2007). Conditioned anti-anthropomorphism. *Comparative Cognition and Behavior Reviews*, **2**, 147–150.

Greenfield, P. M. (1991). Language, tools, and the brain: the ontogeny and phylogeny of hierarchically organized sequential behavior. *Behavioral and Brain Sciences*, **14**, 531–595.

Gumert, M., Kluck, M. & Malaivijitnond, S. (2009). The physical characteristic and usage patterns of stone axe and pounding hammers used by long-tailed macaques in the Andaman Sea region of Thailand. *American Journal of Primatology*, **71**, 594–608.

Hansell, M. & Ruxton, G. (2008). Setting tool use within the context of animal construction behaviour. *Trends in Ecology and Evolution*, **23**(2), 73–78.

Hatley, T. & Kappelman, J. (1980). Bears, pigs, and Plio-Pleistocene hominids: a case for the exploitation of belowground food resources. *Human Ecology*, **8**, 371–387.

Heinrich, J., Heine, S. & Norenzayan, A. (2010). The weirdest people in the world? *Behavioral and Brain Sciences*, **3**, 1–75.

Hernandez-Aguilar, A., Moore, J. & Pickering, T. (2007). Savanna chimpanzees use tools to harvest the underground storage organs of plants. *Proceeding of the National Academy of Sciences USA*, **104**(49), 19210–19213.

Herrmann, E., Wobber, V. & Call, J. (2008). Great apes' (*Pan troglodytes*, *Pan paniscus*, *Gorilla gorilla*, *Pongo pygmaeus*) understanding of tool functional properties after limited experience. *Journal of Comparative Psychology*, **122**(2), 220–230.

Huber, L. & Gajdon, G. (2006). Technical intelligence in animals: the kea model. *Animal Cognition*, **9**, 295–305.

Hunt, G. (1996). Manufacture and use of hook-tools by New Caledonian crows. *Nature*, **379**, 249–251.

Hunt, G., Rutledge, R. & Gray, R. (2006). The right tool for the job: what strategies do wild New Caledonian crows use? *Animal Cognition*, **9**, 307–316.

Ijzendoorn, M. van, Bard, K., Bakermans-Kranenberg, M. & Ivan, K. (2009). Enhancement of attachment and cognitive development of young nursery-reared chimpanzees in responsive versus standard care. *Developmental Psychobiology*, **51**, 173–185.

Inoue-Nakamura, N. & Matsuzawa, T. (1997). Development of stone tool use by wild chimpanzees (*Pan troglodytes*). *Journal of Comparative Psychology*, **111**(2), 159–173.

Keeley, B. (2004). Anthropomorphism, primatomorphism, mammalomorphism: understanding cross-species comparisons. *Biology and Philosophy*, **19**, 521–540.

Kenward, B., Weir, A., Rutz, C. & Kacelnik, A. (2005). Tool manufacture by naive juvenile crows. *Nature*, **433**, 121.

Kummer, H. (1968). *Social Organization of Hamadryas Baboons: A Field Study*. Chicago, IL: University of Chicago Press.

Laden, G. & Wrangham, R. (2005). The rise of the hominids as an adaptive shift in fallback foods: plant underground storage organs (USOs) and australopith origins. *Journal of Human Evolution*, **49**, 482–498.

Leavens, D., Hopkins, W. & Bard, K. (2005). Understanding the point of chimpanzee pointing: epigenesis and ecological validity. *Current Direction in Psychological Science*, **14**(4), 185–189.

Lyn, H., Russell, J. & Hopkins, W. (2010). The impact of environment on the comprehension of declarative communication in apes. *Psychological Science*, **21**(3), 360–365.

Martin-Ordas, G. & Call, J. (2009). Assessing generalization within and between trap tasks in the great apes. *International Journal of Comparative Psychology*, **22**, 43–60.

Martin-Ordas, G., Call, J. & Colmenares, F. (2008). Tubes, tables and traps: great apes solve two functionally equivalent trap tasks but show no evidence of transfer across tasks. *Animal Cognition*, **11**, 423–430.

Matsuzawa, T. (2001). Primate foundations of human intelligence: a view of tool use in nonhuman primates and fossil hominids. In T. Matsuzawa (ed.) *Primate Origins of Human Cognition and Behavior* (pp. 3–25). Tokyo: Springer-Verlag.

McGrew, W. (1974). Tool use by wild chimpanzees in feeding upon driver ants. *Journal of Human Evolution*, **3**, 501–508.

McPherron, S. (2000). Handaxes as a measure of the mental capabilities of early hominids. *Journal of Archaeological Science*, **27**, 655–663.

Mercader, J., Barton, H., Gillespie, J., *et al.* (2007). 4,300-year-old chimpanzee sites and the origins of percussive stone technology. *Proceedings of the National Academy of Sciences USA*, **104**(9), 3043–3048.

Mithen, S. (1996). *The Prehistory of Mind: The Cognitive Origin of Art and Science*. London: Thames and Hudson.

Möbius, Y., Boesch, C., Koops, K., Matsuzawa, T. & Humle, T. (2008). Cultural differences in army ant predation by West African chimpanzees? A comparative study of microecological variables. *Animal Behaviour*, **76**, 37–45.

Morgan, B. & Abwe, E. (2006). Chimpanzees use stone hammers in Cameroon. *Current Biology*, **16**(16), R632–R633.

Moura, A. C. & Lee, P. C. (2004). Capuchin stone tool use in Caatinga dry forest. *Science*, **306**, 1909.

Mulcahy, N. J. & Call, J. (2006). How great apes perform on a modified trap-tube task. *Animal Cognition*, **9**, 193–199.

Nishida, T. & Hiraiwa, M. (1982). Natural history of a tool-using behaviour by wild chimpanzees in feeding upon wood-boring ants. *Journal of Human Evolution*, **11**, 73–99.

O'Connell, S. & Dunbar, R. (2005). The perception of causality in chimpanzees (*Pan spp.*). *Animal Cognition*, **8**, 60–66.

Penn, D. & Povinelli, D. (2007). Causal cognition in human and nonhuman animals: a comparative, critical review. *Annual Review of Psychology*, **58**, 97–118.

Penn, D., Holyoak, K. & Povinelli, D. (2008). Darwin's mistake: explaining the discontinuity between human and nonhuman minds. *Behavioral and Brain Sciences*, **31**, 109–178.

Povinelli, D. (2000). *Folk Physics for Apes: The Chimpanzee's Theory of How the World Works*. Oxford: Oxford University Press.

Premack, D. (2007). Human and animal cognition: continuity and discontinuity. *Proceedings of the National Academy of Sciences USA*, **104**(35), 13861–13867.

Santos, L., Miller, C. & Hauser, M. (2003). Representing tools: how two non-human primate species distinguish between the functionally relevant and irrelevant features of a tool. *Animal Cognition*, **6**, 269–281.

Santos, L., Pearson, H., Spaepen, G., Tsao, F. & Hauser, M. (2006). Probing the limits of tool competence: experiments with two non-tool-using species (*Cercopithecus aethiops* and *Saguinus oedipus*). *Animal Cognition*, **9**, 94–109.

Sanz, C. & Morgan, D. (2009). Flexible and persistent tool-using strategies in honey-gathering by wild chimpanzees. *International Journal of Primatology*, **30**, 411–427.

Sanz, C., Morgan, D. & Gulick, S. (2004). New insights into chimpanzees, tools, and termites from the Congo Basin. *American Naturalist*, **164**(5), 567–581.

Schaik, C. van & Knott, C. (2001). Geographic variation in tool use on *Neesia* fruits in orangutans. *American Journal of Physical Anthropology*, **114**, 331–342.

Schöning, C., Humle, T., Möbius, Y. & McGrew, W. (2008). The nature of culture: technological variation in chimpanzee predation on army ants revisited. *Journal of Human Evolution*, **55**, 48–59.

Seed, A., Emery, N. & Clayton, N. (2009a). Intelligence in corvids and apes: a case of convergent evolution? *Ethology*, **115**, 410–420.

Seed, A., Call, J., Emery, N. & Clayton, N. (2009b). Chimpanzees solve the trap problem when the confound of tool-use is removed. *Journal of Experimental Psychology*, **35**(1), 23–34.

Segall, M., Dasen, P., Berry, J. & Poortinga, Y. (1999). *Human Behavior in Global Perspective: An Introduction to Cross-Cultural Psychology*. 2nd edn. New York: Pergamon Press.

Sharon, G. (2009). Acheulian giant-core technology: a worldwide perspective. *Current Anthropology*, **50**(3), 335–367.

Silva, F. & Silva, K. (2006). Humans' folk physics is not enough to explain variations in their tool-using behavior. *Psychonomic Bulletin and Review*, **13**(4), 689–693.

Silva, F., Page, D. & Silva, K. (2005). Methodological-conceptual problems in the study of chimpanzees' folk physics: how studies with adult humans can help. *Learning and Behaviour*, **53**(1), 47–58.

Souto, A., Bione, C., Bastos, M., *et al.* (2011). Critically endangered blonde capuchins fish for termites and use new techniques to accomplish the task. *Biology Letters*, **7**, 532–535.

Spagnoletti, N., Visalberghi, E., Ottoni, E., Izar, P. & Fragazy, D. (2011). Stone tool use by adult wild bearded capuchin monkeys (*Cebus libidinosus*): frequency, efficiency and tool selectivity. *Journal of Human Evolution*, **61**, 97–107.

Spencer, J., Smith, L. & Thelen, E. (2001). Tests of a dynamic systems account of the A-not-B error: the influence of prior experience on the spatial memory abilities of two-year-olds. *Child Development*, **72**, 1327–1346.

Sugiyama, Y. & Koman, J. (1979). Tool-using and -making behavior in wild chimpanzees at Bossou, Guinea. *Primates*, **20**, 513–524.

Sumita, K., Kitahara-Frisch, J. & Norikoshi, K. (1985). The acquisition of stone-tool use in captive chimpanzees. *Primates*, **26**(2), 168–181.

Taylor, A., Hunt, G., Holzhaider, J. & Gray, R. (2007). Spontaneous metatool use by New Caledonian crows. *Current Biology*, **17**, 1504–1507.

Tomasello, M. & Call, J. (1997). *Primate Cognition*. Oxford: Oxford University Press.

Tomasello, M., Carpenter, M., Call, J., Behne, T. & Moll, H. (2005). Understanding and sharing intentions: the origins of cultural cognition. *Behavioral and Brain Sciences*, **28**, 675–691.

Toth, N. & Schick, K. (1993). Early stone industries and inferences regarding language and cognition. In K. Gibson & T. Ingold (eds.) *Tools, Language and Intelligence: Evolutionary Implications* (pp. 346–362). Cambridge: Cambridge University Press.

Visalberghi, E. & Fragaszy, D. (2006). What is challenging about tool use? The capuchin's perspective. In E. A. Wasserman & T. R. Zentall (eds.) *Comparative Cognition: Experimental Explorations of Animal Intelligence* (pp. 529–552). New York: Oxford University Press.

Visalberghi, E. & Limongelli, L. (1994). Lack of comprehension of cause–effect relations in tool-using capuchin monkeys (*Cebus apella*). *Journal of Comparative Psychology*, **108**, 15–22.

Visalberghi, E. & Tomasello, M. (1998). Primate causal understanding in the physical and psychological domains. *Behavioural Processes*, **42**, 189–203.

Visalberghi, E. & Trinca, L. (1989). Tool use in capuchin monkeys: distinguishing between performance and understanding. *Primates*, **30**, 511–521.

Visalberghi, E., Fragaszy, D., Ottoni, E., *et al.* (2007). Characteristics of hammer stones and anvils used by wild bearded capuchin monkeys (*Cebus libidinosus*) to crack open palm nuts. *American Journal of Physical Anthropology*, **132**, 426–444.

Visalberghi, E., Addessi, E., Truppa, V., *et al.* (2009). Selection of effective stone tools by wild bearded capuchin monkeys. *Current Biology*, **19**, 213–217.

Vonk, J. & Subiaul, F. (2009). Do chimpanzees know what others can and cannot do? Reasoning about "capability." *Animal Cognition*, **12**, 267–286.

Wimpenny, J., Weir, A., Clayton, L., Rutz, C. & Kacelnik, A. (2009). Cognitive processes associated with sequential tool use in New Caledonian crows. *PloS ONE*, **4**(8), e6471.

Wolfheim, J. (1983). *Primates of the World: Distribution, Abundance and Conservation*. Seattle, WA: University of Washington Press.

Wolpert, L. (2003). Causal belief and the origins of technology. *Philosophical Transactions of the Royal Society of London A*, **361**, 1709–1719.

Wynn, T. (1993). Layers of thinking in tool behavior. In T. Ingold & K. Gibson (eds.) *Tools, Language and Intelligence: Evolutionary Implications* (pp. 389–406). Cambridge: Cambridge University Press.

Wynn, T. (2002). Archeology and cognitive evolution. *Behavioral and Brain Sciences*, **25**, 389–438.

Wynne, C. (2007). What are animals? Why anthropomorphism is still not a scientific approach to behavior. *Comparative Cognition and Behavior Reviews*, **2**, 125–135.

3 Chimpanzees plan their tool use

Richard W. Byrne

Centre for Social Learning and Cognitive Evolution and Scottish Primate Research Group, School of Psychology and Neuroscience, University of St. Andrews

Crickette M. Sanz

Department of Anthropology, Washington University

David B. Morgan

Lester E. Fisher Center for the Study and Conservation of Apes, Lincoln Park Zoo
Wildlife Conservation Society, Congo Program

Introduction

To a cognitive psychologist, chimpanzee tool use is exciting because of the opportunity it brings to examine how apes deal with a range of challenging situations that in humans would invoke planning. By *planning* it is meant a special kind of problem solving in which an appropriate course of action for the immediate or distant future is worked out by means of mental computation with brain representations of past or present situations (Miller *et al.*, 1960). These include: a working representation of the current situation that presents a problem; episodic memories of specific past instances and events; and semantic knowledge about how things work or how people behave.

Because a tool is not itself a goal-object, but has meaning and functionality only in regard to achieving a goal, problem solving with tools often makes more of the planning process "visible" than is normally the case (Seed & Byrne, 2010). Because a tool often must be selected to meet specific criteria in order to work, or – more telling still – may have to be made from specific raw materials in a particular way, getting an appropriate tool becomes an extra stage in the planning process. Thus, to approach a cognitive understanding of animal planning, studying tool use is by no means the only approach, but it is certainly a good one. Historically, however, understanding cognition has not been the major driving force in the study of tool use in great apes in primatology: that stemmed instead from anthropology, a subject with a very different agenda.

To oversimplify, anthropology's interest in ape tool use has been predicated on one version of the "silver bullet" theory of human origins – specifically, that using a tool provided the magic ingredient that converted an ancient ape to a person. "Man the tool maker," a phrase attributed to Benjamin Franklin, sums it up – and famously, when Jane Goodall told Louis Leakey of her findings on chimpanzee tool manufacture, his first thought was for the re-definition that it would inevitably prompt some people to suggest – to stop chimpanzees

Tool Use in Animals: Cognition and Ecology, eds. Crickette M. Sanz, Josep Call and Christophe Boesch. Published by Cambridge University Press.

being seen as too human! Additionally, for more pragmatic reasons, the archaeology of early man has been dominated by tools because stone tools are so robust and (usually) identifiable, compared to other hominin traces. Paleoanthropology's apparent fixation with tools is no surprise. But these factors have led to disproportionate focus on the tool, not the process of making or using it (with honorable exceptions: e.g., Wynn, 1993, 2002).

Many species of non-human animals also use tools (Shumaker *et al.*, 2011), whereas few make them. Such a disproportion is consistent with the idea that tool making is much more difficult than tool using; and because humans share tool making with few other species, it is tempting and commonplace to make the assumption that tool manufacture must therefore be cognitively challenging. But we suspect that behind that slightly glib assumption there may be a more interesting but often unspoken logic: that tool making is genuinely challenging because it requires *forethought*, whereas tool use might be driven by the stimulus configuration alone: the perceived problem as physically confronts the animal. As the everyday examples of spiders' webs and wasps' nests remind us, complexity of manufacture need not imply complexity of cognition. To investigate whether cognitive complexity is involved requires a psychological rather than an anthropological approach, and it is from the psychology of human problem solving that the analysis of this chapter is derived.

The psychology of animal problem solving

Most studies of human problem solving have used culturally constructed, "artificial" tasks, such as chess or formal logic, and almost all of them have posed problems in symbolic, often verbal form: the next move in chess; the solution of a puzzle in formal logic; the choice of a dinner menu; and so on (e.g., Newell *et al.*, 1958; de Groot, 1965; Newell & Simon, 1972; Byrne, 1977). This restriction on choice of task poses a problem for attempts to use the comparative method to investigate the evolution of planning: the non-verbal behavior of animals simply cannot be assessed in the same way. In order to understand the evolutionary roots of human planning, therefore, cognitive researchers have a choice: either they can attempt to infer mental processes from observable behavior, or they can deny that planning is even possible without language – and give up. The latter approach is obviously defeatist, and suspiciously self-serving when it comes to promotion of human superiority, although it has been much espoused within experimental psychology (e.g., Macphail, 1985, 1998). However, the more empirical alternative is not an easy option.

Under natural circumstances, it is often ambiguous exactly what an animal's goals are. Retrospectively assuming that their goals must have been to achieve what they are eventually seen to achieve has more than a hint of Dr. Pangloss. Unfortunately, the obvious alternative – experimentally setting goals for the animal – has its own problems. To set a goal for a non-human animal, the experimenter employs a schedule of rewards whose distribution correlates with "success" on the task the experimenter wishes to get the animal to attempt. The idea is that the reward schedule gives the animal an understanding – a mental representation – of the task, which it then tries to solve, reaching its goal by planning. Yet an animal merely operating under the hill-climbing, reward-maximization strategy of reinforcement learning would also show behavior that *looks* representation-based and goal-oriented. Moreover, even if we could be sure that genuinely future-directed planning was happening, the planning itself remains invisible and has to be inferred by its consequences. Human studies normally ask subjects to "think aloud." Under some circumstances, forced verbalization may be misleading; but where subjects find concurrently talking to be a natural behavior, their verbal output may be quite closely linked to the symbolic processes that underlie the planning (Byrne, 1983). Even then, there remains ambiguity. If every verbalized step results in overt behavior, fine; but what of long pauses – is the subject thinking furiously, or dozing off? With non-verbal animals, things get much worse. We are then reliant purely on overt, goal-directed behavior; and when we are lucky enough to witness a sequence of behavior which is goal-directed throughout, we are still left with uncertainty in the end as to whether and to what extent the animal *anticipated* the favorable outcome it has apparently worked "toward."

Beset by so many difficulties, denial of the possibility of planning in non-human animals is understandably tempting: if all planning is language-based, then non-human animals do not plan. On this philosophy, everything that we observe animals doing is the result of two hill-climbing algorithms: (1) genetical evolution by natural selection, which has equipped animals with morphology and behavior that has allowed their ancestors to be successful, and therefore probably will still work for them; and (2) associative learning, which allows correlations in the environment to be passively noticed and remembered, and allows the results of trial-and-error exploration to be recorded so that future behavior becomes more effective than past behavior. These processes are gradual ones, not the most efficient we can imagine – but without language, they are believed by many experimental psychologists to be the best that can exist, and they do work.

The earliest version of this belief, the *tabula rasa* idea of behaviorist learning theory, was firmly set aside by the work of ethologists. But subsequently, learning theorists have shifted to the famous "null hypothesis" of Euan Macphail – that all animals learn in the same way, such that animal learning theory, if correctly employed, can account for everything (Macphail, 1985). According to this canon, the only real differences among species lie in their different motor capacities; the varied limits imposed on them by different perceptual systems; and a range of innate predispositions (constraints on learning: Garcia & Koelling, 1966; Garcia *et al.*, 1966) that guide the same learning process to different endpoints in biologically appropriate circumstances. However, in most complex, real-world situations, persistence of the belief that animal learning theory is sufficient to explain changes in behavior is based more on faith than any testable hypothesis of how that could happen (Byrne & Bates, 2006). Worse, this approach is a bit of a dead end when it comes to explaining how the undeniably special *human* planning abilities arose in evolution: that problem simply becomes another one – how human language arose. Indeed, more cynically, the "null hypothesis" might be seen as a rear-guard action to avoid contemplating the unthinkable: the heretical possibility that animals might plan and think in ways that are recognizably like our own. (For example, see the arguments used by Suddendorf *et al.* (2009) to resist the conclusion, from experimental investigations by Mulcahy and Call (2006) and Osvath and Osvath (2008), that apes are able to anticipate their future use of a tool.) The study of ape tool use has had an important role in opening up the possibility of animal planning and thought to serious scrutiny.

A framework for analyzing tool use

Animal tool use is often just *assumed* to imply "advanced abilities," "cognitive sophistication" or "complexity in behavior" – presumably meaning that the animal is showing signs of abilities such as planning, thinking, anticipation of the future or mental computation. However, deciding whether a case of animal tool use requires planning is not straightforward. Life would be simple if our choice were between a theory that predicted intense deliberation, dramatic leaps of insight and preternatural anticipation of every possible problem, versus one that predicted a gradual increase in competence that depended on continual interaction with the world. The trouble is, even adult humans playing chess (and how much more "cognitively sophisticated" can one get?) show continual improvement with practice. Moreover, leaps of insight are often identified afterwards, when one has forgotten the steps that led to them (Byrne, 1975). In chimpanzees, we might try to identify "insight" where the animal has been inactive for a while and then suddenly acts successfully (e.g., Köhler, 1925) – but that risks a selection bias, if cases of prolonged inaction and then sudden failure are ignored. Just as in the study of human problem solving, there is always a risk that inaction may be torpid dozing rather than intense thought.

Perhaps the involvement of planning in tool use is easier to detect when the tool using is part of a tradition, a cultural product? The greatest achievements of human thought and planning have been built on the cultural legacy of others: they are dependent on social learning, including teaching and imitation. The old idea that cultural learning is how *we* do things, while trial-and-error fumbling is how *animals* do them, is now thoroughly discredited. True, teaching has been pretty elusive to document in animals. In the 17 years since the definitive survey of Caro and Hauser (1992), teaching has been firmly established only in meerkats, babblers and ants (Franks & Richardson, 2006; Thornton & McAuliffe, 2006; Raihani & Ridley, 2008). In none of those three positive cases is there any serious suggestion that the teaching is based on any deep (theory of mind) appreciation by the teacher of the cognitive deficiencies of the learner (Csibra & Gergely, 2009). But in non-Western cultures explicit pedagogy may be less important for cultural learning (Gaskins, 2006), and less glamorous social influences of several kinds have been detected in a huge range of non-human species – with some of the sturdiest evidence of imitation and cultural learning coming from the oddest species: coral reef fish, quail and budgerigars (Hoppitt & Laland, 2008). So, for tool use, is planning an essential part of the explanation of imitation and cultural learning? That depends what is meant by those terms. If imitation means only evoking the same behavior as is seen, an action already in the observer's repertoire, used many times before in other contexts (e.g., Whiten *et al.*, 1996; Stoinski *et al.*, 2001), then it can be explained by a very simple mechanism: response facilitation of matching actions (Byrne, 1994; Hoppitt & Laland, 2008). Experimental demonstrations of cultural transmission within chimpanzee groups may be based on no more than this (Whiten, 2005; Whiten *et al.*, 2007). Some aspects of even human cultures may likewise be based on rather simple learning processes: it is not safe to assume that cultural traditions of tool use require planning.

Rather than making assumptions about particular tasks, what is needed in order to evaluate and compare the mental processes underlying behavior – including tool using and tool making – is an appropriate formalism in which seemingly "clever" actions can be evaluated. Such a formalism must be capable of allowing comparison between humans and other animals "on a level playing field" if we are ever to understand the evolution of the human capacities by means of comparative evidence. Because it is no longer used in attempts to explain human abilities, the associationist formalism preferred by learning-theory psychologists just will not do. However, associationism is still so dominant in animal behavior that it cannot just be ignored: When and why is a cognitive approach preferable?

Whether an animal, or some particular behavior it has shown, should be deemed "cognitive" or "associative" is not an empirically decidable issue. Those are *theoretical frameworks* for analysis of what exists, not *kinds* of existence that can be told apart, even in principle: there is no possible acid test between them. In the 1960s, cognitive psychology as a discipline discarded the whole behaviorist, learning-theory framework as unhelpful. Anyone from the cognitive psychology tradition thus unashamedly tended to talk as if animals had mental representations that governed their behavior (e.g., Tomasello & Call, 1997). Many in the field of animal behavior still hanker for the simplicity of association learning theory (e.g., Heyes, 1993). Learning theory is simpler in that fewer mental entities are proposed – but the danger is that these "simple" explanations of complex phenomena are always constructed post hoc (Byrne & Bates, 2006). Like the wares of snake-oil salesmen, associations can be claimed to have cured everything . . . afterwards! With convenient "help" (e.g., retrospectively asserting what the animal will have noticed or not noticed, and thus which entities "must have been" associated), associationist accounts can be made to fit everything that happened – once we know what it was. Where associationist explanations fail is therefore in their *adequacy*. When it comes to the real world or to the rich captive environments, from which apparently "clever" animal performances are often reported, associationist explanation does not work as a coherent system to produce testable predictions. Moreover, often the post hoc explanations of association theory "over-explain": that is, if they did work as explanation for the species in question, it would be difficult to stop them working to "explain" a great deal of complex behavior in other species – complex behavior that actually never happens.

That is not to deny that rules for making links, including associative links, might form a useful part of a cognitive model; but the explanatory "work" of a cognitive model lies also in its organization, whereas the associationist version of animal learning tends to lack such organization. For instance, Heyes and Ray (2000) proposed a theory of animal imitation to account for imitative learning of chains of actions that the animal observes. They described their model as associationist, as if it flowed naturally from traditional animal learning – yet the associations in the model turn out to be of two, qualitatively different, sorts. "Vertical" links are formed between actions-as-seen and actions-as-done. For instance, an animal that watches itself in a mirror is said to form a vertical association between the perceptual representation of, say, its mouth configuration and the motor program that it is currently executing and which produces that configuration. "Horizontal" links are instead formed between successive movements observed in others. By combining the two, Heyes and Ray suggest that an animal can execute a horizontal sequence by using the vertical links to convert what it sees into what it can do. Nowhere is it specified how the system (the animal's brain) "knows" when to make or use vertical and when horizontal links. But it is evident that simple undifferentiated associations would not do: the model has to have structure. As this case illustrates, by the time an "associationist" account has been specified fully enough to meet the criterion of adequacy, it has become a cognitive model. As such, it has presumably lost some of the "simplicity" and "parsimony" that animal learning theorists yearn for. In this chapter, therefore, we will typically use cognitive descriptions (or everyday language) to spell out what must be involved in the chimpanzee behavior we discuss, in terms of the *mental representations* underlying their actions.

Diagnostics of planning

Consider the associationist alternative to planning: what characteristic attributes of behavior would it predict? To a cognitive theorist, that depends on the mental representation it creates. Radical behaviorists would never talk of mental representations, and the concept remains highly suspect to animal learning theorists, a legacy of behaviorism. However, if we allow a modest level of mental structure, once considered perfectly acceptable to learning theorists (Broadbent, 1961), then we can characterize the results of learning purely by association as characteristically *string-like*. The likely "path," through an imaginary network of possible behavior sequences, is determined by past association frequencies between options, along with direct input from the environment. The external inputs that contribute to determine this path are the *immediately perceptible* features of the environment. (If the task were a navigational one, then paths can be interpreted literally, but the logic applies equally to all behavior.) Endpoints are not *anticipated* in such a system.

Thus, a chimpanzee might see a termite mound and – because of past learning – the sight might trigger a chain of associations that leads from earth mound to (delicious) termite, and also to (useful) sticks, and thence to raw stick-material, triggering the chimpanzee to search for suitable stuff. (As readers will perhaps guess, we have some skepticism as to whether such a network of associations would work efficiently, in the face of the complexity of a real African forest. But let's give it the benefit of the doubt.)

Alternatively, the chimpanzee's behavior might be a result of a mental planning device, which – in principle – can base its decisions on mental representations that encode a structural description of the situation. This structural description might include *remembered* and *inferred* characteristics, as well as the immediately perceptible features in common with the associationist account. Planning might be linear, but could also be *hierarchically* organized: working "down" from a rough or abstract specification, which is successively expanded into a richer structure of motor action. Hierarchical organization permits greater flexibility in several ways. Modularity allows efficiency, when a behavioral sequence assembled in one context is applied to another. Redundant sections of the sequence can be omitted if there is some simple test for their necessity. Iteration to a criterion allows just enough repetition to achieve a detectable result. Hierarchical organization thus generates a distinctive signature (Lashley, 1951; Miller *et al.*, 1960; Dawkins, 1976; Byrne, 2003): The hierarchical structure may be shown up in an analysis of hesitation pauses or behavior at choice points, for instance the omission of optional steps or deployment of alternative processes, depending on the subject's perception of the changing task. Finally, in planning, the desired end may be evoked in advance, so that behavior may sometimes be driven by the mismatch between an *anticipated future state* in the subject's mind and an appreciation of the current state of play, rather than driven always by past and present states that can be perceived or remembered. With association learning, of course, no result is anticipated: action is "pushed" by the operation of automatically triggered, learned responses, rather than "pulled" by the attraction of an attainable goal representation. Equally contrasting with the "fixed action patterns" of early ethology, efficient planning requires comparison of what has been achieved against a pre-specified criterion, to terminate behavior when success has been achieved. In the terms of Miller *et al.* (1960), a *test–operate–test–exit* sequence is a visible indication of using such a criterion. A more elaborate mental representation of a desired future end state has traditionally been called a "schematic anticipation" in human problem solving (Newell & Simon, 1972). Even in the absence of verbal reports, there are potentially ways of detecting an anticipatory schema, both by naturalistic observation and experimental manipulation, especially in tool use: for instance, if tools need to be obtained before moving to the site of use, the criteria for tool selection or construction must come from memory.

In this chapter we evaluate the current data from a single African study site, the Goualougo Triangle of the Republic of Congo, for any of these diagnostic signs of planning in tool use. Is chimpanzee tool use at Goualougo regulated only by immediately perceptible features of the environment, or also by inferred or remembered aspects? Linear and string-like in behavioral organization, or containing signs of flexible, hierarchical organization with optional omission and substitution of subordinate routines? "Pushed" by intrinsic motivation guided by constraints and affordances of the environment, or "pulled" toward a desired future state imagined as a schematic anticipation?

Evidence of planning in Goualougo tool use

Behavior driven by memory and inference

Chimpanzees exhibit several types of tool use to access hidden or encased food items at Goualougo, as at other sites. Extracting and dipping tools are used to gather social insects

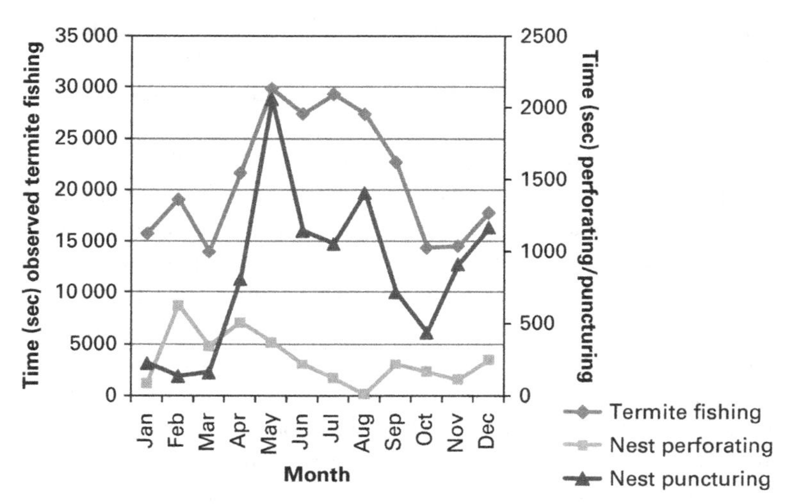

Figure 3.1 Seasonality in tool use observed in termite gathering by chimpanzees of the Goualougo Triangle.

that are not visible until extracted (e.g., *termite fish*) and liquids from recessed areas in trees (e.g., *fluid dip, leaf sponge*). These tools are often fashioned from small twigs, herb stems, leaves or pieces of grass. They may be reused in successive bouts, but do not tend to be conserved for future use by the tool maker or others. This is in contrast to tools used to open insect nests (e.g., *puncture termite nest, pound bee hive*). Those tools tend to be more substantial, being made of stout sticks or branches, and may be left at a particular tool-using site for reuse over consecutive visits. In all these cases, the goal toward which the behavior is directed must be inferred or remembered from past experience.

Parker and Gibson (1977) hypothesized that intelligent tool use in cebus monkeys and great apes evolved as an adaptation for feeding on seasonally available embedded food resources. Indeed, chimpanzee tool use to gain access to nuts, termites and ants shows distinct seasonal peaks (McGrew *et al.*, 1979; Yamakoshi, 1998; Sanz *et al.*, 2010). This most often coincides with the times that these food resources are available, such as the increase in ant dipping by chimpanzees in the Goualougo Triangle that coincides with the seasonal movements of *Dorylus* ants (see Figure 3.1; Sanz *et al.*, 2010), and the wet-season termite fishing on mound-making termites which go deep underground during dry periods (e.g., Uehara, 1982; Goodall, 1986). Seasonality in termite gathering could be related to changes in the structure of the mound, location of termites relative to the surface of the nest in different seasons or the annual reproductive cycle of termites. Evidently, memory and inference are not required to explain seasonal consumption that tracks availability: Parker and Gibson's argument was rather that year-round dependence on an embedded resource would select for morphological adaptations, while occasional but important access to hidden resources might depend on psychological abilities.

In any case, chimpanzees in the Goualougo Triangle seem to have developed tool strategies to overcome some of these difficulties, as they exhibit termite fishing throughout all months of the year. There are peaks in termite fishing and puncturing between May and August, but fishing is still seen in the other months, and at these times the use of perforating tools to open the surface of epigeal nests is most common. This suggests that using perforating tools provides access to this food resource at a time when it would not otherwise be available, resulting in a two-stage process in which each stage requires a quite different tool.

Hierarchical organization of tool-using or tool-making programs

Chimpanzees in the Goualougo Triangle show flexibility in omitting redundant steps in their tool-using sequences, and in deploying alternative processes when necessary to achieve their goal of gathering termites. The two-stage puncturing and fishing task comprises several behavioral "subroutines," each of which consists of specific behavioral elements. Examples of subroutines in this setting include those for the manufacture of a brush-tip tool, for puncturing a termite nest, for perforating a termite nest and for the termite fishing itself. Subroutines are relatively easy to identify as they are seamlessly executed and often separated by natural junctions in tool-using sequences. As an example, we show here the subroutines exhibited by a subadult male chimpanzee, Lewis, during a bout of gathering termites at a subterranean termite nest (November 15, 2006, observations began 08:59; descriptions of subroutines in italic):

9:07 Termite nest puncturing subroutine
The chimpanzee pushes a puncturing stick into the ground. After reaching the desired depth, the tool is removed by pulling upwards with both hands while the chimpanzee stands bipedally. The end of the tool or insertion point is then often smelled or visually inspected. (Humans can detect if a termite nest has been punctured by the smell of termite pheromones that are released when a nest is attacked.) If a nest has not been breached, the chimpanzee continues inserting the puncturing stick in nearby locations.

9:15 Termite fishing subroutine
This involves brush straightening, insertion and extraction of the tool, and gathering the termites from the tool. With the brush-tip fishing probe, it is often necessary to straighten or arrange the brush fibers before each insertion into the tunnel. The probe is inserted into the termite nest, and then extracted with termites clinging to the brush fibers. Termites are gathered into the chimpanzee's mouth using either the pull-through or direct-mouthing technique.

9:27 Using a fishing probe to clear a nest tunnel
After failing to insert the probe, chimpanzee reverses the orientation of a brush-tip tool and uses the blunt end to clear loosened soil from the entrance of the fishing hole. The probe is then reoriented for fishing and the brush straightened before insertion.

9:28 Termite fishing routine
9:28 Using a fishing probe to clear a nest tunnel
9:28 Termite fishing routine
9:32 Using a fishing probe to clear a nest tunnel
9:33 Termite fishing routine
9:45 Termite nest puncturing routine
9:46 Termite fishing routine

9:47 Termite nest puncturing routine
9:56 Termite fishing routine
9:57 Termite nest puncturing routine
9:59 Termite fishing routine
10:00 Termite nest puncturing routine

Structural analysis of the tool-using behavior of chimpanzees in the Goualougo Triangle shows hierarchical organization and flexible use of subroutines in several contexts (Sanz & Morgan, 2009a). For example, puncturing of a termite nest is omitted if termite nest tunnels have already been opened. If a tunnel must be created, then the chimpanzee punctures repeatedly until it succeeds in accessing an active chamber in the termite nest, using a puncturing tool that may need to be manufactured. But if one can be reused from a previous bout that day, it is: manufacture is only employed when necessary. Similarly, brush-tip tool manufacturing steps are omitted when an individual has already created a termite fishing tool at that nest, or brought a tool with them from another nest location. Also, the manufacture steps are bypassed when a youngster receives a manufactured tool from its mother. The flexible use of different subroutines is also seen in honey gathering when individuals alternately use pounding and prying tools to open a hive, and then a dipping probe to extract honey (Sanz & Morgan, 2009b).

Utilization of subroutines in more than one task

Another indication of flexibility is that chimpanzees in the Goualougo Triangle are capable of applying termite fishing knowledge in different settings, and appropriately relating these skills to different tool sets (Sanz & Morgan, 2009a). Nearly identical termite fishing subroutines are generalized across subterranean and elevated termite nest tasks; subtle differences are observed in the length of the fishing probe and in biases toward particular methods of transferring termites from the probe to the mouth in these contexts. Multiple tools can be used in both the subterranean and elevated termite gathering tasks, but it is obligatory to use two tools in gathering termites only at subterranean nests. The entire length of a stout puncturing stick is inserted into the ground to create an access tunnel into the subterranean chambers of a termite nest. After successfully puncturing a nest, a fishing probe is inserted into the tunnel to extract the termites. Besides the obvious differences in the form and function of the puncturing and perforating tools, at an elevated nest the perforating twigs are used only occasionally by chimpanzees to clear debris from an existing termite exit-tunnel during the process of fishing. We predict that further analysis of the composition of tool-using behaviors will reveal that other subroutines are shared between different skilled tasks. Generalization of subroutines across tool tasks allows construction of these skills to be more efficient and less cognitively demanding.

Recovery from interruption at critical points in overall program of action

We have observed that chimpanzees are adept in coping with interruptions during their tool-using sequences. Frequent occurrences of recovery are seen when a mother's termite gathering is interrupted by a youngster's solicitation of her tool. For example, a juvenile female named Malia was having little success fishing at an epigeal termite nest, and threw down her tool; she whimpered and reached out toward her mother, who was fishing nearby. Her mother responded by handing her own brush-tip tool to Malia; she then picked up a discarded tool from the ground and resumed fishing. On other occasions, after transferring their tool to a youngster the mother leaves the termite nest to gather tool materials.

Evaluation of success against pre-set criterion: *test–operate–test–exit*

During the termite nest puncturing subroutine, chimpanzees test to assess their progress toward reaching their anticipated goal of accessing an active termite nest chamber. Chimpanzees repeatedly puncture the ground in a particular location and then pause to inspect the end of the puncturing tool that has been inserted into the nest. The behavior of an adult male, Wallace, shows this clearly (December 29, 2003, 11:57–11:59; and see supplementary video):

Tool action:	Insert/extract puncturing stick (four times)
Test:	Smells/inspects tip of puncturing tool
Tool action:	Insert/extract puncturing stick (three times)
Test:	Smells/inspects tip of puncturing tool
Tool action:	Insert/extract fishing probe
Goal:	Feeds on termites from fishing probe

Termites release specific pheromones (Prestwich 1984; Billen & Morgan, 1998) which can be smelt on the end of the puncturing tool if a termite nest chamber has been breached. After sensing that an active chamber has been located, the chimpanzee inserts a fishing probe into the tunnel to extract the termites.

Anticipation of future needs

At several study sites, chimpanzees have been observed to manufacture tools or to pick up suitable materials in advance of need, sometimes out of sight of the place of use (Goodall, 1964; McGrew, 1974; Boesch & Boesch, 1984). In the Goualougo Triangle, fishing probes are newly manufactured each day, but chimpanzees have been observed transporting a manufactured probe from one termite nest to another. These observations imply that the animals are able to use a mental representation of an adequate tool or appropriate natural materials for a task that is not immediately confronting them (Byrne, 1998). Some researchers would insist that true planning requires the individual to be able to divorce themselves from their current motivational state, and they typically argue that this ability is uniquely human and refer to it as "mental time travel" (Suddendorf & Corballis, 1997). However, this seems to conflate two independent aspects of fore-thought: the ability to imagine feeling different emotions and motivations to now, and the ability to mentally simulate future actions (Craik, 1943; Newell & Simon, 1972). The former is extremely hard to study in non-verbal animals (but see Raby *et al.*, 2007). The latter is evidenced by any clear cases of anticipating needs which are not cued by currently perceptible stimuli, including our observations at Goualougo of construction and transport of tools prior to encountering the problem the tools serve to deal with.

Goualougo chimpanzees show particularly detailed anticipatory planning in their transport of different types of tools necessary for particular tasks. We recorded chimpanzees arriving at termite nests carrying puncturing and/or fishing tools on 45 occasions (Sanz *et al.*, 2004). All tools were transported by adults and appropriate for the type of termite nest being approached. A chimpanzee arrived carrying both a stout stick and fishing probe on seven occasions: all were arrivals at subterranean nests, which necessitate puncturing before fishing. Chimpanzees were never observed to arrive with only a puncturing stick at either a subterranean or epigeal nest; a puncturing tool alone would not be effective at subterranean nests, or appropriate at epigeal mounds where perforating twigs and probes are used. Chimpanzees arrived at termite nests carrying only fishing probes on 38 occasions: 82% of these arrivals were to epigeal mounds and 18% were to subterranean mounds. The significance of the disparity regarding which of the two types of tool is pre-constructed and transported lies in the fact that serviceable wooden puncturing tools are usually found lying around at the mounds; fishing probes, in contrast, usually become damaged in use. The chimpanzees are evidently able to anticipate their likely future needs for *both* of the two necessary steps in the process, and rightly conclude that it is the second step for which a pre-made tool would be valuable.

As researchers have reported from other sites, we have also noted chimpanzees arriving at termite nests with sufficient materials to manufacture multiple tools (Goodall, 1964). This might indicate anticipation of a likely future need for more than one tool, either as a result of a tool becoming inefficient with wear or the likelihood of sharing it with another individual; alternatively, carrying excess material may be accidental. Chimpanzees have been observed to repair tools in mid-use. When using a brush-tip fishing probe, for instance, chimpanzees often pause to repair or maintain the brush (Sanz *et al.*, 2009). This is in contrast to the termite fishing method used by Gombe chimpanzees, who actively remove the frayed end of a termite fishing probe (McGrew *et al.*, 1979). We have also observed chimpanzees modifying puncturing tools during tool-using bouts, sharpening the point (by reducing the diameter of the distal section of the tool, which is accomplished by removing side sections of the tool tip with either their teeth or fingers) for easier insertion into a subterranean termite nest. The length of a bee-hive pounding tool may also be modified mid-use. For example, a pounding tool can be shortened to produce an effective lever. Modification is presumably triggered by the immediate state of the tool and requires no anticipation, although as with tool construction the process of modification may be guided by a mental representation of what an adequate tool should be like.

Summary

Behaviorist learning and cognitive science differ fundamentally as frameworks for explaining behavior, and the main reason that the former has been discarded in human experimental psychology is its inability to make clear advance predictions with action as complex as everyday human activities. Association learning explanations can readily be fashioned after the event; they often seem plausible, but generally lack any demonstration of computational adequacy. If doubts as to adequacy could be set aside, then an associative learning explanation of chimpanzee tool use might differ only subtly from an anticipatory planning account in its observable consequences for behavior. In this chapter we have examined the behavior of chimpanzees at one study site where tool use is particularly prominent for characteristic signs of planning.

In the Goualougo Triangle of northern Republic of Congo, tools are used extensively for extractive foraging, in which the foods that are the goal of the process are not directly visible. Some of these processes involve two steps, for instance puncturing the ground to gain access to subterranean termites, and then fishing for those termites using the access route thus created. Different types of tool are used in the two steps. Although sometimes the resources to fashion suitable tools may be present at the site of use, often tools are constructed in advance and transported to the site, implying anticipation of future need and the ability to use a mental representation of a suitable tool away from the task itself. In the case of two-step tasks, the sturdier puncturing tools are often left lying in serviceable condition at the feeding site; the fact that chimpanzees make and bring a tool for the second stage of the process, a slender fishing tool likely to be unusable when discarded, shows that anticipation extends to the whole complex future sequence. The behavioral sequence itself shows a number of signs of hierarchical planning, rather than the string-like and inflexible organization to be expected from associative learning or use of fixed action patterns. Organization is modular, with smooth and fluid execution within but not between modules, and modules are often used as iterated subroutines. The criterion that stops an iterated sequence is often some type of test, such as the revealing scent of a punctured termite nest, producing a "test–operate–test–exit" pattern of behavior. Subroutines can be shared between different tool-using programs. Flexibility of planning is shown in the omission of any redundant steps, and the smooth recovery from any interruptions.

Whether an association model of learning is also capable of generating elaborate behavior of the types we have reported here, with characteristics mimicking those of planning, is a moot point: but exactly that is adhered to as an article of faith by many animal-learning theorists. Computer simulation of behavioral learning has the potential to demonstrate that such faith is justified, but the experience of human experimental psychology does not offer strong grounds for optimism. In the meantime, we suggest the most parsimonious conclusion is that chimpanzee tool use – and by implication, much else in great ape behavior – is best seen as a planning activity.

References

Billen, J. & Morgan, E. D. (1998). Pheromone communication in social insects: sources and secretions. In R. K. Vander Meer, M. D. Breed, K. E. Espelie & M. L. Winston (eds.) *Pheromone Communication in Social Insects: Ants, Wasps, Bees, and Termites* (pp. 3–33). Boulder, CO: Westview Press.

Boesch, C. & Boesch, H. (1984). Mental map in wild chimpanzees: an analysis of hammer transports for nut cracking. *Primates*, **25**, 160–170.

Broadbent, D. E. (1961). *Behaviour.* London: Methuen.

Byrne, R. W. (1975). Memory in complex tasks. Doctoral thesis, Cambridge University.

Byrne, R. W. (1977). Planning meals: problem-solving on a real data-base. *Cognition*, **5**, 287–332.

Byrne, R. W. (1983). Protocol analysis in problem-solving. In J. S. B. T. Evans (ed.) *Thinking and Reasoning: Psychological Approaches* (pp. 227–249). London: Routledge and Kegan Paul.

Byrne, R. W. (1994). The evolution of intelligence. In P. J. B. Slater & T. R. Halliday (eds.) *Behaviour and Evolution*. Cambridge: Cambridge University Press.

Byrne, R. W. (1998). The early evolution of creative thinking: evidence from monkeys and apes. In S. Mithen (ed.) *Creativity in Human Evolution and Prehistory* (pp. 110–124). London: Routledge.

Byrne, R. W. (2003). Imitation as behaviour parsing. *Philosophical Transactions of the Royal Society of London B*, **358**, 529–536.

Byrne, R. W. & Bates, L. A. (2006). Why are animals cognitive? *Current Biology*, **16**, R445–R448.

Caro, T. M. & Hauser, M. D. (1992). Is there teaching in non-human animals? *Quarterly Review of Biology*, **67**, 151–174.

Craik, K. J. W. (1943). *The Nature of Explanation*. Cambridge: Cambridge University Press.

Csibra, G. & Gergely, G. (2009). Natural pedagogy. *Trends in Cognitive Sciences*, **13**, 148–153.

Dawkins, R. (1976). Hierarchical organisation: a candidate principle for ethology. In P. P. G. Bateson & R. A. Hinde (eds.) *Growing Points in Ethology*. Cambridge: Cambridge University Press.

De Groot, A. D. (1965). *Thought and Choice in Chess*. The Hague: Mouton.

Franks, N. R. & Richardson, T. (2006). Teaching in tandem-running ants. *Nature*, **439**, 153.

Garcia, J. & Koelling, R. A. (1966). Relation of cue to consequence in avoidance learning. *Psychonomic Science*, **4**, 123–124.

Garcia, J., Ervin, F. R. & Koelling, R. A. (1966). Learning with prolonged delay of reinforcement. *Psychonomic Science*, **5**, 121–122.

Gaskins, S. (2006). Cultural perspectives on infant–caregiver interaction. In N. J. Enfield & S. C. Levinson (eds.) *Roots of Human Sociality: Culture, Cognition and Interaction* (pp. 279–298). Oxford: Berg.

Goodall, J. (1964). Tool-using and aimed throwing in a community of free-living chimpanzees. *Nature*, **201**, 1264–1266.

Goodall, J. (1986). *The Chimpanzees of Gombe: Patterns of Behavior.* Cambridge, MA: Harvard University Press.

Heyes, C. M. (1993). Imitation, culture, and cognition. *Animal Behaviour*, **46**, 999–1010.

Heyes, C. M. & Ray, E. D. (2000). What is the significance of imitation in animals? *Advances in the Study of Behavior*, **29**, 215–245.

Hoppitt, W. J. E. & Laland, K. N. (2008). Social processes influencing learning in animals: a review of the evidence. *Advances in the Study of Behavior*, **38**, 105–165.

Köhler, W. (1925). *The Mentality of Apes*. London: Routledge and Kegan Paul.

Lashley, K. S. (1951). The problem of serial order in behaviour. In L. A. Jeffress (ed.) *Cerebral Mechanisms in Behaviour: The Hixon Symposium* (pp. 112–136). New York: Wiley.

Macphail, E. M. (1985). Vertebrate intelligence: the null hypothesis. In L. Weiskrantz (ed.) *Animal Intelligence* (pp. 37–50). Oxford: Clarendon Press.

Macphail, E. M. (1998). *The Evolution of Consciousness*. Oxford: Oxford University Press.

McGrew, W. C. (1974). Tool use by wild chimpanzees feeding on driver ants. *Journal of Human Evolution*, **3**, 501–508.

McGrew, W. C., Tutin, C. E. G. & Baldwin, P. J. (1979). Chimpanzees, tools, and termites: cross cultural comparison of Senegal, Tanzania, and Rio Muni. *Man*, **14**, 185–214.

Miller, G. A., Galanter, E. & Pribram, K. (1960). *Plans and the Structure of Behavior*. New York: Holt, Rinehart and Winston.

Mulcahy, N. J. & Call, J. (2006). Apes save tools for future use. *Science*, **312**, 1038–1040.

Newell, A. & Simon, H. A. (1972). *Human Problem Solving*. New York: Prentice-Hall.

Newell, A., Shaw, J. C. & Simon, H. A. (1958). Elements of a theory of human problem solving. *Psychological Review*, **65**, 151–166.

Osvath, M. & Osvath, H. (2008). Chimpanzee (*Pan troglodytes*) and orangutan (*Pongo abelii*) forethought: self-control and pre-experience in the face of future tool use. *Animal Cognition*, **11**, 661–674.

Parker, S. T. & Gibson, K. R. (1977). Object manipulation, tool use, and sensorimotor intelligence as feeding adaptations in cebus monkeys and great apes. *Journal of Human Evolution*, **6**, 623–641.

Prestwich, G. D. (1984). Defense mechanisms of termites. *Annual Review of Entomology*, **29**, 201–232.

Raby, C. R., Alexis, D. A., Dickinson, A. & Clayton, N. S. (2007). Planning for the future by western scrub-jays. *Nature*, **445**, 919–921.

Raihani, N. J. & Ridley, A. R. (2008). Experimental evidence for teaching in wild pied babblers. *Animal Behaviour*, **75**, 3–11.

Sanz, C. & Morgan, D. (2009a). Complexity of chimpanzee tool using behaviors. In E. V. Lonsdorf, S. R. Ross & T. Matsuzawa (eds.) *The Mind of the Chimpanzee: Ecological and Experimental Perspectives*. Chicago, IL: University of Chicago Press.

Sanz, C. M. & Morgan, D. B. (2009b). Flexible and persistent tool-using strategies in honey-gathering by wild chimpanzees. *International Journal of Primatology*, **30**, 411–427.

Sanz, C., Morgan, D. & Gulick, S. (2004). New insights into chimpanzees, tools and termites from the Congo Basin. *American Naturalist*, **164**, 567–581.

Sanz, C., Call, J. & Morgan, D. (2009). Design complexity in termite-fishing tools of chimpanzees (*Pan troglodytes*). *Biology Letters*, **5**, 293–296.

Sanz, C., Schöning, C. & Morgan, D. (2010). Chimpanzees prey on army ants with specialized tool set. *American Journal of Primatology*, **71**, 1–8.

Seed, A. M. & Byrne, R. W. (2010). Animal tool use. *Current Biology*, **20**, R1032–R1039.

Shumaker, R. W., Walkup, K. R. & Beck, B. B. (2011). *Animal Tool Behavior: The Use and Manufacture of Tools by Animals*. Baltimore, MD: Johns Hopkins University Press.

Stoinski, T. S., Wrate, J. L., Ure, N. & Whiten, A. (2001). Imitative learning by captive western lowland gorillas (*Gorilla gorilla gorilla*) in a simulated food-processing task. *Journal of Comparative Psychology*, **115**, 272–281.

Suddendorf, T. & Corballis, M. C. (1997). Mental time travel and the evolution of the human mind. *Genetic, Social and General Psychology Monographs*, **123**, 133–167.

Suddendorf, T., Corballis, M. C. & Collier-Baker, E. (2009). How great is great ape foresight? *Animal Cognition*, **12**, 751–754.

Thornton, A. & McAuliffe, K. (2006). Teaching in wild meerkats. *Science*, **313**, 227–229.

Tomasello, M. & Call, J. (1997). *Primate Cognition*. New York: Oxford University Press.

Uehara, S. (1982). Seasonal changes in the techniques employed by wild chimpanzees in the Mahale Mountains, Tanzania, to feed on termites (*Pseudacanthotermes spiniger*). *Folia Primatologica*, **37**, 44–76.

Whiten, A. (2005). The second inheritance system of chimpanzees and humans. *Nature*, **437**, 52–55.

Whiten, A., Custance, D. M., Gomez, J.-C., Teixidor, P. & Bard, K. A. (1996). Imitative learning of artificial fruit processing in children (*Homo sapiens*) and chimpanzees (*Pan troglodytes*). *Journal of Comparative Psychology*, **110**, 3–14.

Whiten, A., Spiteri, A., Horner, V., et al. (2007). Transmission of multiple traditions within and between chimpanzee groups. *Current Biology*, **17**, 1038–1043.

Wynn, T. (1993). Layers of thinking in tool behavior. In K. R. Gibson & T. Ingold (eds.) *Tools, Language and Cognition in Human Evolution* (pp. 389–406). Cambridge: Cambridge University Press.

Wynn, T. G. (2002). Archaeology and cognitive evolution. *Behavioral and Brain Sciences*, **25**, 389–437.

Yamakoshi, G. (1998). Dietary responses to fruit scarcity of wild chimpanzees at Bossou, Guinea: possible implications for ecological importance of tool use. *American Journal of Physical Anthropology*, **106**, 283–295.

Part II

Comparative cognition

4 Insight, imagination and invention: Tool understanding in a non-tool-using corvid

Nathan J. Emery
Queen Mary University of London

Introduction

The ability of animals to use tools has tended to represent a hallmark of intelligent behavior; i.e., those species that use tools are thought to be smarter than those species that do not. However, many species have been described as tool users, including species that we do not traditionally endow with complex cognition (Beck, 1980). For example, sea otters float with flat rocks on their chests onto which they break shellfish; green herons lower bait or lures (feathers, flowers or insects) onto the surface of the water to attract fish, which they then catch. Tool manufacture, rather than tool use *per se*, may be a finer-grade distinction in terms of intellectual capacity.

Evidence supporting a relationship between tool use and intelligence can be seen in analyses of species of birds and mammals who demonstrate true tool use (i.e., use of objects detached from the environment as tools, such as probes or hammers), have relatively larger brains than species which either do not use tools, or which only demonstrate proto-tool use (i.e., use of objects *in situ* to facilitate the function of an action, such as dropping shells onto a rock to break them; Lefebvre *et al.*, 2002; Reader & Laland, 2002). It not clear whether these are meaningful relationships due to issues with the data used (see below) or the result of some other factor, such as the diet of tool users being better than non-tool users, which has the resultant effect of causing increases in brain size.

But is there any evidence that tool users are actually smarter than non-tool users? In this chapter I will discuss a series of experiments on a member of the corvid family of birds, rooks (*Corvus frugilegus*). Rooks have not been observed using tools in the wild and do not spontaneously use objects as tools in captivity. After examining whether rooks could solve a causal reasoning task that had been adapted for non-tool-using species (two-trap-tube task; Seed *et al.*, 2006), we wanted to further investigate what this species may have understood about how tools worked, what made a functional tool and even whether certain materials could be manipulated to make them into functional tools. This chapter will describe these tasks in detail, but extend the discussion on one remarkable result and its implications for our understanding of insight, imagination and invention.

Tool Use in Animals: Cognition and Ecology, eds. Crickette M. Sanz, Josep Call and Christophe Boesch. Published by Cambridge University Press.

Are tool users smarter than non-tool users?

Tools and brains

Theories as to why the brain, and subsequently intelligence, evolved are still the subject of great debate. Sociality, technology, environmental complexity, extractive foraging, diet, climatic variability and the physical environment have all been proposed as drivers for an increase in brain size and subsequent leaps in intellect during the evolution of primates (Gibson, 1986; Dunbar, 1992; Sterelny, 2003; Potts, 2004), and presumably other animals displaying complex cognition, such as corvids, parrots, elephants and cetaceans (Emery, 2006; van Horik *et al.*, 2011).

The argument is that greater computing power requires a more sophisticated processor, which is reflected in the size of the brain over that required for basic bodily functions, control of movement, etc. This is why *relative brain size* is often used as a proxy for neural processor. This may not necessarily be the most appropriate measure (Healy & Rowe, 2007); indeed, studies in primates have found that overall brain size is a more informative measure when predicting general intelligence (Deaner *et al.*, 2007). Chittka and Niven (2009) even question whether a large brain adds any computational complexity over that seen in much smaller brains, such as in invertebrates. Other problems regarding brains that are often overlooked are that: brains are often not measured correctly using modern stereological techniques; data sets tend to be an accumulation of brain volumes from different sources that either used different methods to process or measure the brains; and the data sets are usually incomplete (i.e., missing species), which may skew subsequent correlations. In particular, the Stephan data set on primate brains (and brain parts) that is used over and over again is 30 years old; each species is often based on only one individual and not always the same sex for each species; the brains were processed using very old methods and have not been verified to check that they represent the brain areas they are said to demark (Stephan *et al.*, 1981).

The problems in the collection and use of socioecological data are equally as problematic. For example, the data on tool use are based on the frequency of reports in the literature by amateur ornithologists and field primatologists. In the majority of cases, tool use was not the primary purpose of the observation and the frequency of tool use by the individual species (i.e., whether it is habitual or one-off) is not clear. Although collection of data is entirely dependent on research effort, there seems to be little or no effort to control the conditions in which the data were reported. For example, it may be very difficult to report tool use in a nest- or canopy-dwelling species in which clear observations of its behavior are impossible.

These technical issues therefore suggest that studies on brain–behavior relationships may be useful as a first step in determining where to look for cognitive differences, but cannot answer the question of whether tool users are smarter than non-tool users.

Physical cognition

What about evidence from laboratory tasks? There are three lines of evidence that suggest tool use is not necessarily related to being smarter: (1) species that use tools in the wild which have problems with tool-related physical cognition tasks; (2) species that use tools that perform better on tool-related physical cognition tasks when a tool does not have to be used; and (3) species that do not use tools in the wild which perform well on tool-related physical cognition tasks.

First, tool-using species do not necessarily solve tool-related physical problems. In a series of physical cognition tests, Povinelli (2000) described the difficulties chimpanzees had with tasks testing their understanding of traps, such as trap-tube and trap-table problems. These tasks focus on causal reasoning (i.e., what is the effect of performing some action when there is an obvious obstacle that will prevent an alternative effect, such as pulling food over a hole will cause the food to fall into the trap, whereas pulling the food away from the hole will result in gaining access to it) and an understanding of gravity. Chimpanzees require around 80–100 trials to learn to avoid the trap in simple versions of this task (Limongelli *et al.*, 1995; Povinelli, 2000; however see Mulcahy & Call, 2006), but do not appear to transfer knowledge between different trap problems (Martin-Ordas *et al.*, 2008).

Tool-using New Caledonian crows (NCCs), in some cases, do not fare much better. When given a choice of *Pandanus* leaf tool to use to extract meat from a hole (the tool was already located in the hole) – either oriented with barbs in an upright position (functional) or with barbs facing downwards or barbs not present (non-functional), many of the crows tested did not choose the functional tool or did not flip the non-functional tool to make it functional (Holzhaider *et al.*, 2008). However, in a similar setup, the crows did flip a non-functional stick tool (with a functional lateral extension poking outside the hole) to become a functional tool, suggesting that perhaps the NCC does not attend to barbs when using this particular tool.

Studies on undergraduate students have revealed surprising inadequacies in their tool-choice behavior. For example, when presented with the rope-and-banana problem – in which they are asked to choose between a series of images of bananas and ropes with differing degrees of connection between the two objects, to pull on a rope to bring a banana toward them – only half of the subjects select the image of the rope tied around the banana. Some subjects choose the image of the rope lying under the banana, whereas other subjects only choose images of the rope that touched the banana, either on top of or next to it (Silva *et al.*, 2008). Silva and Silva (2006) also found that undergraduate students made irrational decisions on trap-table tasks, by not necessarily choosing a tool on the side of the table without a trap (actually only a painted trap) compared to the side with the functional trap.

Second, tool-using species may perform better on tool-related physical tasks when they do not have to use a tool. As stated earlier, chimpanzees do not necessarily display proficient understanding of the trap-tube problem. However, when certain constraints on how tools can be used in the tube are removed, such as increasing the width of the tube, allowing the apes to pull the food toward them rather than push it away, then the chimpanzees' performance improves (Mulcahy & Call, 2006). A perhaps surprising constraint on physical reasoning is tool use itself. Chimpanzees asked to solve the trap-box problem (like the two-trap-tube problem, see below) were given two versions; one in which they could use a tool to move the food, or a second in which they could move the food directly using their finger. Chimps that used their finger learned the task much faster than chimps that had to use a tool (Seed *et al.*, 2009), suggesting that tool use adds an additional level of cognitive load onto the task, when compared to the task not requiring tool use.

Finally, non-tool-using species may also demonstrate success on tool-related physical tasks. Non-tool-using species have been tested for their comprehension of tool-related physical tasks. For example, when given the choice between different hook-shaped tools or between pieces of cloth with food placed on top or to the side of the cloth, cotton-top tamarins and vervet monkeys made choices based on relevant (e.g., shape) rather than irrelevant (e.g., color) features of the tools (Hauser, 1997; Hauser *et al.*, 1999; Santos *et al.*, 2003, 2006).

Although string pulling in birds may not strictly be classified as tool use (St Amant & Horton, 2008), it does require a complex series of object manipulations akin to tool use. It has been described for a number of non-tool-using birds, such as ravens, keas and African gray parrots, and has even been proposed as evidence for insight (Heinrich, 1995; Pepperberg, 2004; Heinrich & Bugnyar, 2005; Werdenich & Huber, 2006). However, pulling up a string with food attached, although not a natural behavior for birds, is part of most perching birds' behavioral repertoire, as it replicates the actions used in eating (e.g., standing on the food, reaching down and pulling off pieces of the food), and so would not fulfill various criteria for insight (see below).

Finally, non-tool-using rooks have been tested for their understanding of physical concepts, such as contact and causal reasoning, using the expectancy violation procedure (Bird & Emery, 2010) and various trap-problems, such as the two-trap-tube task (Helme *et al.*, 2006; Seed *et al.*, 2006; Tebbich *et al.*, 2007; also see Liedtke *et al.*, 2011 for similar studies in non-tool-using keas, but with a lack of success). For example, rooks look longer, more frequently and with a longer first-look duration at images that are impossible (i.e., violate physical reality, such as an inanimate object floating in mid-air) than images that are possible, such as the same object resting on a flat surface (Bird & Emery, 2010). This method has been used to great success in human babies to test for their implicit understanding of physical concepts like contact; the patterns of the rooks' looking behavior are identical to six-month-old babies.

An adapted version of the trap-tube task (two-trap-tube task; Seed *et al.*, 2006) with two traps, one functional (i.e., traps the food) and one non-functional (i.e., is either baseless so the food falls through or the base has been raised so that food passes over the trap), and an inserted tool to eliminate the need for tool use has been used to test for causal reasoning, while attempting to eliminate explanations for responses based on perceptual discriminations and associative learning using transfer tests. Seven out of eight rooks tested on the two-trap-tube task learned to avoid the food falling into the functional trap within 40–70 trials and transferred their performance to a different trap configuration within ten trials. One rook was found to transfer immediately (or in one case, by trial 2) when previously rewarded non-functional traps were pitted against one another (such as a trap without a base and a trap with a raised base on the same tube), but with an external manipulation that made one of the non-functional traps functional. One example was lowering the whole tube, so that the trap without a base was now provided with a base (i.e., the wooden platform it had been lowered onto) and a second example was adding rubber bungs into either end of the tube, but allowing a stick tool to pass through), so that pulling the food toward the previously rewarded raised-base trap would now trap the food (Seed *et al.*, 2006). The performance of tool-using NCCs on trap problems, such as the two-trap-tube task (Taylor *et al.*, 2009), is certainly not any better than found in non-tool-using rooks, so this and other evidence presented briefly here does not appear to differentiate tool users from non-tool users (Emery & Clayton, 2009). So what is the extent of a non-tool-using animal's understanding of tools?

What does a non-tool-using corvid know about how tools work?

Although rooks have not been observed using tools, either in the wild or in captivity, they do interact with their physical environment, either in terms of building nests or extractive foraging. Rook nests are found at the top or near the top of deciduous trees, and are maintained throughout the lifetime of the pair who originally built the nest. The nests are built using a combination of long sticks, mud, grasses and moss, alongside other natural and artificial materials. Any damage to the nest is repaired. The nests have to withstand extremes in weather conditions, including high winds, rain and snow. Although study of the cognition of nest building is in its infancy (Healy *et al.*, 2008), we claim that if any species uses cognition to build its nest, then rooks would be a strong candidate.

Additional support for the notion that rooks may understand their physical environment has recently been found regarding their innate knowledge of contact using an adapted expectancy violation paradigm described above (Bird & Emery, 2010).

Training, transfers and triangulation

In a recent study (Bird & Emery, 2009a), four rooks were trained to use stones as tools in a platform-releasing task (see Figure 4.1). In this task a tasty treat was placed on a

(a)
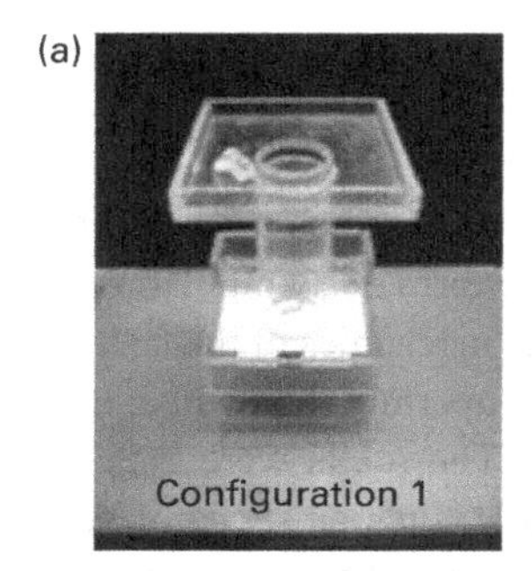

(b)
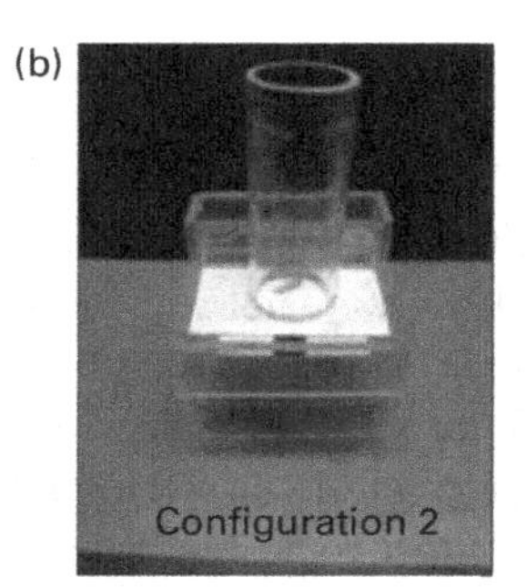

(c)
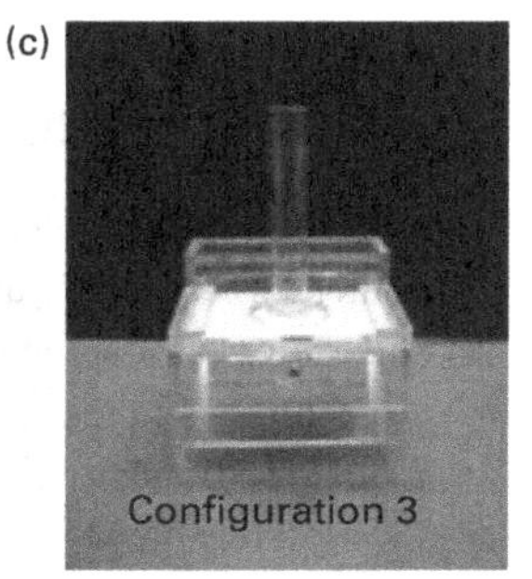

Figure 4.1 Photographs of the tube platform tasks: (a) original training-tube with a platform located next to the aperture on which a stone was placed (Configuration 1); (b) wide-aperture tube (Configuration 2); and (c) narrow-aperture tube (Configuration 3). Photographs by Chris Bird/Nathan Emery.

collapsible platform under a vertical tube. Dropping an object, such as a stone, into the tube caused the platform to release the food.

The birds rapidly learned this behavior, first taking only five trials to consistently nudge stones placed at the aperture of the tube to cause the platform to deposit its reward, and eventually picking up stones at the base of the tube or elsewhere in the aviary to put into the tube.

The rooks therefore rapidly learned the affordances of the task: that dropping a stone of a certain shape and size would release the platform. We were not particularly interested in how the rooks learned the original task mechanism (accident [the most likely occurrence], social learning, individual trial-and-error, innovation, ghost controls, etc.), only that they learned how the task worked. We have never claimed that the rooks originally dropped stones into tubes using insight, contrary to some suggestions (von Bayern *et al.*, 2009). Our claim for insight refers to a specific form of tool-related behavior (see below).

Our test that the birds understood the task affordances was to determine if they (1) dropped other objects of equal weight into the tube; (2) refrained from choosing objects that were of an insufficient weight; and (3) that they chose objects that would fit into the tube when the tube aperture was decreased in width, but which would function as appropriate tools.

In all cases the rooks made the correct choice of tool on the first trial that they were presented with the manipulated task. For example, the rooks were initially provided with large stones as tools (and were subsequently rewarded when dropped), but were then given a choice of either the same-sized stones or smaller stones. When the tube aperture was wide (i.e., the original size), the birds predominantly chose the large stones (albeit with some choices of the small stones). When the aperture of the tube was decreased, the birds (after one trial) only chose small stones. In the first trial the birds chose a large stone, tried it, found that it did not fit into the tube; then, on the second and subsequent trials, only chose small stones.

The rooks could also manipulate the stones in order that they could fit into the narrow-aperture tube. When presented with either large, round stones or long, thin stones that were the same length as the large stones, but narrow, the birds rotated the stones into the correct orientation in order to fit into the narrow-aperture tube. They did this on the first trial and three-quarters of the subjects did this without trying to fit the stone into the tube first.

Functional tool choice

These results suggest that the rooks appreciated that some stones were functional and others were not, or at least, were not without some form of manipulation. These results also suggest that the birds may have understood the mechanics of the task and responded to changes in these affordances with changes in the aperture of the tube. But how did the rooks respond to changes in the available tools? The rooks were then provided with completely new objects that could be used as tools, such as sticks of various lengths, widths and weights. For example, a light, thin stick would fit into a narrow-aperture tube, but would not have sufficient weight to trip the platform, whereas a thick, heavy stick would have sufficient weight, but would not fit into a narrow-aperture tube.

The birds performed different actions on the different sticks, depending on the size of the aperture. If the aperture size was wide, and the birds were presented with a heavy stick, they would drop the stick into the tube and collect the worm. If the birds were presented with a long, light stick, they would insert the stick into the tube, but push down onto the stick at the same time, adding the necessary force that would trip the platform without weight alone. If the birds were presented with either a wide, heavy stick or a long, thin stick, they chose the appropriate tool depending on the aperture of the tube (wide stick for wide aperture, long stick for narrow aperture). We do not know what information the rooks may have been using to drive their decision making; however, it is likely that the rooks chose the correct stick (light and thin or heavy and thick) the first time they were presented with them, based on a combination of their visual appearance, size and the shape of the aperture. Their subsequent actions with the sticks (i.e., pushing down on the light stick or dropping the heavy stick) could be achieved through rapid feedback of the weight of the stick, also combined with previous experience and then feedback from the consequences of using the stick for the first time (i.e., that the light stick was not heavy enough to displace the platform). This information could then have driven behavior on subsequent trials.

The rooks were then presented with a choice of a functional tool or a non-functional tool, such as either a functional long stick and a non-functional large stone or a functional small stone and a non-functional short, wide stick, both with a narrow-aperture tube. All birds chose the appropriate functional tool during the first trial for the functional stone versus non-functional stick choice, and three out of four of the birds chose correctly on the first trial for the functional stick versus non-functional stone choice (and the fourth bird chose correctly on the second trial).

Tool modification

What may turn a non-functional object, such as a branch with side twigs and leaves attached, into a functional tool, such as a stick probe, is the ability to modify the object in an efficient manner. This can be the result of removing items from the main object, such as stripping off twigs, or sculpting the object so that it transforms into a different shape, such as molding the end of a broken twig into a hook.

Tool-using capuchins were presented with a tube problem in which the stick tools they could use to push the food out of the tube needed to be modified in one of three ways. The sticks were presented: in a bundle that was too large to fit into the tube without being dismantled; in three short pieces that needed to be attached together to form a longer stick to reach the food; or as two cross pieces that were attached to the stick which had to be pushed out so that the stick could fit into the tube (Visalberghi & Limongelli, 1996). Capuchins were tested to see whether they could modify the tools to push the food out of the tube, but more importantly, whether they modified the tools before trying to insert them into the tube. If they modified the stick before an attempt, this suggested that they had formed a representation of a functional tool. Although capuchins quickly solved the three tasks, they did not show any improvement over time and made a number of significant errors, whereas chimpanzees and children did not make these errors.

Although rooks do not make tools, we were interested, based on our earlier findings, as to whether rooks would modify natural stick tools with side twigs that could not fit into the tube, by removing the twigs and hence turning the stick into a functional tool. We found that the rooks did not remove the side twigs before inserting the stick into the tube. However, this appeared to be because the side branches were too cumbersome to remove (without hands), and so the birds used leverage from the tube to break off the twigs. It is important to note that the rooks inserted the longest part of the branch into the tube before modification, although this could have been because they were attempting to trigger the platform without modification or to increase stability for modification. Sticks required 1–4 modifications (based on the number of side twigs) and the rooks made 1–7 modifications per stick, which were 100% successful as tools if modified.

Sequential tool use

Humans (and in some cases, chimpanzees) often use multiple tools to achieve a goal, either using one tool to support, modify or even make another tool (metatool use; Matsuzawa, 1991), or using one tool in order to reach a second tool (sequential tool use). In the animal tool-use literature, these two terms are often used interchangeably in error (unfortunately including by ourselves: Bird & Emery, 2009a). So, metatool use is often used to refer to sequential tool use, in which an animal can use one tool (that is non-functional in terms of accessing the goal object) to reach a second tool (from a choice) that is functional. NCCs have been tested on both two-step (Taylor *et al.*, 2007) and three-step (Wimpenny *et al.*, 2009; Taylor *et al.*, 2010) sequential tool-use problems and have demonstrated great proficiency on this task. The test itself, especially in more complex forms with multiple steps, could be thought of as a planning test equivalent to the classic Tower of Hanoi (or London) task in human psychology (Tower of Hanoi: move a stack of discs largest at the base and smallest at the top from the left peg to the right peg via a middle peg, making sure that the discs finish stacked largest at the base, smallest at the top and do so in the smallest number of moves; Shallice, 1982). But the test does not necessarily have to be solved using goal-directed behavior or planning (as suggested by Wimpenny *et al.*, 2009). Certainly, it is difficult to propose a more complex mechanism than chaining (i.e., reinforcing individual actions in a sequence to form a complex behavior) if only one choice of tool is available or if the initial choice is between one tool that has previously been rewarded and one that has never been rewarded (as in Taylor *et al.*, 2007; see also Clayton, 2007). Other problems with previous tasks raised by Wimpenny *et al.* (2009) were whether the inaccessible tool was located close to the reward and so increasing the chance that the subject would retrieve it by chance or if the complexity of the task was increased in a stepwise fashion, thus increasing the opportunity for learning.

In our study of sequential tool use, we tried to eliminate these problems by providing the rooks with three tubes; two side tubes with wide apertures containing either a large stone on the platform or a small stone and a central tube with a narrow aperture containing a worm (Figure 4.2). The rooks were then presented with a large stone, which they could either insert into the tube containing the large stone, the tube containing

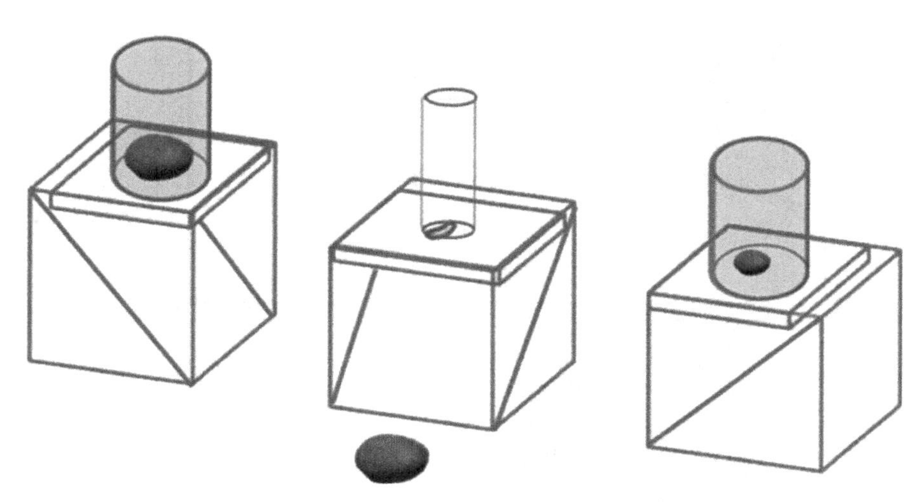

Figure 4.2 Drawing of the sequential tool-use task (Bird & Emery, 2009a), using three tube platforms; the first tube holds a large stone, the second tube holds a small stone and the middle tube holds a worm. The subject is provided with a large stone. Drawing by Chris Bird.

the small stone or the tube containing the worm. All four subjects spontaneously solved the task on the first trial, inserting the large stone into the tube with the small stone, and then inserted the small stone into the narrow-aperture tube to release the worm. Three out of four subjects demonstrated 100% success (mean 96.7 ± 2.6% success across all subjects). It is difficult to explain the rooks' success, especially on the first trial, in terms of chaining, trial and error, training or chance retrieval, but we cannot categorically ascribe planning without the addition of control conditions, such as removing the reward to determine whether the stone dropping was goal-directed.

"Necessity is the mother of invention"

One significant difference between chimpanzees and NCCs is whether they create and/or use hooks; NCCs do, chimpanzees do not. Wild NCCs create hooks from the ends of branches that have been stripped of all side twigs; the broken end is sculpted until it is clean and shaped into a hook (Hunt, 1996; Hunt & Gray, 2004). In addition, a captive female crow (Betty) was given a tube task in which meat was placed into a bucket at the bottom of the tube and then given a hook made from wire. However, when the dominant male (Abel) stole this tool and only a straight piece of wire remained, Betty created hooks by bending the wire (Weir *et al.*, 2002). This remarkable finding has proven difficult to replicate in other individuals, suggesting that Betty may have either done this before when previously kept in captivity, or that she may have been just an exceptional crow.

We therefore gave rooks hook-shaped tools made from wooden sticks with a V-shape taped to the bottom of the stick (Figure 4.3). The hook tool was placed next to a novel task modeled on Weir *et al.* (2002), in which a small card bucket with a handle was placed at the bottom of a vertical tube. A worm was placed inside the bucket. The rooks were very successful in using the hook to pull the bucket out of the tube (mean 90.8 ± 4.3% correct),

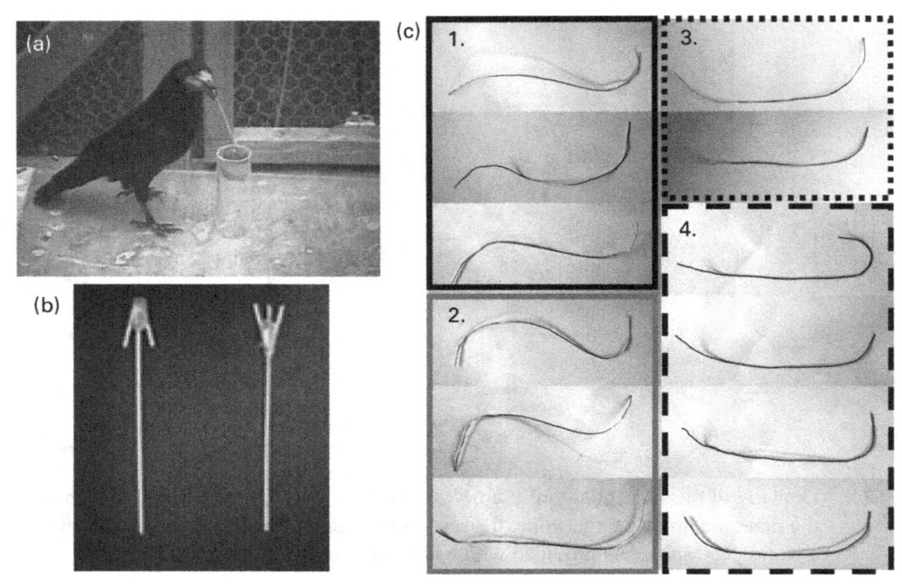

Figure 4.3 Photograph of (a) the rook using a pre-made wooden hook tool to retrieve a bucket containing food from a vertical tube; (b) wooden hook-shaped tools (V-shaped and ∧–shaped tools); and (c) hook tools produced by the rook subjects. Birds 1 (black border) and 2 (gray border) form one pair; birds 3 (small dash border) and 4 (large dash border) form a second pair. Photographs by Chris Bird/Nathan Emery.

with three out of four birds successful on the first trial and the final bird successful on the second trial. All the birds inserted the hooked end into the tube first, even though they had previously been rewarded for inserting sticks (without hooks) into tubes (see earlier).

We then provided the rooks with a choice of tools: either the V-shaped tool they had previously used or a novel ∧–shaped tool (Figure 4.3b). The two tools were similar (i.e., made of sticks and tape), except that the hook end was either functional or non-functional. All subjects inserted the V-shaped end into the tube significantly more than the ∧–shaped end, although this may have been due primarily to their previous success with this tool.

Due to the rooks' success with wooden hook tools, we decided to determine whether the non-tool-using rooks would create their own hook tools. We therefore provided the rooks with a straight piece of wire (17 cm long) and the same tube/bucket apparatus for ten trials. *All four rooks modified the wire and successfully retrieved the food.* Three out of four subjects were successful on their first trial and the fourth subject was successful on the fourth trial. It should be noted that birds were only successful in creating hooks on four out of ten ($n = 2$) or three out of ten ($n = 2$) trials. An analysis of the hooks themselves found that successful tools were $100 \pm 8.5°$ angled, whereas unsuccessful hooks were $75 \pm 4.5°$ angled. This was in contrast to Betty's hooks, which were all less than 75° angled (Weir *et al.*, 2002). Of particular interest, and something that as yet we cannot explain, the subjects' tools were constructed of two designs, either a curved hook or an S-shaped hook (i.e., a hook at both ends; Figure 4.3) and these designs were conserved within pair bonds, such that pair A favored design A and pair B favored design B. How this information may have passed between the partners or whether this was just a case of random occurrence is not yet clear.

The fact that the rooks created a tool (hook) from a novel material (wire) on the first trial on which it was presented, when they do not use tools in the wild and with only limited experience of a functionally similar, but not perceptually similar, hook tool, provides a reasonable case for insight (see later for a definition). We discuss this proposal, evidence supporting the proposal and why we think that alternative explanations are not sufficient to explain the behavior below.

Aesop's Fable realized

Of the many Aesop's Fables that feature a corvid is the story of "The Crow and the Pitcher." In the fable, a thirsty crow comes across a pitcher of water, but the level of the water is too low for the crow to reach. A number of stones are lying around the base of the pitcher. The crow hits upon the idea of placing the stones into the pitcher until the water level rises enough for it to drink and quench its thirst. This story is frequently used to suggest insight or the moral of the story: "Necessity is the mother of invention."

Tasks based on manipulating water have only recently been attempted with animals. Mendes *et al.* (2007) tested orangutans on an experimental design based on the fable, but instead of water being present in a pitcher, or in their case a vertical tube, the tube was empty except for a peanut. The apes appeared to spontaneously add water to the tube by taking a mouthful from a nearby water source and spitting it into the tube (Mendes *et al.*, 2007). In control conditions, in which the peanut was accessible without adding water to the tube, the orangutans did not collect water, suggesting that their behavior was goal-directed. Mendes *et al.* (2007) suggested that this was evidence for insightful behavior and challenged corvid researchers to test for similar behavior in crows in order to emulate the original fable.

We therefore adapted the original fable (rather than use Mendes *et al.*'s [2007] design, as crows cannot carry water in their mouths like apes), using a vertical tube containing some water, but with a worm floating on the surface (housed in a "boat" to avoid the worm sinking). The birds were presented with this novel task – novel as they had not been presented with a vertical tube containing water previously – and next to the tube was a pile of large stones (a similar size to the stones they had been previously rewarded for dropping into vertical tubes in Bird & Emery, 2009a). The rooks spontaneously dropped the appropriate number of stones into the water to raise the worm to within reach (Bird & Emery, 2009b; Figure 4.4). As the birds had been previously rewarded for dropping stones into tubes, albeit tubes not filled with water or requiring a number of stone drops in succession before receiving a reward, we do not suggest that this behavior is an example of insight (see below). The rooks' behavior is likely based on generalizing from their previous experience and responding to immediate feedback as the worm steadily approached them without moving into reach (although the rooks did not reach for the worm after each stone drop).

In subsequent tests we examined whether the rooks made the most efficient stone choices, i.e., did they choose the larger stones that would result in the greatest displacement

Figure 4.4 Photograph of a rook placing stones into a vertical tube containing water and a floating worm. Photograph by Chris Bird/Nathan Emery.

of water? Although the rooks spontaneously dropped stones into the tube, when given a choice between large and small stones, they tended to deposit small stones and then eventually (after 5–10 trials) switched preference to large stones. This does not mean that the rooks did not necessarily understand the distinction between the two sizes (Taylor & Gray, 2009), only that both stones had been previously rewarded in earlier tube tasks and both also led to a reward in the Aesop's Fable task. A choice between a functional (i.e., large stone) and non-functional (i.e., polystyrene ball) may have resulted in different responses (see Cheke *et al.*, 2011 for such controls and results in Eurasian jays).

In a final experiment, the rooks were presented with two vertical tubes, one containing sand and the other containing water. Again, surprisingly, the birds preferred to add stones to the tube containing sand, a substrate that could not be displaced. As previously, the birds learned relatively rapidly (5–10 trials) to add stones to the water tube. It is not clear why the rooks initially chose to deposit stones into the sand-filled tube.

Who's afraid of insight?

Our assertion that wire-bending, at least in rooks, may be an example of insight has met with criticism from many quarters (Kacelnik, 2009; Lind *et al.*, 2009; Shettleworth, 2010; Ed Wasserman & Clive Wynne, reported in Kloc, 2009), often invoking arguments based on an unclear and anthropomorphic definition of insight which is not particularly useful in supporting behavioral criteria that can be tested in non-human animals.

This folk psychological or anthropomorphic concept of insight often refers to the "Aha-moment," in which, after a period of confusion, a solution to a problem appears as if by magic. This is not particularly satisfactory for an empirical approach. Other problems arise when discussing apparent examples of insight by animals. The classic case is the study of chimpanzees by Wolfgang Kohler on the island of Tenerife in the early part of the twentieth century (Kohler, 1927). In these studies, Kohler provided his apes with various problems in which a tasty treat was out of reach, and the subjects were provided with various objects that could be used in order to either bring the chimp closer to the food or the food closer to the chimp. In one case, a banana was placed inside a cage and the chimp was provided with sticks that could fit together. The chimp eventually combined the sticks into a longer tool that could be used to pull the banana closer. In another case the problem facing the chimps was a banana hanging from a piece of string, too high for them to reach without assistance. After some effort trying to reach the banana without the use of tools and then using individual items, such as standing on a box or reaching with a stick, a chimp eventually moved the box under the banana, stood on the box and reached the banana (more complicated variants included stacking boxes or using a stick to knock the banana onto the floor). Kohler (1927) interpreted the chimp's actions as the result of insight – that the chimp *eventually* visualized the problem as a whole (or *gestalt*) and put the various individual solutions (e.g., stick, box in relation to the banana, etc.) together mentally to form the final solution.

However, this supposed demonstration of insight has been rightly criticized because of the important role of trial-and-error learning and previous experience in forming the chimp's eventual success. Birch (1945), for example, tested young chimps with similar problems and found that the chimps required experience of the individual components of the task in order to demonstrate success.

Even pigeons have been shown to demonstrate behavior resembling insight in a similar problem-solving scenario. Epstein *et al.* (1984) trained a number of pigeons to produce the behavioral components of Kohler's test. Some birds were "trained to push the box towards a green spot . . . which was placed at random positions along the base of the chamber wall(s). Pushing was extinguished in the absence of the green spot." The pigeons were then "trained to climb onto the box and peck the banana . . . pecking it [the box] was never reinforced" (Epstein *et al.*, 1984: 61). So, both pushing the box and pecking the hanging banana were reinforced. When both banana and box were presented, three pigeons first appeared "confused," looking repeatedly between the box and the banana, which had both been reinforced, but then each subject eventually pushed the box toward the banana until it was directly underneath, then climbed onto the box and pecked the banana. Pigeons that had only been trained on one action (pushing box or pecking banana) did not produce this sequence of actions and hence solve the problem. Epstein *et al.* (1984) suggested that because they had not trained the pigeons to push the box toward the banana (only toward the green spot, which was then extinguished), this could be seen as a case of insight, in the same manner as Kohler's chimps, based on a functional generalization between two previously reinforced actions (push and peck) in relation to two objects (box and banana).

Could this really be a demonstration of insight and could the same behavioral principles explain the rooks' wire-bending? Before attempting to answer this question, it is necessary to evaluate the definition of insight being used here. A clearer description based on behavioral criteria has been provided by Thorpe, who defined insight as "the sudden production of a new adaptive response not arrived at by trial behavior or as the solution of a problem by the sudden adaptive reorganization of experience" (Thorpe, 1964: 110).

The important terms to consider here are *sudden, new, adaptive* and *reorganization of experience*. For an action to be considered the result of insight, it must be spontaneous (i.e., not the result of explicit training or trial and error), novel (i.e., not performed before), functional (i.e., solve the problem and be goal-directed) and built from previous, untrained, similar behavior (i.e., not produced from copying earlier learned responses, but adapting previous behavior into new actions).

These behavioral criteria can be applied to any action that is proposed as insightful. For example, in the case of using water as a tool by orangutans described earlier (Mendes *et al.*, 2007), we cannot be certain that the behavior was insightful because, although it was spontaneous, functional and built from previous behavior, we do not know whether the behavior itself was novel (i.e., moving water from one container to another via spitting). More information on the previous behavior of the orangutans in this case would provide stronger support for insight using Thorpe's definition; for example, how often do the orangutans spit water during their day-to-day lives outside the context of the experiment?

What about Epstein's pigeons? Their behavior was not spontaneous (it was based on training), not functional (no reward), not novel (pecking and pushing – a form of peck – are both in a pigeon's repertoire) and was not adapted from previous similar behavior (rather than previous same behavior), so could not be considered as insight in any sense.

Finally, what about rooks? Wire-bending was spontaneous (performed on the first trial), novel (wire-bending, hook-making or even tool making are not within the behavioral repertoire of rooks and not possible outside the context of the experiment, in contrast to the spitting orangutans who have constant access to water), functional (brought the food to within reach) and adapted from previous behavior (using wooden hook tools to raise buckets containing food and inserting sticks into tubes). We are therefore confident that the rooks' actions constitute an unambiguous case of insightful behavior.

We will consider these ideas in detail:

(1) Rooks do not use tools in the wild or in captivity during their daily lives. When rooks are presented with objects that *can* function as tools, rooks will use them in an appropriate context, such as pulling food up on a piece of string (Seed, Emery & Clayton, unpublished observations), pulling a string with another rook to pull in a wooden shelf holding food (Seed *et al.*, 2008), pulling a stick to drag food out of a clear Perspex tube (Helme *et al.*, 2006; Seed *et al.*, 2006; Tebbich *et al.*, 2007) and inserting sticks/dropping stones into tubes to release a platform or raise water (Bird & Emery, 2009a). However, these are examples of basic tool use, not insight. The action of bending wire is not something that is part of rooks' normal behavioral repertoire. Wire-bending is making a tool from a novel material. Although straight wire could potentially be used as a functional tool, and indeed both rooks (Bird & Emery, 2009a) and NCCs (Weir *et al.*, 2002) were relatively successful in using the straight wire, they were more successful when the wire was bent into a hook.

(2) Rooks were hand-raised. Their developmental and experimental history was well documented and controlled. We are certain that the rooks had no experience of wire

in an experimental context and the only experience outside of experiments was of the wire that formed their aviaries.

(3) The rooks had no previous experience of wire hooks. The first time the rooks saw a piece of wire bent into a hook was when they had formed one themselves. The rooks also could not have formed reward associations between wire and food. It is an open question as to why they inserted the wire into the tube, although it is likely that the rooks generalized their previous use (and reinforcement) of sticks; however, this cannot explain how they came to bend the wire into hooks.

(4) The only experience rooks had of hooks were wooden sticks with a V-shape taped to one end. The rooks were also given some trials with a choice between a functional V-shaped wooden hook and a non-functional ∧–shaped wooden hook. These hooks do not look like the wire hooks created by the rooks (see Figure 4.3). It is possible that the rooks formed a mental image of a functional hook tool and then used this template to form a hook from the novel material.

(5) The rooks' behavior was spontaneous, produced on the first trial in which they experienced the novel material. Therefore, there was no potential for trial-and-error learning.

(6) The rooks did not appear confused at any time during a trial and we did not observe a change in state from "confusion to solution" (i.e., an "Aha-moment'). In both Epstein *et al.*'s (1984) and Kohler's (1927) studies, at some point during proceedings, the pigeons and chimpanzees appeared to be in a confused state, i.e., they looked frequently between the objects available to them without acting or they repeatedly attempted to reach the banana that was out of reach. The rooks did not do this. They tended to look at the tube from different angles and they did not attempt to reach the food without inserting and bending the wire.

(7) The rooks could not have accidentally created a hook. If hook-making was accidental, then surely the rook would have attempted to pierce the worm with the straight piece of wire at the start of every trial, rather than concentrate on manipulating the end of the wire closest to the opening of the tube.

(8) The finished hook was not located next to the food. Once the hook was created, the rook left the tool *in situ* and examined the task again (Figure 4.5). The rook then removed the tool from the tube, flipped it over, inserted the hooked end into the tube, manipulated the hook until it passed under the handle of the bucket and then pulled the tool upwards until the bucket and reward were removed from the tube (Figure 4.5).

It is very easy to claim that wire-bending in rooks is *just the same* as the pseudo-problem solving displayed by Epstein's pigeons and just leave it at that. However, they are very different problems, applied in completely different contexts that actually make such simple exertions non-parsimonious and frankly implausible. Hopefully, this attempt to dissect both Epstein's results with pigeons and Bird and Emery's results with rooks provides some structure on which to make reasoned arguments about insight (as a form of mental operation) rather than dissenters crying anthropomorphism and associative learning without any scientific structure to their criticisms.

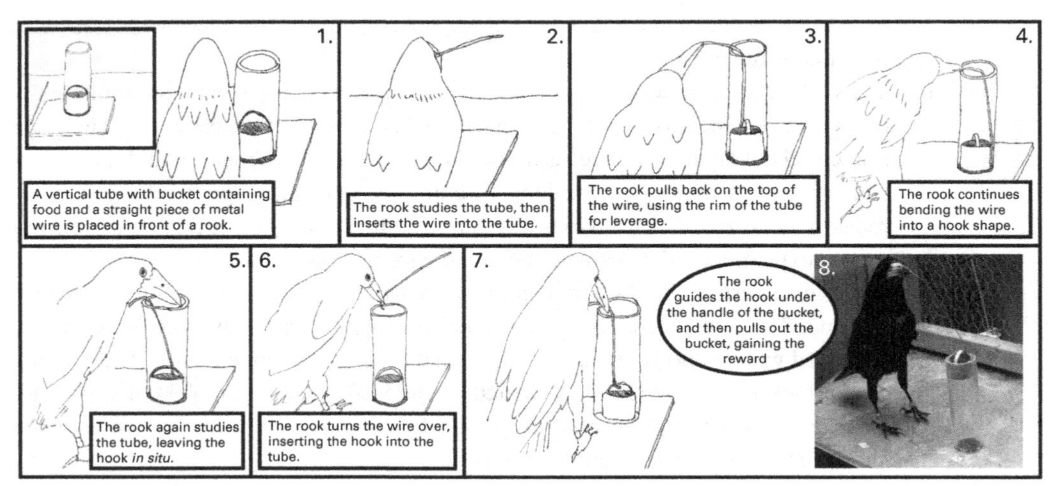

Figure 4.5 Diagram redrawn from the original video displaying a typical successful wire-bending trial. (1) The rook retrieves the wire (a novel material) – this behavior has not been previously reinforced. (2) The rook studies the tube without pecking at the worm and then inserts the wire into the tube – insertion of sticks has been previously reinforced, albeit with wooden sticks. (3–4) The rook bends the wire using the top of the tube – this behavior has never been previously reinforced and is not part of the rooks' behavioral phenotype (i.e., rooks are not tool users). (5) The rook studies the worm/bucket, but again does not peck at it – behavior is not reinforced. (6) The rook removes the hooked end of the wire and turns it over so the hooked end is now in the tube near to the bucket – perceives that the functional end is not in a functional position, so removes the wire (behavior is not reinforced) and reinserts (remove and reinsert have never been reinforced, but insertions have been previously reinforced). (7) The rook guides the hooked end under the bucket handle and pulls up the bucket until it is out of the tube – this behavior has been previously reinforced with wooden sticks. (8) Photograph of a rook using a recently created hook tool to pull a bucket out of a tube. Drawings by Nathan Emery. Photograph by Chris Bird/Nathan Emery.

The how and why of insight in rooks

If we assume that our analysis is correct – that some part of rooks' actions toward certain tools is representative of insight – then why rooks and how could this psychological process be manifest in the rook brain?

There are a number of potential reasons for why rooks. First, although rooks have not been demonstrated to use or make tools in the wild, field studies on rook behavior have not specifically addressed this issue, so it is possible that tool use has just never been reported, but it is present. This seems very unlikely, as rooks are one of the most common European and Asian birds. Amateur and professional birdwatchers have dedicated thousands of hours of observation to this species and fervently submit short notes to ornithological journals reporting unusual behaviors, including potential examples of tool use.

Second, insight and other aspects of physical cognition (folk physics) are the result of non-tool-related behaviors which involved object manipulation, such as extractive foraging. The best candidate is perhaps nest building (Healy *et al.*, 2008; Walsh *et al.*, 2010), which requires individuals to form artificial structures from large numbers of objects, such as sticks, mosses and grasses, that can withstand huge fluctuations in temperature, wind and weather elements, as well as predators, with the function of protecting offspring. Rook nests, in particular, are built at the top of deciduous trees and so are more exposed than most nests. Rook pairs return to the same nest site each breeding season and repair their nests.

Is there something special about rook nests? For example, do rooks build "better" nests than other birds? How are their nests built, maintained and repaired? Does nest structure improve over time? How is "folk physics" involved in nest construction? Answers to these questions from detailed observations of nest-building behavior, simulations of nests under various climatic scenarios and psychological tests on the possible mechanisms of construction behavior should help answer these questions.

Third, the common ancestor of the *Corvus* genus (or perhaps all corvids) had already evolved technical intelligence and so the behaviors described in this chapter are more common than we perhaps realize. These behaviors may have evolved in response to increased opportunities for object manipulation and extractive foraging in the *Corvus* (or corvid) common ancestor.

Fourth, the rook mind is constructed from a series of adaptive specializations (domain-specific modules) *and* general processes (domain-general; Figure 4.6). Emery and Clayton (2004) suggested a potential "cognitive tool kit" in corvids and

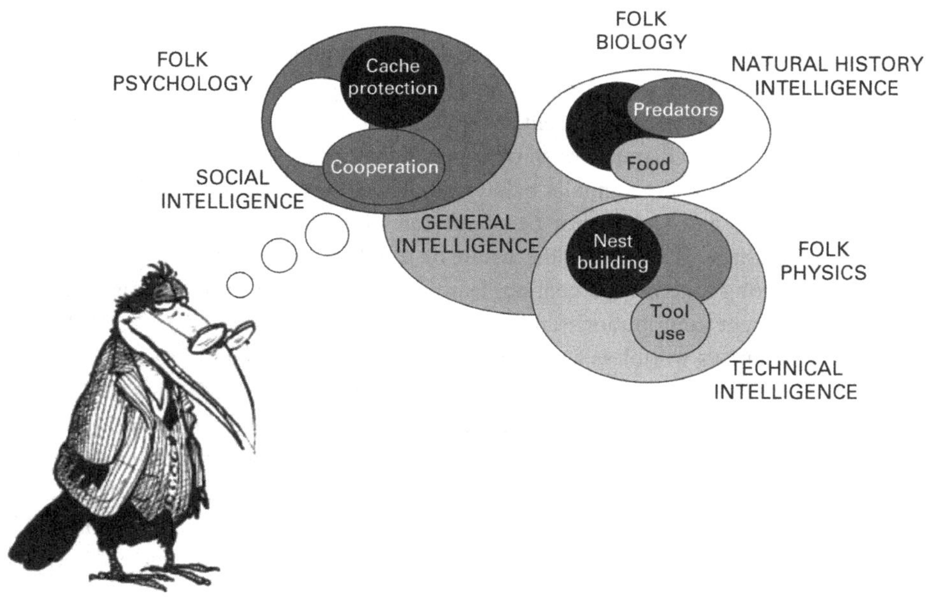

Figure 4.6 Diagram representing domain-general and domain-specific (physical, social and natural history intelligence) cognitive architecture in rooks. The domain-specific modules specialize in solving specific socioecological problems using problem-specific processing (e.g., face recognition is important for social cognition), but they also overlap because they also use domain-general processes, such as memory, concepts, causal and analogical reasoning, planning and imagination. The degree of overlap is perhaps more pronounced in *Homo sapiens*.

apes that could be utilized across domains, comprising imagination, prospection, causal reasoning and flexibility. We suggest that as rooks do not use tools in the wild, and yet appear to understand how tools work when tested in captivity, this could be evidence that the cognition underlying their response to tools is domain-general (i.e., independent of the ecological context) and not domain-specific (i.e., dependent on the ecological context). Hence, the rook mind has refined its cognitive architecture from solving one set of physical problems (e.g., nest building) to solving another set of physical problems outside the domain for which it originally evolved (e.g., tool use). Such cognitive architecture may require such general processes as causal reasoning and insight.

How might such a system be realized in the rook brain? Potential models of the cognitive architecture of insight could be based on the associative-cybernetic model (de Wit & Dickinson, 2009) or global workspace theory (Shanahan & Baars, 2005). Both models include a means for simulating alternative actions and scenarios (mental trial and error) without having to act on the simulations, something akin to imagination or insight. The development of such models in terms of insight and imagination *per se* would be a significant step in our understanding of invention and the creative process, but at present such models in this context are unrealized.

Conclusion

When interpreting an animal's behavior in cognitive terms, especially when those terms have been borrowed from human psychology (or worse, human folk psychology), a cautious approach is essential (Shettleworth, 2010). However, many critics are possibly too quick to dismiss much of animal behavior as "simple" associative learning without considering the animal's behavior in its entirety. Associative learning is anything but simple, and should not be contrasted directly with cognition as if they were polar opposites. Behavior is the result of a collection of different psychological processes working together: perception, learning and cognition. Without discussing or testing the different contributions and forms that these processes take, the "killjoy" hypotheses become as worthless as the "anthropomorphic" terms they are suggested to replace.

In the case of insight described here, Epstein *et al.*'s (1984) studies on pigeons are important, perhaps overlooked, contributions to comparative psychology, but they have tended to be wielded without careful consideration of what arguments they are being used to support. Hopefully, the analysis of the wire-bending study of Bird and Emery (2009a) presented in this chapter may dispel some erroneous notions that the rooks' behavior is a rather simple case of previous experience and conditioning (Lind *et al.*, 2009). By using a clear definition of insight (Thorpe, 1964), we can hopefully distinguish between real candidates for insightful behavior rather than based on our folk psychological notions of insight as the "Aha-moment" (see Chapter 1 for an up-to-date discussion of modern psychological theories of insight). This may as well be a magical process and certainly one that cannot be analyzed at any behavioral level. In getting us closer to the most parsimonious explanation for the processes underlying a particular behavior, we may advocate a liberal dose of Occam's aftershave (Emery & Clayton, 2008) after the brutal wielding of the razorblade of "killjoy explanations."

Acknowledgments

The experiments described in this chapter formed part of Chris Bird's PhD thesis. I thank him profusely for his sterling efforts in increasing our understanding of the rook mind. I would also like to thank the three editors, Crickette Sanz, Josep Call and Christophe Boesch, for initially inviting me to the meeting and for their patience in waiting for this chapter to finally appear. I would also like to thank the other meeting participants for a stimulating and enjoyable conference. My research was funded by the Royal Society, BBSRC and University of Cambridge and I was supported by a Royal Society University Research Fellowship.

References

Beck, B. (1980). *Animal Tool Behavior: The Use and Manufacture of Tools by Animals*. New York: Garland.

Birch, H. G. (1945). The relation of previous experience to insightful problem-solving. *Journal of Comparative Psychology*, **38**, 367–383.

Bird, C. D. & Emery, N. J. (2009a). Insightful problem solving and creative tool modification by captive nontool-using rooks. *Proceedings of the National Academy of Sciences USA*, **106**, 10370–10375.

Bird, C. D. & Emery, N. J. (2009b). Rooks use stones to raise the water level to reach a floating worm. *Current Biology*, **19**, 1410–1414.

Bird, C. D. & Emery, N. J. (2010). Rooks perceive support relations similar to six-month old babies. *Proceedings of the Royal Society of London B*, **277**, 147–151.

Cheke, L. G., Bird, C. D. & Clayton, N. S. (2011). Tool use and instrumental learning in the Eurasian jay (*Garrulus glandarius*). *Animal Cognition*, **14**, 441–455.

Chittka, L. & Niven, J. (2009). Are bigger brains better? *Current Biology*, **19**, R995–R1008.

Clayton, N. S. (2007). Animal cognition: crows spontaneously solve a metatool task. *Current Biology*, **17**, R894–R895.

de Wit, S. & Dickinson, A. (2009). Associative theories of goal-directed behaviour: a case for animal–human translational models. *Psychological Research*, **73**, 463–476.

Deaner, R. O., Isler, K., Burkart, J. M. & van Schaik, C. P. (2007). Overall brain size, and not encephalization quotient, best predicts cognitive ability across non-human primates. *Brain, Behavior and Evolution*, **70**, 115–124.

Dunbar, R. I. M. (1992). Neocortex size as a constraint on group size in primates. *Journal of Human Evolution*, **20**, 469–493.

Emery, N. J. (2006). Cognitive ornithology: the evolution of avian intelligence. *Philosophical Transactions of the Royal Society of London B*, **361**, 23–43.

Emery, N. J. & Clayton, N. S. (2004). The mentality of crows: convergent evolution of intelligence in corvids and apes. *Science*, **306**, 1903–1907.

Emery, N. J. & Clayton, N. S. (2008). Imaginative scrub-jays, causal rooks and a liberal application of Occam's aftershave. *Behavioural and Brain Sciences*, **31**, 134–135.

Emery, N. J. & Clayton, N. S. (2009). Tool use and physical cognition in birds and mammals. *Current Opinion in Neurobiology*, **19**, 27–33.

Epstein, R., Kirschnit, C. E., Lanza, R. P. & Rubin, L. C. (1984). "Insight" in the pigeon: antecedents and determinants of an intelligent performance. *Nature*, **308**, 61–62.

Gibson, K. R. (1986). Cognition, brain size and the extraction of embedded food resources. In J. G. Else & P. C. Lee (eds.) *Primate Ontogeny, Cognition and Social Behaviour* (pp. 93–105). Cambridge: Cambridge University Press.

Hauser, M. D. (1997). Artifactual kinds and functional design features: what a primate understands without language. *Cognition*, **64**, 285–308.

Hauser, M. D., Kralik, J. & Botto-Mahan, C. (1999). Problem solving and functional design features: experiments in cotton-top tamarins. *Animal Behaviour*, **57**, 565–582.

Healy, S. D. & Rowe, C. (2007). A critique of comparative studies of brain size. *Proceedings of the Royal Society of London B*, **274**, 453–464.

Healy, S., Walsh, P. T. & Hansell, M. (2008). Nest building by birds. *Current Biology*, **18**, R271–R273.

Heinrich, B. (1995). An experimental investigation of insight in common ravens (*Corvus corax*). *Auk*, **112**, 994–1003.

Heinrich, B. & Bugnyar, T. (2005). Testing problem-solving in ravens: string-pulling to reach food. *Ethology*, **111**, 962–976.

Helme, A. E., Clayton, N. S. & Emery, N. J. (2006). What do rooks (*Corvus frugilegus*) understand about physical contact? *Journal of Comparative Psychology*, **120**, 288–293.

Holzhaider, J. C., Hunt, G. R., Cambell, V. M. & Gray, R. D. (2008). Do wild New Caledonian crows (*Corvus moneduloides*) attend to the functional properties of their tools? *Animal Cognition*, **11**, 243–254.

Hunt, G. R. (1996). Manufacture and use of hook-tools by New Caledonian crows. *Nature*, **379**, 249–251.

Hunt, G. R. & Gray, R. D. (2004). The crafting of hook tools by wild New Caledonian crows. *Proceedings of the Royal Society of London: Biology Letters*, **271**, 88–90.

Kacelnik, A. (2009). Tools for thought or thought for tools? *Proceedings of the National Academy of Sciences USA*, **106**, 10071–10072.

Kloc, J. (2009). Invoking the magic of the mind. *Seed Magazine*. http://seedmagazine.com/content/article/invoking_the_magic_of_the_mind/.

Kohler, W. (1927). *The Mentality of Apes*. New York: Vintage Books.

Lefebvre, L., Nicolakakis, N. & Boire, D. (2002). Tools and brains in birds. *Behaviour*, **139**, 939–973.

Liedtke, J., Werdenich, D., Gajdon, G. K., Huber, L. & Wanker, R. (2011). Big brains are not enough: performance of three parrot species in the trap-tube paradigm. *Animal Cognition*, **14**, 143–149.

Limongelli, L., Visalberghi, E. & Boyzen, S. T. (1995). The comprehension of cause–effect relations in a tool-using task by chimpanzees (*Pan troglodytes*). *Journal of Comparative Psychology*, **109**, 18–26.

Lind, J., Ghirlanda, S. & Enquist, M. (2009). Insight and learning. *Proceedings of the National Academy of Sciences USA*, **106**, E76.

Martin-Ordas, G., Call, J. & Colmenares, F. (2008). Tubes, tables and traps: great apes solve two functional equivalent trap tasks but show no evidence of transfer across tasks. *Animal Cognition*, **11**, 423–430.

Matsuzawa, T. (1991). Nesting cups and meta-tools in chimpanzees. *Behavioural and Brain Sciences*, **14**, 570–571.

Mendes, N., Hanus, D. & Call, J. (2007). Raising the level: orangutans use water as a tool. *Biology Letters*, **3**, 453–455.

Mulcahy, N. J. & Call, J. (2006). How great apes perform on a modified trap-tube task. *Animal Cognition*, **9**, 193–199.

Pepperberg, I. M. (2004). "Insightful" string-pulling in grey parrots (*Psittacus erithacus*) is affected by vocal competence. *Animal Cognition*, **7**, 263–266.

Potts, R. (2004). Paleoenvironmental basis of cognitive evolution in great apes. *American Journal of Primatology*, **62**, 209–228.

Povinelli, D. J. (2000). *Folk Physics for Apes*. New York: Oxford University Press.

Reader, S. M. & Laland, K. N. (2002). Social intelligence, innovation and enhanced brain size in primates. *Proceedings of the National Academy of Sciences USA*, **99**, 4436–4441.

Santos, L. R., Miller, C. T. & Hauser, M. D. (2003). Representing tools: how two non-human primate species distinguish between the functionally relevant and irrelevant features of a tool. *Animal Cognition*, **6**, 269–281.

Santos, L. R., Pearson, H. M., Spaepen, G. M., Tsao, F. & Hauser, M. D. (2006). Probing the limits of tool competence: experiments with two non-tool-using species (*Cercopithecus aethiops* and *Saguinus oedipus*). *Animal Cognition*, **9**, 94–109.

Seed, A. M., Tebbich, S., Emery, N. J. & Clayton, N. S. (2006). Investigating physical cognition in rooks. *Current Biology*, **16**, 697–701.

Seed, A. M., Clayton, N. S. & Emery, N. J. (2008). Cooperative problem solving in rooks (*Corvus frugilegus*). *Proceedings of the Royal Society of London B*, **275**, 1421–1429.

Seed, A. M., Call, J., Emery, N. J. & Clayton, N. S. (2009). Chimpanzees solve the trap problem when the confound of tool-use is removed. *Journal of Experimental Psychology: Animal Behavior Processes*, **35**, 23–34.

Shallice, T. (1982). Specific impairments of planning. *Philosophical Transactions of the Royal Society of London B*, **298**, 199–209.

Shanahan, M. P. & Baars, B. J. (2005). Applying global workspace theory to the frame problem. *Cognition*, **98**, 157–176.

Shettleworth, S. J. (2010). Clever animals and killjoy explanations in comparative psychology. *Trends in Cognitive Sciences*, **14**, 477–481.

Silva, F. J. & Silva, K. M. (2006). Humans' folk physics is not enough to explain variations in their tool-using behaviour. *Psychonomic Bulletin and Review*, **13**, 689–693.

Silva, F. J., Silva, K. M., Cover, K. R., Leslie, A. L. & Rubalcaba, M. A. (2008). Humans' folk physics is sensitive to physical connection and contact between a tool and reward. *Behavioural Processes*, **77**, 327–333.

St Amant, R. & Horton, T. E. (2008). Revisiting the definition of animal tool use. *Animal Behaviour*, **75**, 1199–1208.

Stephan, H., Frahm, H. & Baron, G. (1981). New and revised data on volumes of brain structures in insectivores and primates. *Folia Primatologia*, **35**, 1–29.

Sterelny, K. (2003). *Thought in a Hostile World*. New York: Blackwell.

Taylor, A. H. & Gray, R. D. (2009). Animal cognition: Aesop's fable flies from fiction to fact. *Current Biology*, **19**, R731–R732.

Taylor, A. H., Hunt, G. R., Holzhaider, J. C & Gray, R. D. (2007). Spontaneous metatool use in New Caledonian crows. *Current Biology*, **17**, 1504–1507.

Taylor, A. H., Hunt, G. R., Medina, F. S. & Gray, R. D. (2009). Do New Caledonian crows solve physical problems through causal reasoning? *Proceedings of the Royal Society of London B*, **276**, 247–254.

Taylor, A. H., Elliffe, D., Hunt, G. R. & Gray, R. D. (2010). Complex cognition and behavioural innovation in New Caledonian crows. *Proceedings of the Royal Society of London B*, **277**, 2637–2643.

Tebbich, S., Seed, A. M., Emery, N. J. & Clayton, N. S. (2007). Non tool-using rooks (*Corvus frugilegus*) solve the trap tube task. *Animal Cognition*, **10**, 225–231.

Thorpe, W. H. (1964). *Learning and Instinct in Animals*. London: Methuen & Co. Ltd.

van Horik, J., Clayton, N. S. & Emery, N. J. (2011). Convergent evolution of cognition in corvids, apes and other animals. In J. Vonk & T. Shackleford (eds.) *The Oxford Handbook of Comparative Evolutionary Psychology* (pp. 80–101). New York: Oxford University Press.

Visalberghi, E. & Limongelli, L. (1996). Acting and understanding: tool use revisited through the minds of capuchin monkeys. In A. E. Russon, K. A. Bard & S. T. Parker (eds.) *Reaching Into Thought: The Minds of the Great Apes* (pp. 57–79). Cambridge: Cambridge University Press.

von Bayern, A. M. P., Heathcote, R. J. P., Rutz, C. & Kacelnik, A. (2009). The role of experience in problem solving and innovative tool use in crows. *Current Biology*, **19**, 1965–1968.

Walsh, P. T., Hansell, M., Borello, W. D. & Healy, S. D. (2010). Repeatability of nest morphology in African weaver birds. *Biology Letters*, **6**, 149–151.

Weir, A. A. S., Chappell, J. & Kacelnik, A. (2002). Shaping of hooks in New Caledonian crows. *Science*, **297**, 981.

Werdenich, D. & Huber, L. (2006). A case of quick problem-solving in birds: string pulling in keas, *Nestor notabilis*. *Animal Behaviour*, **71**, 855–863.

Wimpenny, J. H., Weir, A. A. S., Clayton, L., Rutz, C. & Kacelnik, A. (2009). Cognitive processes associated with sequential tool use in New Caledonian crows. *PLoS ONE*, **4**, e6471.

5 Why is tool use rare in animals?

Gavin R. Hunt
Department of Psychology, University of Auckland

Russell D. Gray
Department of Psychology, University of Auckland

Alex H. Taylor
Department of Psychology, University of Auckland

Introduction

Tool use is widespread in the animal kingdom. It has been reported in taxa ranging from insects to primates (see reviews in Beck, 1980; Bentley-Condit & Smith, 2010; Shumaker *et al*., 2011). However, although it is taxonomically widespread, tool use is relatively rare. The rarity of tool use is surprising given the potential evolutionary advantages that a species can gain. Tools can be used to extract rich food sources such as termites and wood-boring larvae that would otherwise be extremely difficult to obtain. Given the obvious advantages of tool use, an equally obvious question is why tool use is seen in very few species.

A glance across the species that use objects as tools rules out any simple association between the presence or absence of tool use and level of cognitive ability. Tool use is seen in insects, marine invertebrates and fish, as well as in birds and mammals. Indeed, Jane Goodall (1970) recognized that the evolutionary processes underpinning tool use across the animal kingdom will be very different. Beck (1980) emphasized that there was no simple correlation between the presence of tool use and cognitive abilities. Hansell and Ruxton (2008) recently proposed another possible explanation for the rarity of tool use in animals – that tool use was rare simply because of the lack of ecological contexts in which it was advantageous (we call this the lack-of-utility hypothesis). However, we will show here that an "excess of opportunity" clearly contradicts the lack-of-utility hypothesis because in evolutionary terms tool use appears to be potentially much more useful than its frequency in the animal kingdom indicates. Given its potential usefulness, why is tool use so rare?

In this chapter we propose that (1) the rarity is explained by constraints associated with the "dynamic mechanical interactions" (St Amant & Horton, 2008) inherent in tool use, and (2) that these constraints are different depending on the processes and mechanisms underpinning tool use. In most invertebrates and fish, relatively low cognitive ability means that tool use only evolves as stereotyped behavior. That is, it is usually ubiquitous

Tool Use in Animals: Cognition and Ecology, eds. Crickette M. Sanz, Josep Call and Christophe Boesch. Published by Cambridge University Press.

within species and shows little flexibility or learning. The requirement for stereotyped tool use to evolve from suitable pre-existing behavior severely constrains its evolution. In birds and mammals, relatively high cognitive ability means that tool use can not only evolve as stereotyped behavior, but also develop as flexible behavior based on a significant learning component. The development of flexible tool use is likely to have three main constraints. The first constraint is the need to learn that an object can be used to facilitate the solving of a problem that would be otherwise difficult to solve without a tool. The second is the working memory capacity to deal with the greater sequential and relational complexity that is involved in a tool task compared to a comparable non-tool task. The third constraint is the practical difficulty of manipulating an object in a controlled way to achieve a specific goal.

Definition of tool use

A clear and accurate definition of what constitutes tool use is essential for any attempt to answer the question of why this behavior is rare in animals. Without identifying the essential characteristic that constitutes tool use we cannot carry out a meaningful investigation of what constraints might be acting on the behavior. Beck's (1980) widely used definition of tool use over the last 30 years has been updated in the second edition of his book (Shumaker *et al.*, 2011: 10). The updated definition states that tool use is:

> the external employment of an unattached or manipulable attached environmental object to alter more efficiently the form, position, or condition of another object, another organism, or the user itself, when the user holds and directly manipulates the tool during or prior to use and is responsible for the proper and effective orientation of the tool.

Shumaker *et al.* stated that the changes made to Beck's original definition were heavily influenced by the new definition of tool use recently proposed by St Amant and Horton (2008). However, unlike St Amant and Horton's definition, it fails to clearly identify the intrinsic factor that distinguishes tool use as a special behavioral category. Narrowing in on this intrinsic factor, as we believe St Amant and Horton have done, would (1) allow the same fundamental behavior to be compared across the multitude of ways that tool use can occur, and (2) make it far easier to establish the processes and mechanisms that give rise to tool use and as a consequence enable inferences about how tool use might be constrained. The definition proposed by St Amant and Horton (2008: 103) states that tool use is:

> the exertion of control over a freely manipulable external object (the tool) with the goal of (1) altering the physical properties of another object, substance, surface or medium (the target, which may be the tool user or another organism) via a dynamic mechanical interaction, or (2) mediating the flow of information between the tool user and the environment or other organisms in the environment.

St Amant and Horton's definition identifies "dynamic mechanical *interactions*" (DMIs) as the intrinsic factor that distinguishes tool use as a special category of behavior, thus also helping to eliminate much of the ambiguity in previous definitions. Active bait fishing, which Beck (1980) and Shumaker *et al.* (2011) classified as tool use, is an appropriate object behavior with which to illustrate the application of St Amant and Horton's definition. Bait fishing is practiced by only a handful of bird species and involves a bird placing food or a lure on the water surface near itself to attract fish in the area (Ruxton & Hansell, 2011). The bird then stands still and waits for fish to come and eat or inspect the object. Active bait fishing is not classified as tool use by St Amant and Horton (2008) because it does not involve any DMI, a crucial point that both Shumaker *et al.* (2011) and Ruxton and Hansell (2011) appear to dismiss. However, if bait was directed at visible fish there would be a dynamic interaction in the context of tool use (St Amant & Horton, 2008). In this case it would then be valid to compare bait fishing with other kinds of tool use. The bait fishing example and the focus on DMIs is not simply a question of being pedantic. Rather, it crucially enables a more rigorous standardization of what constitutes tool use, which in turn provides the framework for meaningful comparative study.

If tool use is based on DMIs, non-tool object manipulation involves the establishment of "static" relationships between objects (St Amant & Horton, 2008). An example of static object use in construction behavior is nest building by birds. Hansell and Ruxton (2008) argued that there is no difference, cognitively or otherwise, between tool use and construction behavior such as nest building. However, as St Amant and Horton (2008) make clear, the material to build a nest is not used in a dynamic interaction like a stick is used, for example, to extract an insect from a hole. Rather than being used to alter the physical properties of another object, nest material is simply used in combination with other objects to create a structure. Non-tool object manipulation also includes objects manipulated dynamically without the intent of a goal-directed interaction with another object. An example would be a badger throwing material behind itself when digging a hole. We define such dynamic object manipulation as object-related mechanical *actions* (ORMAs) and will propose that they are a crucial prerequisite for the evolution of stereotyped tool use. The bait fishing by birds described above also involves only "static" object manipulation because birds simply place bait on the water without actively directing it at fish.

Another example of how the focus on DMIs has clarified what is or is not tool use is the relatively common behavior of active anting. This behavior is performed by many species of birds, especially Passeriformes such as starlings (*Lamprotornis* spp.), and involves holding invertebrates such as ants and caterpillars in the bill and wiping them on the skin or feathers (the precise reason why this is done is unclear). Active anting was considered a controversial case of tool use by Beck (1980), but it is now clearly classified as tool use based on St Amant and Horton's definition (Bentley-Condit & Smith, 2010; Shumaker *et al.*, 2011).

Tool manufacture is much rarer than tool use in the animal kingdom (Shumaker *et al.*, 2011), possibly because it increases the complexity of the tool-using process. In our chapter we specifically investigate why tool use – as opposed to tool manufacture – is rare, and follow St Amant and Horton's definition of tool use (goal 1 of the definition, above).

Patterns in the spread of tool use across the animal kingdom

The comparative method is a powerful tool in evolutionary biology. A look at the spread and frequency of tool use across the animal kingdom might reveal interesting associations leading to hypotheses for the rarity of tool use. Beck (1980) stated that "The evolution of tool behavior appears to represent a complex of parallelism and convergence that is resistant to a simple phyletic analysis." From his catalog of tool use that was similar to the current list of species that are known to use tools (Bentley-Condit & Smith, 2010; Shumaker *et al.*, 2011), he considered it was safe only to conclude "that apes use and manufacture tools more than monkeys, that apes and monkeys use and manufacture tools more than mammals, and that birds and mammals use tools more than other classes." We take a more optimistic view about the degree of meaningful information that a comparative analysis of tool use can reveal. To carry out such an analysis we first identified the main relevant taxonomic groupings from invertebrates to non-human primates. We then made a rough estimation of the frequency of the independent occurrence of tool use in each group (Table 5.1). To do this we estimated when tool use across species probably came from one or many independent innovative events. Independent occurrence could be at any taxonomic level (e.g., order, family, genus or species), but was restricted to one such occurrence per species (excluding agonistic tool use in primates and active anting in passerines).

We split invertebrates into insects, chelicerates (e.g., spiders) and marine species (tool use is not known in invertebrates outside these three groups). We also split both birds (passerines and non-passerines) and mammals (primates and others) into subgroups that distinguish both important divergence in relative brain size and variation in the frequency of species reported to use tools (Lefebvre *et al.*, 2004; Striedter, 2005; Bentley-Condit & Smith, 2010; Shumaker *et al.*, 2011). Finally, both Passeriformes (perching birds) and primates are similar in that they each have a type of tool use that is relatively widespread among species and that appears to be rather stereotyped. In passerines this behavior is active anting, and in primates it is agonistic aimed throwing or dropping of objects in defense. Given that both of these behaviors are thought to have a common historical origin (Potter, 1970; van Schaik *et al.*, 1999), we separated the independent occurrence of these two cases of tool use in passerines and primates from all other independent occurrences. We then calculated a simple ratio for the relative frequency of independent occurrence using the estimated number of known species in a taxonomic group. We stress that the data in Table 5.1 are intended to only provide a general indication of the taxonomic patterns associated with tool use, and that the estimates and ratios are only approximate values. We have also not controlled for observer effort in documenting the presence of tool use. This probably means that the frequency of tool use in those taxonomic groups where there has been relatively little observer effort (e.g., fish and insects) is higher than shown in Table 5.1. However, Nicolakakis and Lefebvre (2000) showed that variation in research effort had no significant effect on data showing differential innovation rates in birds. Therefore, it is possible that variation in research effort has only had minimal influence on the large differences in the frequency of tool use between birds/mammals and invertebrates/fish, shown in Table 5.1.

Table 5.1 Summary of tool use in the animal kingdom.

Taxonomic group	Est. number of independent occurrences	Dominant tool use	Extent of tool use within species	Est. number of known species in group	Independent occurrence ratio
Invertebrates: Chelicerata	2	Subsistence	Probably widespread	77 000	1:38 000
Invertebrates: Myriapoda	None	–	–	13 000	–
Invertebrates: insects	13	Subsistence	Probably widespread	2 000 000	1:153 846
Invertebrates: marine	5	Self-care	Probably widespread	40 000	1:8 000
Fish	2	Subsistence	Probably widespread	28 000	1:14 000
Amphibians	None	–	–	6350	–
Reptiles	None	–	–	8200	–
Birds: non-passerines	29	Subsistence	Usually one to several individuals	5000	1:172
Birds: passerines	1	Self-care; active anting	Unknown	3600	1:3600
Birds: passerines	33	Subsistence	Usually one to several individuals	3600	1:109
Mammals: non-primates	18	Self-care; subsistence	Usually one to several individuals	5000	1:278
Mammals: primates	1	Self-care: agonistic	Probably widespread	420	1:420
Mammals: primates	37	Subsistence	Habitual in some populations	420	1:11

Note

In column 2 we estimated when tool use across species probably came from one or many independent innovative events. Independent occurrence could be at any taxonomic level (e.g., order, family, genus or species), but was restricted to one such occurrence per species. However, we did separate out agonistic tool use in primates and active anting in passerines from other independent occurrences in these two groups. We included tool use by captive animals, but not when it was associated with controlled experiments. In insects, one independent occurrence often covered many species; for example, sand throwing in antlions and wormlions, and transporting food and weaving in nest construction in ants. The estimate of the number of known species in each taxonomic group was obtained from Wikipedia. The following reviews were consulted in addition to individual papers: Beck (1980); Pierce (1986); Lefebvre *et al.* (2002), Bentley-Condit and Smith (2010); Shumaker *et al.* (2011).

Tool use is reported in most of the main taxonomic groups, with notable exceptions being Myripoda (millipedes and centipedes), amphibians and reptiles (Table 5.1). Shumaker *et al.* (2011) cite a reported case of tool use in amphibians that involves horned frogs flicking dirt onto their backs with their legs to hide themselves. However, as in the case of bait fishing in birds discussed earlier, the flicking of dirt in this case involves a static relationship between objects (an ORMA), not a dynamic one (DMI). It is therefore not tool use under St Amant and Horton's (2008) definition. Other than tool use for self-care in active anting by passerines and the agonistic use of objects by primates, the majority of vertebrate species use tools for subsistence. Subsistence is also the main reason why insects are reported to use tools, but marine invertebrates use tools for self-care. The frequency of the independent occurrence of tool use is relatively low in invertebrates and fish compared to that in vertebrates. In vertebrates, non-passerine birds and non-primate mammals have the lowest frequency, and passerines the next lowest. The highest frequency, as expected, is in primates.

It is interesting to speculate on why tool use is absent from certain important taxonomic groups. There is no obvious explanation for the lack of tool use in amphibians and reptiles, or for that matter in other non-tool-using groups such as Myripoda. Given the low frequency of independent occurrence in invertebrates and fish, it might not be surprising that tool use is absent or unknown in Myripoda, amphibians and reptiles where species numbers in each of these groups are relatively small.

The final interesting pattern evident in Table 5.1 is that tool use in invertebrates/fish is usually widespread within species, but tool use in birds/mammals is not usually widespread (excluding active anting and agonistic tool use, respectively). If Hansell and Ruxton's lack-of-utility hypothesis is correct, tool use is therefore generally more useful for a far greater number of individuals within species in invertebrates/fish (with relatively lower cognitive abilities) than it is in birds/mammals (with relatively higher cognitive abilities). However, individuals with increased cognitive abilities (e.g., primates) should be more innovative and thus discover many uses for tools. Therefore, the contrast in the degree of dissemination of tool use between invertebrates/fish and birds/mammals suggests constraints other than utility prevent tool use in individual birds and mammals.

Different processes and mechanisms underlie tool use

The data in Table 5.1 support an important observation made by previous authors about tool use. Based only on the ratios in the last column, the independent occurrence of tool use clearly separates two distinct groups of animals: invertebrates/fish and birds/mammals. This division was noted by Alcock (1972), who saw the split as indicating a difference in the processes underpinning tool use – leading to stereotyped tool use in invertebrates and fish and flexible tool use in birds and mammals. This dichotomy is undoubtedly an oversimplification, but we think that it provides an accurate general description of the different processes behind tool use across animals.

Tool use in invertebrates and fish is generally consistent with stereotyped behavior for several reasons. First, as we mentioned above it is usually widespread within species.

Second, the adult behavior develops with little or no social contact (e.g., sand throwing in antlions). Third, the distribution of any independent occurrence of tool use is often at the genus level (e.g., objects used to transport liquids in *Aphaenogaster* spp. ants) and probably even occurs across closely related genuses (e.g., tool use to construct breeding burrows in Sphecinae digger wasps and sand throwing in Myrmeleontidae antlions). Fourth, the tool use occurs in a highly specific context with relatively little variation across individuals. Last, Haidle (2010) showed in a cognigram of digger wasp tool use that the working memory involved in invertebrate tool use is minimal. Digger wasps use various objects such as small pebbles to compact the tops of their nest burrows in which they seal their eggs, which are laid in paralyzed prey. The complete oviposition process involves stimulus-driven, instinctive behavior and any interruption of it means the female must repeat all or part of the process.

The situation in birds and primates is strikingly different. Active anting in certain bird species and agonistic aimed throwing and/or dropping in primates appear to be widespread and more consistent with stereotyped tool use. However, flexible tool use practiced by virtually all individuals is the exception and may only occur in two non-human species: New Caledonian crows (*Corvus moneduloides*) and chimpanzees (*Pan troglodytes*) (Savage, 2005). Chimpanzees' tool use is strikingly diverse across many contexts, but there is still high variation across individuals and populations, even within the same context, such as foraging (McGrew, 2010). Tool use in the wild by the New Caledonian crow is restricted to foraging and, contrary to speculation by Hansell and Ruxton (2008), is an integral part of their lifestyle and not restricted to periods of seasonal food shortages. We have observed tool use by crows in all months of the year on the island of Maré and have also collected tools from birds throughout the year on mainland Grande Terre (Hunt & Gray, 2002; our unpublished data). Also, the supply of large wood-boring grubs which crows commonly extract with stick tools is available year round because the grubs almost certainly feed for at least two years before pupating (Hunt, 2000). Nevertheless, tool use by the crows still varies considerably across individuals and populations in the diversity and types of foraging tools made and used (Hunt & Gray, 2002, 2003).

An indication that tool use in birds and mammals is mostly flexible is that unlike stereotyped tool use, a particular independent occurrence of tool use is rarely seen in more than one species. Indeed, innovative behavior is commonly seen in birds and primates (Lefebvre *et al.*, 1997; Reader & Laland, 2002), but is rarely described in invertebrates and fish. The ontogeny of tool use in New Caledonian crows and chimpanzees also contrasts with that in invertebrates and fish. In the wild, juveniles' tool use in crows and chimpanzees proceeds through distinct developmental stages over considerable periods of time in close association with their parent(s), when obvious individual learning of tool skills occurs (Matsuzawa, 1994; Inoue-Nakamura & Matsuzawa, 1997; de Resende *et al.*, 2008; Holzhaider *et al.*, 2010a).

Tool use in highly encephalized octopi is potentially very interesting because these invertebrates use tools in several ways (Shumaker *et al.*, 2011). For example, they use water jets in defense and for shelter construction, and use objects to cache themselves from predators and to aid food extraction. Octopi are renowned for their behavioral flexibility and have some of the relatively largest brains among invertebrates (Packard,

1972; Mather, 2008). Their brain structure also appears to have similarities with brain organization in mammals (Hochner *et al.*, 2006; Hochner, 2008). However, it is unknown if tool use in octopi is flexible or stereotyped.

One important point to emphasize is that a mix of stereotyped and flexible tool use exists across both mammals and birds, and even occurs in individual species (e.g.; both agonistic behavior and extractive foraging in chimpanzees). Such a mix of tool use with likely different underlying processes does not appear to exist in invertebrates and fish. The degree to which flexible tool use occurs alongside stereotyped tool use in taxonomic groups is also associated with relative brain size and cognitive abilities. The degree is highest in Corvidae and primates, groups that consist of animals with relatively large brains among birds and mammals, respectively (Reader & Laland, 2002; Lefebvre *et al.*, 2004; Striedter, 2005). If tool use in birds and mammals is predominantly flexible, then this kind of tool use develops more frequently across species than stereotyped tool use evolves.

The excess-of-opportunity problem

In the preceding two sections, we pointed out important associations between cognitive ability and the frequency of tool use across the animal kingdom. These associations provide explanations for the variation in the frequency of tool use between distinctly different groups of species. Hansell and Ruxton (2008) proposed that ecological utility explains the presence or absence of tool use within taxonomic groupings. In this and the following section we show (1) that there is ample opportunity for tool use to evolve in the important area of food extraction, at least in vertebrates, but such opportunities remain mostly unexploited with tools (we call this the excess-of-opportunity problem), and (2) that there are important constraints other than utility on the presence of tool use (see next section). Animals have also rarely evolved tool-like morphological adaptations to more easily exploit rich sources of difficult-to-extract food. In the case of wood-living insects, exceptions include an elongated finger in both the Madagascan aye-aye (Erickson, 1991) and the Australian striped possum (Rawlins & Handasyde, 2002), an elongated top mandible in several species of Hawaiian *Hemignathus* finches (James & Olsen, 2003) and an elongated tongue in woodpeckers. These highly specialized morphological extensions function as tools for extracting rich food resources from wood.

We address point 1 above in this section using three examples in which tool use appears to be underutilized. Our point is not that tool use is required for an animal to obtain its *absolute* daily requirements during foraging. It is that there are unexploited ecological opportunities where tool use could confer a selective advantage over conspecifics who do not use tools.

Termites are a highly rich food source found throughout the tropical and subtropical regions (Eggleton *et al.*, 1994; Paoletti *et al.*, 2003; Figure 5.1). For example, the soldiers of *Syntermes aculeosus*, eaten year round by South America Indians, are of extremely high nutritional value (Paoletti *et al.*, 2003). They are especially rich in proteins, essential amino acids and minerals such as iron and calcium, and also contain a range of other nutrients and essential fatty acids. The chemical makeup of the proteins in these termites is equivalent to that of meat, and their iron content is far higher than that of meat. Termites are difficult and

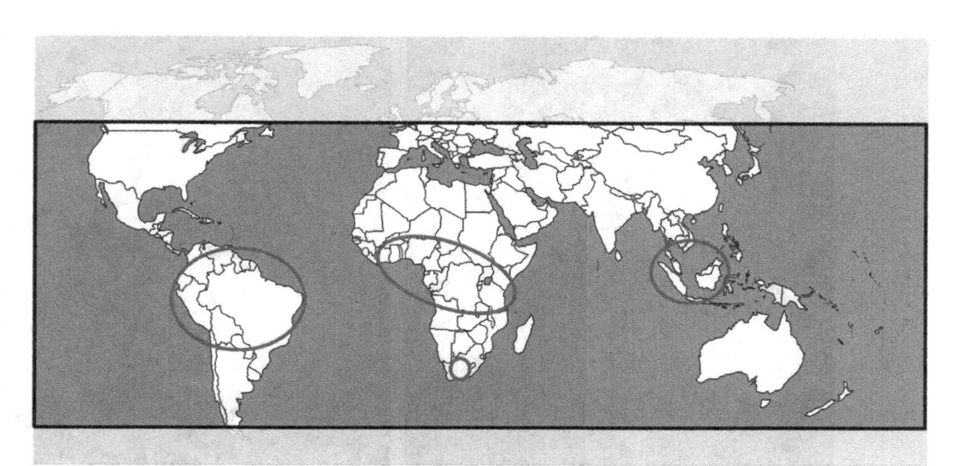

Figure 5.1 Worldwide distribution of termites indicated by the regions within the rectangle. Areas within the circles are where the highest generic richness occurs. Figure adapted from Eggleton *et al.* (1994).

energetically costly to exploit without the use of tools because this requires considerable physical effort to break apart the soil or wood in which they live. However, termites can be efficiently exploited with narrow fishing probes or wands, as used by South American Indians to catch *S. aculeosus*. Surprisingly, only one non-human species (chimpanzees) exploits termites with tools, and this occurs in only a very small region of the worldwide termite distribution. Given that termites are an extremely rich food source, we would have expected other species (e.g., primates) to have evolved the capacity to exploit them efficiently with tools throughout their range, but this is not the case. Furthermore, not only are chimpanzees the only species to exploit termites with tools, termite fishing is absent in some of their populations without any obvious ecological explanation (Whiten *et al.*, 1999). Therefore, tool use to obtain termites contradicts the lack-of-utility hypothesis proposed by Hansell and Ruxton (2008).

Large wood-boring grubs, particularly of the family Cerambycidae, are found both inside and outside tropical/subtropical regions; they therefore have a greater distribution than termites. Wood-boring grubs are also a highly rich food source (Beggs & Wilson, 1987; Rutz *et al.*, 2010). Rutz *et al.* (2010) estimated that only three average-sized larvae would satisfy the entire daily energy requirements of a New Caledonian crow. Like termites, wood-boring grubs are difficult and energetically expensive to exploit without tools because this involves digging into dead, but still relatively hard, wood. Nevertheless, the kaka (*Nestor meridionalis*), a New Zealand parrot, spends a good deal of its feeding time (ca. 35%) excavating large Cerambycidae grubs from dead wood using only its bill (Beggs & Wilson, 1987). Kakas must engage in considerable energetically expensive digging to extract a grub (Figure 5.2). It takes an estimated 81 minutes for a kaka to extract a grub in its pupal chamber with its exit hole cut. Grubs in their pupal chambers are both more restricted in their movements and closer to the surface of the wood compared to younger ones deeper in the wood. Nevertheless, tool use to extract these large grubs is more efficient because only a small hole into a grub's chamber needs to be made to extract it.

Figure 5.2 (a) A kaka using its bill to excavate a hole to capture a large wood-boring Cerambycidae grub (image: Tony Wills). (b) The kaka with the extracted larva in its bill (image: Tony Wills). (c) A New Caledonian crow using a tool to extract the same kind of grub (image: Gavin Hunt). Kaka images from: http://commons.wikimedia.org/wiki/File:Kaka_feeding_01.jpg.

A tool inserted into the chamber can then be used to "fish" or manipulate the grub out of the hole, rather than having to excavate a much larger hole to grasp it directly (Hunt, 2000).

Given the costs in terms of both time and energy of wood excavation, it seems likely that tool use would be useful to kakas, yet they have not evolved it. Instead, only two species habitually exploit wood-boring grubs with tools: woodpecker finches (*Cactospiza pallida*) in the Galápagos Islands (Eibl-Eibesfeldt, 1961) and New Caledonian crows in the south-west Pacific (Hunt, 2000). New Caledonian crows are highly proficient at using tools to extract the large Cerambycidae grubs (*Agrianome fairmairei*) that specialize in feeding on dead wood from candlenut trees (*Aleurites moluccana*) (Hunt, 2000; Bluff *et al.*, 2010) (Figure 5.2). Rutz *et al.* (2010) found from blood and feather samples that *A. fairmairei* constituted a large part of New Caledonian crows' diet in areas where candlenut trees grow. As crows can extract *A. fairmairei* from rotten wood using only their bills, it remains unclear exactly what percentage of the larvae that they eat are extracted with tools. However, that tools are commonly found left at extraction sites suggests that larvae are often obtained with tools (Hunt, 2000; Bluff *et al.*, 2010). Candlenut trees are widely distributed in both the old and new tropics, presumably in association with wood-boring larvae (Elevitch & Manner, 2006). The rich larval food source associated with candlenut trees is therefore distributed in areas where a range of *Corvus* and primate species live. We stress that candlenut trees are only one of many tree species whose dead wood is eaten by wood-boring larvae. That only New Caledonian crows and woodpecker finches habitually exploit wood-boring larvae with tools again contradicts the lack-of-utility hypothesis.

In the first two examples offered here we focused on rich, widespread food sources that were rarely exploited with tools. A comparison of similar foraging niches also suggests that tool use is potentially more useful than its rarity indicates. The New Caledonian crow is a relatively small *Corvus* species and is restricted to living in the southwest Pacific on tropical islands. These birds probably evolved in forest with limited access to carrion and other meat because of the historic lack of mammals and large reptiles such as snakes. Consequently, they forage for mostly small invertebrate prey in vegetation, in both traditional trees and *Pandanus* spp. trees. They can also drive their bills into dead wood in a woodpecker-like fashion to try and excavate grubs (Hunt, 2000); there are no woodpeckers in New Caledonia. Living in this tropical forest niche they evolved the complex tool-using lifestyle that they have today, which is likely to be adaptive compared to a non-tool-using lifestyle. The Mariana crow (*Corvus kubaryi*), now virtually restricted to the small island of Rota, has a very similar foraging niche to that of the New Caledonian crow (Morton *et al.*, 1999; Plentovich *et al.*, 2005). They are relatively small crows (dos Anjos *et al.*, 2009) and probably evolved on small tropical islands in the west Pacific with a similar food supply to that available to New Caledonian crows. For example, they also forage mostly for small invertebrate prey in vegetation that includes *Pandanus* spp. trees (Tomback, 1986). In fact, they even rip leaves in the center of *Pandanus* spp. trees to try and access prey there (pers. commun., Sarah Faegre). As New Caledonian crows demonstrate, tool use is advantageous when foraging in *Pandanus* spp. trees as it enables extraction of prey lodged in the densely packed leaf bases. Thus, the use of tools in foraging could potentially be advantageous to Mariana crows, yet this species does not use tools.

The examples above using termites and wood-boring grubs demonstrate that rich food sources which are exploited much more efficiently with tools remain little used. If the only limiting factor on tool use is utility, over evolutionary time these two food sources should have come to be exploited much more with tools than is the case. The Mariana crow example suggests that there may also be many potential niches where animals could do better in evolutionary terms by using tools compared to conspecifics that do not use tools. It therefore seems unlikely that Hansell and Ruxton's hypothesis is correct, because it claims that tool use is absent only because it has no ecological benefit. In the next section we propose ways besides lack of utility in which tool use can be constrained.

Constraints on tool use other than utility

Pre-existing behavior is necessary for stereotyped tool use

Alcock (1972) speculated that most of the animal tool use known at the time could have evolved from the novel use of "pre-existing behaviour patterns." He also stated that "These behaviour patterns may have been performed in a way that more or less accidentally involved the use of objects as tools in special situations." If this process does occur, it importantly implies a necessarily close match in the underlying activity of the pre-existing behavior and the subsequent tool use. We agree with Alcock that non-tool behavior can plausibly evolve into tool use from a change in context. However, Alcock appears to suggest that tool use could evolve by chance from an inherited pre-existing behavior without phenotypic change. We argue here that stereotyped tool use evolves from an inherited behavior pattern, but only through phenotypic evolution. We also argue that there is a specific, very limited category of pre-existing behavior that can potentially become tool use via this process. This would be ORMAs that can become very similar dynamic mechanical *interactions* (DMIs) (*sensu* St Amant & Horton, 2008), within the same underlying activity. Identifying these specific ORMAs before any tool use evolves may be extremely difficult. For example, the use of sticks in nest building by birds (an ORMA) has been suggested as pre-existing behavior that led to the use of tools by woodpecker finches (Alcock, 1972). However, the extremely low frequency of tool use in birds that build nests out of twigs suggests that this generally stereotyped object use has a very low potential to evolve into tool use. This illustrates that the evolution of stereotyped tool use requires more than an ORMA, it also requires the appropriate ecological context and associated phenotypic change. In birds it would appear that nest-building behavior (collection of material, making the nest) is not closely associated with contexts in which a phenotypic change to the behavior could evolve tool use (e.g., for extracting food in foraging).

Below are six examples of stereotyped tool use from a wide range of taxa that seem to be associated with plausible pre-existing behaviors as defined above. In each of the six cases the DMIs involved in the tool use are associated with plausible pre-existing ORMAs.

(1) *Flicking sand at insects by antlions and wormlions*. Antlions and wormlions dig conical pits in sand by flicking out material. They hide at the base of these pits and flick sand in the direction of small insects such as ants, which they detect on the rim of the pit or inside it (Guillette *et al*., 2009). The sand flicking to help bring insects within reach (DMI) has the obvious pre-existing behavior of flicking sand randomly to maintain the pits (ORMA) (Alcock, 1972).

(2) *Water spitting by fish*. Archer fish shoot water jets at insects on low branches to knock them off so they fall on the water surface (DMI) (Figure 5.3a). These fish also leap out of the water to try and catch insect prey on branches (Figure 5.3b). Leaping out with an

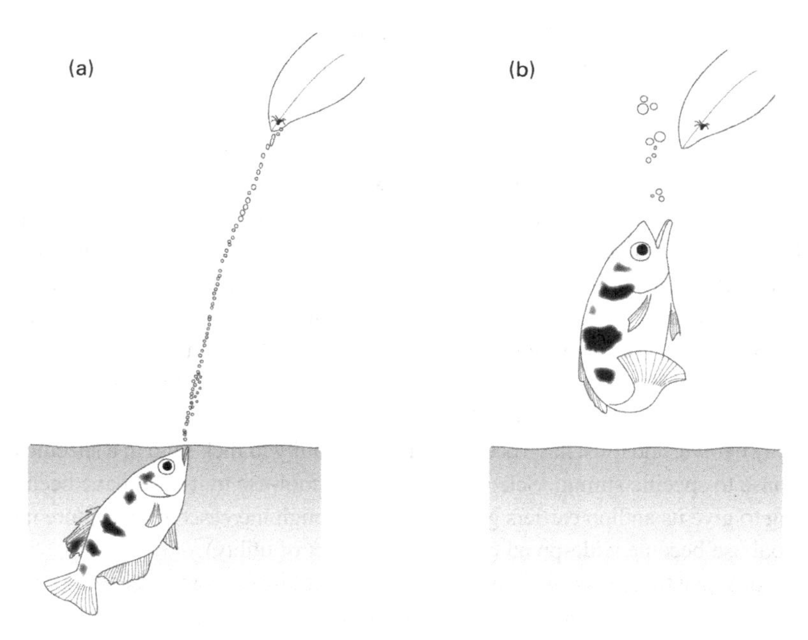

Figure 5.3 An archer fish illustrating a plausible pre-existing behavior leading to tool use. (a) An archer fish using a jet of water as a tool to try and knock an insect off a leaf, rather than leaping up to capture it. (b) An archer fish leaping out of the water to try and directly capture an insect with its mouth. Some jumping archer fish coincidentally cause water to be projected ahead of them when they jump (the drawing is taken from a photograph of an actual jump). If the projected water originates from the mouth, then this would provide a plausible pre-existing behavior for the spitting.

open mouth may also incidentally throw water from the mouth at the insect (ORMA), and this effect could have evolved into aimed shooting of water (Alcock, 1972).

(3) *Throwing rocks at ostrich eggs by Egyptian vultures*. The vultures throw rocks at ostrich eggs to break them (DMI), behavior that can develop without social learning (Thouless *et al*., 1989). They also use the same stereotyped actions to throw eggs on the ground to break them. Alcock (1972) suggested that rock throwing might have started in displacement activity in the presence of ostrich eggs that the vultures could not pick up (ORMA).

(4) *Compacting nest entrances by digger wasps*. Digger wasps seal their nesting burrows in the ground with soil, then compact it with objects such as stones (DMI). Plausible pre-existing behavior in this case could have been the placement of objects on top of the filled-in burrow to cache the entrance (ORMA), behavior also seen in digger wasps (Brockmann, 1985).

(5) *Throwing sand at snakes by California ground squirrels*. The squirrels throw sand with their feet at snakes near their burrows in defensive behavior (DMI) (Coss *et al*., 1993). A possible pre-existing behavior could be sand throwing in displacement activity in response to a predator threat outside excavated burrows (ORMA).

(6) *Agonistic dropping or throwing down of material onto intruders by primates*. Many primate species engage in this agonistic aimed dropping or throwing (DMI) (Hall, 1963; Kortlandt & Kooij, 1963; van Schaik *et al*., 1999; Shumaker *et al*., 2011). This

tool use is suggested to have evolved from displacement activity in response to a threat when tree material was broken off and dropped or thrown in a non-aimed way (Hall, 1963; Kortlandt & Kooij, 1963) (ORMA).

If stereotyped tool use initially evolves as a phenotypic change to a very similar pre-existing behavior (i.e., an ORMA), two main factors combined would affect the frequency with which it initially evolves. These factors are: (1) the frequency of pre-existing ORMAs in animals (phylogenetic constraint); and (2) the opportunity for ORMAs to become tool use after a phenotypic change and persist in the population (utility constraint). Antlion tool use provides a practical demonstration of the above two factors. These animals had the likely pre-existing behavior of flicking sand in pitfall trap construction (factor 1: the effect of phylogeny). A change of context from flicking sand randomly to flicking it in the direction of prey would have led to tool use. The tool-use trait of aimed flicking obviously was able to initially evolve, and must have been based on the ability to flick sand in a specific direction in response to specific stimuli. Selection for the new tool-use trait must have been sufficiently strong to give its antlion carriers greater fitness through increased prey capture rates because the tool use became widespread (factor 2: the effect of utility).

We propose that the evolution of stereotyped tool use is contingent on a complex set of conditions, making it highly unlikely to be a common event. This would explain the relatively very low frequency of the independent occurrence of tool use in invertebrates and fish if, as seems likely, tool use in these animals evolves predominantly from such a process. In contrast, the relatively higher frequency in birds and mammals is probably a consequence of the predominantly learning-based process by which tool use is suggested to develop in these two groups (Alcock, 1972).

Cognitive demands on flexible tool use

In this and the following section we deal with flexible tool use, which does not initially evolve via a novel phenotypic trait from pre-existing behavior. Implicit in St Amant and Horton's (2008) definition is that tool use is relationally more complex than object-related non-tool use in a comparable situation. A tool user not only manipulates an object, but does so in a dynamic, goal-orientated way to affect another object. An ORMA might also be a dynamic action with a particular purpose (e.g., flicking sand in pit maintenance by antlions), but without the goal of affecting another object. We first suggest three ways in which the development of flexible tool use in the wild is constrained by cognitive demands. Call (Chapter 1) also proposes that flexible tool use is based on multiple cognitive skills, which he suggests are to do with cognitive, motivational and sensorimotor aspects. We take a slightly different approach and focus on cognitive, memory and manipulation requirements. Finally in this section, we discuss recent findings in woodpecker finches and rooks (*Corvus frugilegus*) that at first glance appear to provide support for Hansell and Ruxton's lack-of-utility hypothesis.

Constraints

Conceptual knowledge about the use of objects as tools

Frey (2008: 1951) suggested that the ability of humans to use tools in context-dependent ways does not depend solely on sensorimotor processes, but also on "conceptual knowledge about objects and their functions, the actor's intended goals and interpretations of prevailing task demands." Similarly, we propose that the initial innovation of flexible tool use depends on the much more basic conceptual knowledge that an object can be used in a DMI to affect a target (i.e., as a tool). Whether initial tool use by an individual occurs by chance or more intentional behavior, this conceptual knowledge is still required for tool use to be sustained in the wild. We propose that such understanding requires a "cognitive leap" that is extremely difficult for a non-tool-using animal in the wild to make.

Why is a "cognitive leap" necessary and so difficult? We propose that it is very difficult for flexible tool use in the wild to develop from the alternative of trial-and-error learning. This is because of the set of conditions that must be met to enable successful tool use. First, a naive individual would need to be in a situation in which tool use was possible and advantageous (e.g., the availability of extractable food and the potential tools to obtain it). Second, the individual must have sufficient control over an object so that it could be used to tackle the specific problem at hand (i.e., to bring food within reach). Third, it would need to successfully use an object in a completely novel way to solve the problem (i.e., obtain the food). Last, it must rapidly learn *all* the preceding steps of the first successful trial in order to repeat the behavior at another time. A "cognitive leap" that involved an ability to identify and exploit accidental causal interactions quickly would be a much more likely way of initiating tool use. This ability could be underpinned by a range of cognitive skills, such as rapid learning and use of perceptual-motor feedback, causal reasoning and possibly, but not necessarily, mental scenario building (Taylor *et al.*, 2010, 2012a, b).

The requirement to be able to use a familiar object in a novel way may be especially demanding because of functional fixedness (German & Defeyter, 2000; German & Barrett, 2005), particularly if it means using it in a very different context. Behavior that is strongly context-specific, such as nest building and object play, may be especially hard to transfer to a novel context. Our work with New Caledonian crows shows that this effect might also constrain the diversity of tool use in a species by limiting its potential application. As we described earlier, these birds have complex tool skills in the wild and can solve an impressive array of tool-related problems in captivity (e.g., Taylor *et al.*, 2007, 2009a, 2011a, 2012a). However, in the wild they only use tools in a foraging context and only as probes to directly obtain food. The crows need to be in a highly scaffolded setting to grasp the concept of tool use outside their usual foraging techniques. For example, without previous experience of using stones as tools, New Caledonian crows failed to solve the Aesop's Fable test (Taylor *et al.*, 2011a). The Aesop's Fable task involves picking up stones and dropping them into a water-filled tube to raise the water level to get a floating reward. Rooks solved this task, but probably only because they had previous experience of dropping stones into a tube in a different setup (Bird & Emery, 2009b; Taylor & Gray, 2009). However, once the New Caledonian crows learned via accidental stone dropping into the tube that this action raises the water level, they readily dropped stones to obtain the floating food and attended to the functional properties of novel objects while stone dropping. Therefore, even for a tool-using species in a highly scaffolded laboratory setting, being able to conceptualize that a familiar object can be used as a novel tool in a novel way is cognitively demanding. That innovating novel tool behavior is cognitively demanding has been clearly demonstrated with young children

(Cutting *et al.*, 2011). Children as young as two years old are able to use hooks as tools and make inferences about the function and design of artifacts (Brown, 1990; Casler & Kelemen, 2005). However, Cutting *et al.* (2011) found that 24 children of 4–5 years old usually failed on their first trial to innovate relatively straightforward modifications to non-functional tools (bending and unbending pipe cleaners) unless they first received a demonstration (mean failure rate in two tasks was 79%).

As we mentioned above, a captive environment appears especially conducive for initiating tool use in non-tool-using species. Thus both captivity generally, and controlled experimental conditions in particular, appear to provide the scaffolding that facilitates the conceptualization of tool use in certain non-tool-using and tool-using species. This effect may be similar to Alcock's (1972) suggestion that moving into novel environments helps facilitate tool use in animals. Similarly, the concept of tool use related to foraging may develop more frequently in harsher or more competitive environments (Alcock, 1972). The association of tool use with more difficult environmental conditions would provide a similar frequency distribution to that predicted by Hansell and Ruxton's lack-of-utility hypothesis. However, we stress that although the advantages of tool use in harsher conditions may be greater, it does not mean that they are absent in more benign environments.

The cognitive difficulty of increased problem–solution distance

The sequential complexity of tool tasks varies enormously. Even the sequential structure of a simple extraction task using a stick would be cognitively challenging for a first-time tool user (see above). Humans have the most sequentially complex tool tasks in which tools can be used to make tools. Haidle (2010) described the sequential complexity of a tool task as the "problem–solution distance," and used it as a measure of the cognitive complexity, flexibility and decision making involved. The measure indicates the amount of working memory required. It also reflects the relational complexity of the task, defined as the number of related dimensions or sources of variation that can be processed in parallel by working memory (Halford *et al.*, 1998). The relational complexity of a task is considered to be positively correlated with both the processing requirements necessary to carry it out and the amount of working memory needed (Halford *et al.*, 1998). Therefore, the relational complexity of a tool-use task is probably closely positively correlated with its problem–solution distance. Haidle (2010) constructed cognigrams to illustrate the problem–solution distance of tool use for a range of species. The working memory required for stereotyped tool use (digger wasps) was minimal, but increased across species for tool use in dolphins (sponge use for nose protection), chimpanzees (tool set to extract termites), Oldowan humans (stone tools to cut meat) and finally to spear making and use in prehistoric humans. Read (2008) also proposed that important technological advances in human evolution were associated with increases in working memory capacity and the associated greater ability for relational complexity (e.g., greater problem–solution distances). Therefore, the increased problem–solution distance of flexible tool use would appear to place additional cognitive demands on the tool user via working memory capacity compared to non-tool use in a comparable task.

New Caledonian crows provide support for the idea that working memory capacity is important for flexible tool use. This support comes from the association between their problem-solving abilities with tools and their relatively large associative forebrain regions

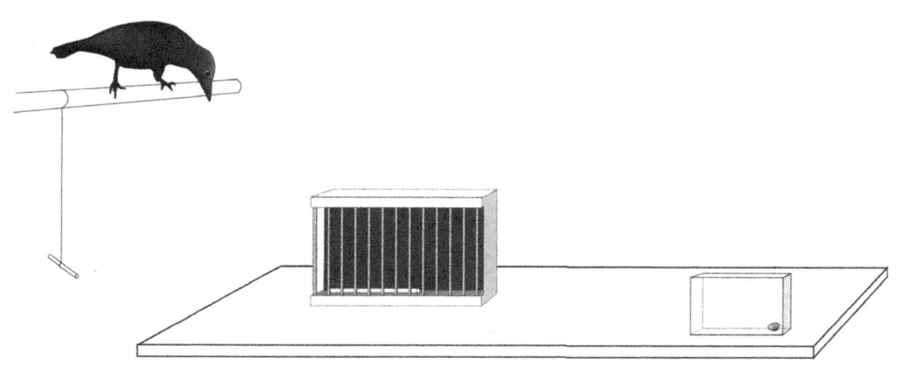

Figure 5.4 A New Caledonian crow and the three-stage metatool task (figure from Taylor *et al.*, 2010).

(Cnotka *et al.*, 2008; Mehlhorn *et al.*, 2010). The crows have recently solved a novel three-stage metatool problem with a considerable problem–solution distance (Taylor *et al.*, 2010). The task involved three distinct stages (Figure 5.4): (1) obtaining a short stick by pulling up a string; (2) using the short stick as a metatool to extract a long stick from a toolbox; and finally (3) using the long stick to extract food from a hole. Three crows with experience of only using string and tools to access food directly successfully solved the problem, innovating two behaviors in the process. Their performance was consistent with the transfer of an abstract, causal rule: "Out-of-reach objects can be accessed using a tool." In other words, the crows appeared to use cognition more complex than basic associative learning mechanisms. If, as seems likely, the sequential three-stage task involved an element of planning and decision making, then it required more working memory resources than would be needed for only associative learning (Haidle, 2010).

The performance of the crows in Taylor *et al.* (2010), together with their large associative brain regions (Mehlhorn *et al.*, 2010), suggest a possibly enhanced working memory capacity. A vital component of working memory is the "central executive" that is represented by one or more cognitive functions (e.g., decision making, planning) (summarized in Haidle, 2010). A cognigram of the crows' behavior in Taylor *et al.* (2010) implicates a range of executive functions such as inhibition, task-relevant decision making, attention to a short-term goal and a plan of action (Figure 5.5). New Caledonian crows have relatively large brains among birds, mostly due to a larger nidopallium and mesopallium (Mehlhorn *et al.*, 2010). The nidopallium is heavily involved in both spatial and non-spatial working memory, and its connections with the highly associative mesopallium may be a crucial cognitive link involving components of working memory (Diekamp *et al.*, 2002). The size of the mesopallium is correlated with innovative and flexible behavior in birds generally (Timmermanns *et al.*, 2000), and is involved in diverse associative functions and the production of complex learned motor sequences (Mehlhorn *et al.*, 2010). The genus *Corvus*, which consists of crows and ravens, provides an ideal opportunity to test hypotheses about the links between relational complexity of behavior, working memory and innovation capacity, and tool use.

Praxic skills

An absence of tool use in the wild does not imply that a species lacks the manipulative ability to use tools. Nor does it mean that a species lacks the cognitive skills to cope with the

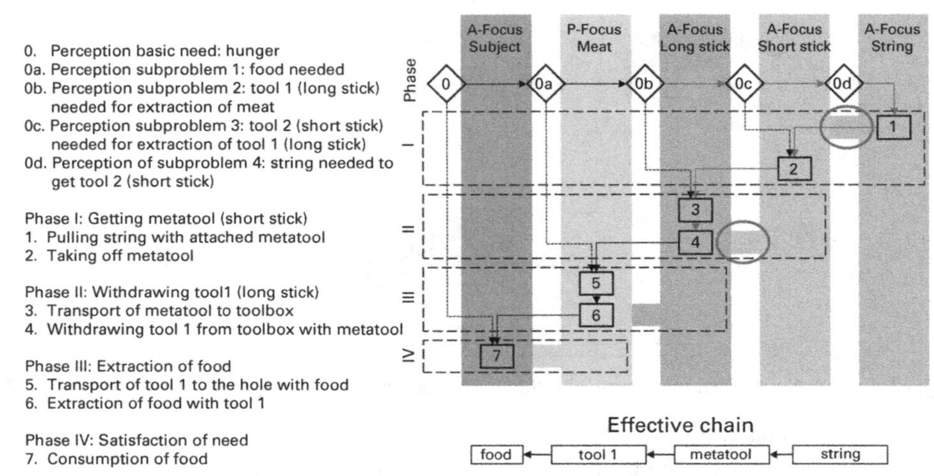

Figure 5.5 Cognigram of the New Caledonian crow behavior in Taylor *et al.* (2010) for the innovation group birds (kindly produced by Miriam Haidle). The diamonds and associated colored vertical bars indicate a new focus of attention (A-Focus = a focus actively controlled by the animal; P-Focus: a passive focus not actively controlled). Connections between foci indicate when one foci affects another foci. The four phases (rectangular boxes across more than one foci) indicate the integration of closely associated activities with a common intermediate aim. The boxes within each phase indicate operational steps or activities. Solid-line arrows (black and gray) show the direction of the primary chain of activities. The gray arrows indicate the novel sequence of perceptions and actions required of the crows to solve the problem. This required the new principal of metatool use and the introduction of two new behaviors with effects to put it into practice (string pulling to obtain the short tool, then its use to extract the long tool; shown by circles). Dashed-line arrows show the direction of additional problem solving. For further details on interpreting cognigrams, see Haidle (2010).

increased problem–solution distances involved in flexible tool use. Some non-tool-using species such as rooks have considerable skills at solving problems with tools in a captive setting (Bird & Emery, 2009a). However, the practical and cognitive hurdles of using a tool for the first time in the wild may place important constraints on the innovation of tool use. An inept attempt may be unsuccessful and therefore unlikely to be repeated. Although non-tool-using birds such as rooks can manipulate objects as tools, their skill at fine manipulation is nowhere near that of expert tool users such as New Caledonian crows. For example, when "fishing" for large grubs the crows can use the tool tip for two different types of precision manipulation (Hunt, 2000). They first use the tool tip to irritate the grub, touching its head or body; they then hold the tip at the mandibles so the now-aggressive grub will latch onto the tool. This kind of precision manipulation of the working end of a tool is not evident in rook tool use (compare video footage of rooks' tool use in Bird & Emery, 2009a to that of New Caledonian crows' tool use in Taylor *et al.*, 2009a). Although juvenile New Caledonian crows are initially inefficient at manipulating tools (Holzhaider *et al.*, 2010a), their tool use develops within a highly scaffolded setting that motivates them to persevere and thus improve their skills. Non-tool-using animals in the wild do not have the advantage of such scaffolding to enhance any initial tool use.

New Caledonian crows may have specific adaptations to enable the proficient use of tools. Indeed, Lefebvre *et al.* (2002) suggested that the larger telencephalon of "true" tool-using birds compared to "proto" tool-using ones might be mostly related to the "subtle coordination of visual and somatosensory information" required for tool use. Hunt and Gray (2007) raised the possibility that New Caledonian crows have adaptive specialization for tool behavior (e.g., precision manipulation of stick tools) evolved through the genetic assimilation of learned behavior (i.e., the Baldwin effect). They also stressed that genetic assimilation of behavior does not eliminate learning. This would explain why New Caledonian crows and woodpecker finches have an inherited disposition specifically for basic stick-like tool use, but still require a considerable amount of time and practice to learn how to successfully use and make a basic tool (woodpecker finches: Tebbich *et al.*, 2001; New Caledonian crows: for stick tools in captivity, see Kenward *et al.*, 2006; for pandanus tools in the wild, see Holzhaider *et al.*, 2010a).

Experienced crows usually align the tool with their bills, with individuals strongly lateralized for the side of the head along which they hold a tool (Hunt, 2000; Hunt & Gray, 2004; Rutledge & Hunt, 2004). Their strongly lateralized tool use is likely to be adaptive because it probably increases coordination skills and problem-solving abilities (Magat & Brown, 2009). A crow extracting food from a hole is capable of doing so using precision movements of the tool tip in much the same way as it would use its bill (Hunt, 2000; Hunt & Gray, 2004). The proficiency with which New Caledonian crows hold and use tools suggests that they might incorporate tools (or the tool tip) into peripersonal space as some primates are reported to do (Maravita & Iriki, 2004). That is, they may use a tool as a physical extension of the bill. Precision tool use by New Caledonian crows probably required behavioral (e.g., lateralized tool holding), morphological (e.g., bill sensitivity and shape, specialized vision for tool manipulation) and neural (e.g., for the complex task of monitoring a tool's working end, lateralized processing) adaptations for tool manipulation after initial tool use developed. New Caledonian crows have recently been shown to have morphological adaptations that facilitate tool use; however, it is unclear whether these evolved before or after the initial appearance of the behavior (Troscianko *et al.*, 2012). Such adaptations would facilitate the development of tool use while at the same time increasing its precision and problem-solving effectiveness. Therefore, it might not be surprising that even an adult non-tool user (e.g., a rook) with previous experience at manipulating objects unrelated to tool use (e.g., nest building) would have considerable difficulty in using a stick tool for the first time in a precise and efficient goal-directed interaction. It would also be interesting to investigate if species like rooks that do not use tools in the wild are capable of the context-dependent tool use exhibited by New Caledonian crows (Taylor *et al.*, 2011b).

Fine motor control of a tool by a human hand is cognitively demanding because of constantly monitoring the spatial and dynamic movements of the tool tip in relation to another object in a predictive way (Wolpert & Flanagan, 2001). There is some experimental support from non-humans that the practical difficulty of flexible tool use is also cognitively challenging. Chimpanzees do not perform well on the trap paradigm when they use tools, but perform much better when they can use their fingers (Seed *et al.*, 2009). Rooks also performed well on trap-tube tests when tool use was not required (Seed *et al.*, 2006; Tebbich *et al.*, 2007). In contrast, New Caledonian crows perform well with trap tasks when they use tools (Taylor *et al.*, 2009a, 2009b). Three out of six crows solved the initial trap-tube task, then successfully transferred to a functionally similar, but perceptually different, trap-table task (Taylor *et al.*, 2009a). Great ape species, including chimpanzees, failed to transfer between similar versions of the trap-tube and trap-table tasks when using tools (Mulcahy & Call, 2006). Seed *et al.* (2009) suggested that tool use might have been cognitively demanding for chimpanzees because of: (1) the need for greater attention; (2) the need to sometimes assess functional properties visually; and (3) decision making about how to use and hold tools.

Do recent findings with rooks and finches support the lack-of-utility hypothesis?

Teschke *et al.* (2011; see also Chapter 7) propose that flexible tool use in species with tool-using lifestyles may not require any special cognition. The authors found no significant differences in the performances of woodpecker finches and their close relatives, (non-tool-using) small tree finches in both physical cognition and general learning tests that did not involve tool use. From this they concluded that woodpecker finches probably had the cognitive abilities for their current level of tool use before the behavior evolved. The implication if this is correct is that all the other Galápagos finches can use tools like woodpecker finches, but lack of ecological need prevents them doing so. A major concern we have is that the above claim is based on experiments that do not involve tool use and therefore cannot test for cognition specific to this behavior. We argued earlier that initial tool use requires a "cognitive leap." One implication of this is that a species with a tool-using lifestyle should need cognitive as well as manipulative adaptations to facilitate and maintain the development of tool use in individuals. A likely cognitive adaptation would be one that facilitated a conceptual grasp of the usefulness of tools and how they work. Our work with New Caledonian crows shows that when individuals of a species with a tool-using lifestyle are able to use tools rather than make a choice between which object to pull, as in the (non-tool) cane task (Teschke *et al.*, 2011; Chapter 7), they perform significantly better (Taylor *et al.*, in preparation). Thus we disagree that Teschke *et al.* (2011) showed woodpecker finch tool use in its current form does not require any special cognitive capacity.

Bird and Emery (2009a; Chapter 4) argue that hand-raised rooks solved tool tasks "insightfully" by using generalized intelligence (i.e., finding solutions by mental scenario building before first attempting a task). The implication if this claim is correct is that "clever" birds and mammals that do not use tools in the wild might still have the concept of a tool and can effectively use one, but like most Galápagos finch species have no ecological need for tools. Bird and Emery's claim of "insight" needs to be treated very cautiously (Kacelnik, 2009; Shettleworth, 2009). First, individuals of many non-tool-using species are reported to use tools only in captivity (Bentley-Condit & Smith, 2010; Shumaker *et al.*, 2011). Thus captivity appears to often provide scaffolding to facilitate the use of objects as tools, which is usually lacking in the wild. As Boesch emphasized (Chapter 2), the cognitive demands that tool users face in the wild may be very different to those in controlled situations in captivity. Second, if imagination-based cognitive mechanisms were widespread in birds and mammals we would expect flexible tool use to be common in the wild, not rare, to take advantage of the excess of opportunity for tool use. Last, apparently insightful behavior can often be explained by relatively simple cognitive mechanisms (Shettleworth, 2010; Taylor *et al.*, 2012). The term "insight" is also considered not to be scientifically rigorous, because it does not identify the cognition responsible for behavior that appears insightful (Kacelnik, 2009; von Bayern *et al.*, 2009; Taylor & Gray, 2009). Thus, if the cognitive mechanisms are not experimentally identified, any claim of insight is highly speculative and open to alternative explanations. Von Bayern *et al.*'s (2009) novel attempt to deconstruct apparently insightful behavior using New Caledonian crows demonstrated the difficulty of pinpointing the cognition involved. Although they showed that innovation of stone dropping was possible if birds only had knowledge of the functional properties of the task, the authors stressed that further experimental work was still essential to determine the exact cognition responsible for the innovative actions (as did Shettleworth, 2009 in her commentary on the paper).

Emery (Chapter 4) bases the claim of insight in rooks on the wire-bending experiment in Bird and Emery (2009a). We agree that the rooks' behavior in this experiment is very impressive, even if insight was not involved. Indeed, rooks appear to have an ability to learn quickly; Taylor and Gray (2010) proposed previous experience combined with rapid learning as a plausible alternative explanation for rooks' seemingly insightful behavior when they solved the Aesop's Fable task (Bird & Emery, 2009b). Next we provide alternative explanations for two crucial aspects of the wire-bending experiment. These aspects are to do with the assertion that the rooks intentionally (1) bent the end of the wire that protruded from the top of the tube because they realized by visually inspecting the apparatus that a bent end was more efficient than a straight end; and (2) flipped the newly modified tool to place the bent end into the tube first so they could use it to extract the bucket containing food. An alternative explanation for bending the wire over the edge of the tube is that the bending is a consequence of the "pulling" action that the rooks used to remove tools from tubes. As the supplemental video footage in Bird and Emery (2009a) clearly shows, rooks do not have the precision tool manipulation skills of New Caledonian crows. A crow would most likely have first used the straight wire in its bill as a tool and found that it did not work (as Betty did; Weir *et al.*, 2002). However, the non-tool-using rook first pulled on the top of the wire, causing it to bend over the top of the tube before continuing to manipulate it. Evidence that this initial pull against the side of the tube was the method the rooks naturally used to try and remove tools is that all of the 12 successful tools in the experiment were bent at both ends to at least some degree (Figure 4.3c). If the rooks had used insight from the very first trial, it must be explained why they failed to extract the bucket on 28 of the 40 trials in which the wire had been bent. Also, even 4–5-year-old children, who have far greater conceptual knowledge about tools and potential for insightful behavior than rooks, usually failed (22 of 24 children) on the first trial to bend a pipe cleaner to solve the same task that the rooks were given (Cutting *et al.*, 2011).

An alternative explanation for the rooks' subsequent flipping of the bent tools comes from Cook's behavior. On its first trial, Cook removed the wire by holding the bent end, then quickly reinserted it. After seeing that the extraction was unsuccessful Cook, perhaps unsurprisingly, reinserted the only available tool to try again. A close look at Cook's behavior in the first trial raises another possible reason for why the bent end was inserted first. At the start of the trial Cook placed the straight wire into the tube by inserting the end it was holding rather than inserting the distal end (as experienced New Caledonian crows would usually do). Thus Cook's flipping of the tool may have just been a consequence of its preference for inserting the proximal end first. Interestingly, non-tool-using New Zealand keas also appear to prefer to hold the working ends of stick tools when initially inserting them into holes (Auersperg *et al.*, 2011).

Dissemination processes for flexible tool use

The occurrence of flexible tool use across individuals within a species is usually highly restricted compared to stereotyped tool use. It is unlikely that utility alone could be responsible for this clear association between the degree of spread of tool use within species and the process underlying tool use. Many cases of occasional tool use in free-living animals are to do with the use of sticks or stones in foraging. Such tool use has disseminated within species in both passerines (woodpecker finches, New Caledonian crows) and mammals (elephants, Andaman long-tailed macaques, bearded capuchins, orangutans, chimpanzees), which suggests that it is particularly useful in a range of situations. The woodpecker finch also shows that flexible tool use underpinned by a specific learning ability can be widespread within species outside the main tool-using groups of primates and *Corvus* species (Tebbich *et al.*, 2001, 2002). So, why is flexible tool use at least similar to that of woodpecker finches so rare? According to Hansell and Ruxton's lack-of-utility hypothesis, the wide application of the proficient use of sticks or stones for foraging within species is rarely useful. We propose that constraints other than utility restrict the spread of such tool use within species.

Implicit in flexible tool use is that learning plays an important role in its ontogeny. This is true for New Caledonian crows and woodpecker finches, which can develop basic tool use without social contact from conspecifics (Tebbich *et al.*, 2001; Kenward *et al.*, 2005; Hunt *et al.*, 2007). Therefore, once flexible tool use is innovated, the cognitive constraints we proposed earlier would continue to affect the chance of transmission to other individuals. The individual(s) in a species who innovate tool use may do so because they have more appropriate cognitive skills for this behavior. A reduced level of these skills in the other individuals that need to acquire rather than innovate tool use may act to constrain its transmission. However, individual-level constraints do not explain why patterns in the spread of flexible tool use within species differ between birds and primates.

Flexible tool use that is widespread within non-human primate species is often contextually diverse and not confined to foraging (McGrew, 2010). It also usually has a "patchy" distribution both locally and geographically within species, a characteristic which has been used as evidence of cultural variation (McGrew, 1992; Whiten *et al.*, 1999, 2001; van Schaik & Knott, 2001; van Schaik *et al.*, 2003; Lycett *et al.*, 2010). However, genetic (Langergraber *et al.*, 2011) or ecological (Byrne, 2007) effects might also have had a role in causing such geographical "patchiness" in primate behavior. In contrast, flexible tool use in woodpecker finches and New Caledonian crows is highly context-specific and much more spatially predicable. Another potentially important difference between birds and primates is that an inherited disposition specifically for the use of stick-like tools to extract food has been found in the finches and crows (Tebbich *et al.*, 2001; Kenward *et al.*, 2005; Hunt *et al.*, 2007). As yet, there is no evidence in non-human primates of an inherited disposition for a specific type of tool use. In fact, chimpanzees' poor performance with stick tools in the trap-tube problem compared to that when using their hands suggests the absence of an avian-like inherited disposition (Seed *et al.*, 2009). Some primate species, though, probably have a general disposition for tool-like actions that precede the development of tool use, such as hitting a surface with an object (Schiller, 1952; Fragaszy & Adams-Curtis, 1997; Takeshita *et al.*, 2005; Chapter 10). The contextually diverse tool use in both chimpanzees and orangutans (Chapter 9) suggests that any inherited disposition for tool use in the great apes is more to do with the generalized use of objects as tools.

The differences in the characteristics of flexible tool use between birds and non-human primates might be caused by different processes and mechanisms of transmission. Birds and primates likely have a differential capacity generally for social learning; both social learning and tool use are correlated with relative brain size in primates, but in birds only tool use appears to be correlated with relative brain size (Reader & Laland, 2002). Although the transmission of tool skills in chimpanzees seems to occur mostly via vertical transmission from mother to juvenile (Matsuzawa *et al.*, 2001; Lonsdorf, 2006; Humle *et al.*, 2009; Chapter 8), the increased social learning capacity of primates and their complex group living probably allows tool use to spread both vertically and horizontally without any inherited learning ability for specific kinds of tool use. In contrast to chimpanzees, tool-using New Caledonian crows' social system appears to minimize the possibility for the horizontal transmission of tool information (Holzhaider *et al.*, 2010b, 2011). This is because the crows in primary forest generally live and forage in small units consisting only of immediate family (mated pair and recent offspring). Juveniles usually stay in close association with both their parents up until at least the following breeding season, at which time they have mostly developed their tool skills (Holzhaider *et al.*, 2010a, 2011). Finally, New Caledonian crows (Singh *et al.*, in preparation) and non-human primates (Chapter 6) appear to focus more on the physical aspects of tool use than the intentional actions, although primates most likely pay greater attention to social information than do the crows.

Thus we propose different models for the dissemination of tool use within bird and non-human primate species. In the bird model, limited social learning combined with an inherited disposition for a specific kind of tool use facilitates vertical transmission and thus the spread of adaptive tool use within species (in contrast to stereotyped tool use, flexible tool use is not necessarily adaptive). Minimizing horizontal transmission also facilitates the standardization of tool skills and therefore their potential enhancement (Sterelny, 2006). The highly context-specific tool use and probable diversification and cumulative change of tool designs in New Caledonian crows are consistent with this bird model (Hunt & Gray, 2003). In the primate model, lack of a disposition for a specific type of tool use and an increased reliance on low-level social learning (providing both horizontal and vertical transmission of tool skills) facilitates greater contextual diversity of tool use within species, consisting of both adaptive and non-adaptive tool use. However, the primate model would limit the standardization, spread and cumulative complexity of any particular tool skill. The contextually diverse tool use in both orangutans and chimpanzees and the often patchy distribution of any particular tool-use variant (Whiten *et al.*, 1999, 2001; van Schaik *et al.*, 2003; Chapter 9) is consistent with the primate dissemination model. The mother–dependent juvenile relationship in chimpanzees and orangutans compared to the biparental care of dependent New Caledonian crows may also play a role in the patchiness of tool variants in the two great ape species. This is because modeling suggests that it is very difficult for single-parent social transmission to maintain cultural conditions over the long term (Enquist *et al.*, 2011).

A complex range of factors would influence the effectiveness of the above dissemination models. These factors would include those that affected the amount and/or quality of both social and individual learning associated with tool use, such as social conformity, tolerance of conspecifics, motivation to watch, learn and experiment, level of parental care and access to tools and tool behavior of experienced individuals (Beck, 1980; Laland *et al.*, 2000; Reader & Laland, 2003; Price *et al.*, 2009; Holzhaider *et al.*, 2010a, 2011). Our work on New Caledonian crows illustrates how the successful generational transmission of tool skills operates in an avian species with a special disposition for stick-like tools. Although we found that the species in primary forest habitat is one of the least social corvids because of relatively few social relationships, the relationships they have are high quality (Holzhaider *et al.*, 2011). The high-quality relationships occur in a social system of small family units, exceptionally long periods of parental care and sometimes delayed dispersal of independent offspring (Holzhaider *et al.*, 2011; Hunt *et al.*, 2012). This social system facilitates the vertical transmission of tool skills because adults scaffold juveniles' learning in their first year of life when tool skills mostly develop (Holzhaider *et al.*, 2010a, 2011). Hunt *et al.* (2012) raised the possibility that the unusually long period of juvenile dependence in the New Caledonian crow might be adaptive to allow juveniles to learn complex tool skills in a close family situation.

Conclusion

Hansell and Ruxton (2008) proposed a simple explanation for the rarity of tool use in animals: that tool use was not often useful. We found that their lack-of-utility hypothesis is fundamentally flawed. One crucial problem with it is that the opportunity for tool use appears to be far greater than the frequency of tool use in animals suggests. We illustrated this excess-of-opportunity problem with two examples of rich food resources (termites and wood-boring grubs) that can be exploited with tools but remain mostly unexploited. We propose that there are constraints more important than utility on the presence of tool use, and that they are different depending on whether tool use is "stereotyped" or "flexible." Stereotyped tool use evolves via new phenotypic traits from a contextual change to pre-existing object manipulation. This kind of tool use is constrained by two main factors: (1) the frequency of pre-existing behaviors that could potentially become tool use; and (2) the opportunity for these pre-existing behaviors to become tool use after a phenotypic change, and to persist in the population. Lack of utility is unlikely to have played a dominant role in the relatively low frequency of stereotyped tool use in animals. Flexible tool use develops via learning-based behavior. We argue that flexible tool use is initially difficult to develop in the wild because of three related cognitive constraints: (1) conceptual knowledge that an object–object interaction can provide an advantageous additional step in a sequence of goal-directed actions; (2) the greater problem–solution distance and relational complexity inherent in tool use and its associated increased working memory requirements; and (3) the practical difficulty of holding and using a tool effectively. We also propose that the different characteristics of tool use within species between birds and non-human primates are dependent on whether inherited dispositions are specific or more general and the degree and type of social learning.

Acknowledgments

We thank the organizers of the "Tool use" workshop and this volume for their invitation to participate. We also thank the two reviewers for their comments on a submitted draft of our chapter.

References

Alcock, J. (1972). The evolution of the use of tools by feeding animals. *Evolution*, **26**, 464–473.

Auersperg, A. M. I., Huber, L. & Gajdon, G. K. (2011). Navigating a tool end in a specific direction: stick-tool use in kea (*Nestor notabilis*). *Biology Letters*, **7**, 825–828.

Beck, B. B. (1980). *The Use and Manufacture of Tools by Animals*. New York: Garland STPM Press.

Beggs, J. R. & Wilson, P. R. (1987). Energetics of South Island kaka (*Nestor meridionalis meridionalis*) feeding on the larvae of kanuka longhorn beetles (*Ochrocydus huttoni*). *New Zealand Journal of Ecology*, **10**, 143–147.

Bentley-Condit, V. K. & Smith, E. O. (2010). Animal tool use: current definitions and an updated comprehensive catalog. *Behaviour*, **147**, 185–221.

Bird, C. D. & Emery, N. J. (2009a). Insightful problem solving and creative tool modification by captive nontool-using rooks. *Proceedings of the National Academy of Sciences USA*, **106**, 10370–10375.

Bird, C. D. & Emery, N. J. (2009b). Rooks use stones to raise the water level to reach a floating worm. *Current Biology*, **19**, 1410–1414.

Bluff, L. A., Troscianko, J., Weir, A. A. S., Kacelnik, A. & Rutz, C. (2010). Tool use by wild New Caledonian crows *Corvus moneduloides* at natural foraging sites. *Proceedings of the Royal Society of London B*, **277**, 1377–1385.

Brockmann, H. J. (1985). Tool use in digger wasps (Hymenoptera: Sphecinae). *Psyche*, **92**, 309–329.

Brown, A. L. (1990). Domain-specific principles affect learning and transfer in children. *Cognitive Science*, **14**, 107–133.

Byrne, R. W. (2007). Culture in great apes: using intricate complexity in feeding skills to trace the evolutionary origin of human technical prowess. *Philosophical Transactions of the Royal Society of London B*, **362**, 577–585.

Casler, K. & Kelemen, D. (2005). Young children's rapid learning about artifacts. *Developmental Science*, **8**, 472–480.

Cnotka, J., Güntürkün, O., Rehkämper, G., Gray, R. D. & Hunt, G. R. (2008). Extraordinary large brains in tool-using New Caledonian crows (*Corvus moneduloides*). *Neuroscience Letters*, **433**, 241–245.

Coss, R. G., Gusé, K. L., Poran, N. S. & Smith, D. G. (1993). Development of antisnake defences in California ground squirrels (*Spermophilus beecheyi*): II. Microevolutionary effects of relaxed selection from rattlesnakes. *Behaviour*, **124**, 137–162.

Cutting, N., Apperly, I. A. & Beck, S. R. (2011). Why do children lack the flexibility to innovate tools? *Journal of Experimental Child Psychology*, **109**, 497–511.

de Resende, B. D., Ottoni, E. B. & Fragaszy, D. M. (2008). Ontogeny of manipulative behavior and nut-cracking in young tufted capuchin monkeys (*Cebus apella*): a perception-action perspective. *Developmental Science*, **11**, 828–840.

Diekamp, B., Gagliardo, A. & Güntürkün, O. (2002). Nonspatial and subdivision-specific working memory deficits after selective lesions of the avian prefrontal cortex. *Journal of Neuroscience*, **22**, 9573–9580.

dos Anjos, L., Debus, S. J. S., Madge, S. C. & Marzluff, J. M. (2009). Family Corvidae (crows). In J. del Hoyo, A. Elliott & D. A. Christie (eds.) *Handbook of the Birds of the World* (pp. 494–640). Barcelona: Lynx Edicions.

Eggleton, P., Williams, P. H. & Gaston, K. J. (1994). Explaining global termite diversity: productivity or history? *Biodiversity and Conservation*, **3**, 318–330.

Eibl-Eibesfeldt, I. (1961). Uber den Werkzeuggebrauch des Spechtfinken *Camarhynchus pallidus* (Scalter und Salvin). *Zeitschrift fur Tierpsychologie*, **18**, 343–346.

Elevitch, C. R. & Manner, H. I. (2006). *Aleurites moluccana* (kukui), ver. 2.1. In C. R. Elevitch (ed.) *Species Profiles for Pacific Island Agroforestry*. Holualoa: Permanent Agriculture Resources.

Enquist, M., Strimling, P., Eriksson, K., Laland, K. & Sjostrand, J. (2010). One cultural parent makes no culture. *Animal Behaviour*, **79**, 1353–1362.

Erickson, C. J. (1991). Percussive foraging in the aye-aye, *Daubentonia madagascariensis*. *Animal Behaviour*, **41**, 793–801.

Fragaszy, D. M. & Adams-Curtis, L. E. (1997). Developmental changes in manipulation in tufted capuchins (*Cebus apella*) from birth through 2 years and their relation to foraging and weaning. *Journal of Comparative Psychology*, **111**, 201–211.

Frey, S. (2008). Tool use, communicative gesture and cerebral asymmetries in the modern human brain. *Philosophical Transactions of the Royal Society of London B*, **363**, 1951–1957.

German, T. P. & Barrett, H. C. (2005). Functional fixedness in a technologically sparse culture. *Psychological Science*, **16**, 1–5.

German, T. P. & Defeyter, M. A. (2000). Immunity to functional fixedness in young children. *Psychnomic Bulletin & Review*, **7**, 707–712.

Goodall, J. van Lawick (1970). Tool-using in primates and other vertebrates. *Advances in the Study of Behaviour*, **3**, 195–249.

Guillette, L. M, Hollis, K. L. & Markarian, A. (2009). Learning in a sedentary insect predator: antlions (Neuroptera: Myrmeleontidae) anticipate a long wait. *Behavioural Processes*, **80**, 224–232.

Haidle, M. N. (2010). Working-memory capacity and the evolution of modern cognitive potential. *Current Anthropology*, **51**, S149–S166.

Halford, G. S., Wilson, W. H. & Phillips, S. (1998). Processing capacity defined by relational complexity: implications for comparative, developmental, and cognitive psychology. *Behavioral and Brain Sciences*, **21**, 803–865.

Hall, K. R. L. (1963). Tool-using performances as indicators of behavioural adaptability. *Current Anthropology*, **4**, 479–494.

Hansell, M. & Ruxton, G. D. (2008). Setting tool use within the context of animal construction behaviour. *Trends in Ecology and Evolution*, **23**, 73–78.

Hochner, B. (2008). Octopuses. *Current Biology*, **18**, R897–R898.

Hochner, B., Shomrat, T. & Fiorito, G. (2006). The octopus: a model for a comparative analysis of the evolution of learning and memory mechanisms. *Biological Bulletin*, **210**, 308–317.

Holzhaider, J. C., Hunt, G. R. & Gray, R. D. (2010a). The development of pandanus tool manufacture in wild New Caledonian crows. *Behaviour*, **147**, 553–586.

Holzhaider, J. C., Hunt, G. R. & Gray, R. D. (2010b). Social learning in New Caledonian crows. *Learning and Behavior*, **38**, 206–219.

Holzhaider, J. C., Hunt, G. R., Sibley, M. D., *et al.* (2011). The social system of New Caledonian crows. *Animal Behaviour*, **81**, 83–92.

Humle, T., Snowden, C. T. & Matsuzawa, T. (2009). Social influences on ant-dipping acquisition in the wild chimpanzees (*Pan troglodytes verus*) of Bossou, Guinea, West Africa. *Animal Cognition*, **12**, S37–S48.

Hunt, G. R. (2000). Tool use by the New Caledonian crow *Corvus moneduloides* to obtain Cerambycidae from dead wood. *Emu*, **100**, 109–114.

Hunt, G. R. & Gray, R. D. (2002). Species-wide manufacture of stick-type tools by New Caledonian crows. *Emu*, **102**, 349–353.

Hunt, G. R. & Gray, R. D. (2003). Diversification and cumulative evolution in tool manufacture by New Caledonian crows. *Proceedings of the Royal Society of London B*, **270**, 867–874.

Hunt, G. R. & Gray, R. D. (2004). Direct observations of pandanus-tool manufacture and use by a New Caledonian crow (*Corvus moneduloides*). *Animal Cognition*, **7**, 114–120.

Hunt, G. R. & Gray, R. D. (2007). Parallel tool industries in New Caledonian crows. *Biology Letters*, **3**, 173–175.

Hunt, G. R., Lambert, C. & Gray, R. D. (2007). Cognitive requirements for tool use by New Caledonian crows (*Corvus moneduloides*). *New Zealand Journal of Zoology*, **34**, 1–7.

Hunt, G. R., Holzhaider, J. C. & Gray, R. D. (2012). Prolonged parental feeding in New Caledonian crows. *Ethology*, **118**, 423–430.

Inoue-Nakamura, N. & Matsuzawa, T. (1997). Development of stone tool use by wild chimpanzees (*Pan troglodytes*). *Journal of Comparative Psychology*, **111**, 159–173.

James, H. F. & Olsen, S. L. (2003). A giant new species of Nukupuu (Fringillidae: Drepanidini: *Hemignathus*) from the island of Hawaii. *Auk*, **120**, 970–981.

Kacelnik, A. (2009). Tools for thought or thoughts for tools? *Proceedings of the National Academy of Sciences USA*, **106**, 10071–10072.

Kenward, B., Weir, A. A. S., Rutz, C. & Kacelnik, A. (2005). Tool manufacture by naive juvenile crows. *Nature*, **433**, 121.

Kenward, B., Rutz, C., Weir, A. A. S. & Kacelnik, A. (2006). Development of tool use in New Caledonian crows: inherited action patterns and social influences. *Animal Behaviour*, **72**, 1329–1343.

Kortlandt, A. & Kooij, M. (1963). Protohominid behaviour in primates. *Symposium of the Zoolological Society London*, **10**, 61–88.

Laland, K. N., Odling-Smee, J. & Feldman, M. W. (2000). Niche construction, biological evolution, and cultural change. *Behavioral and Brain Sciences*, **23**, 131–146.

Langergraber, K. E., Boesch, C., Inoue, E., *et al.* (2011). Genetic and "cultural" similarity in wild chimpanzees. *Proceedings of the Royal Society of London B*, **278**, 408–416.

Lefebvre, L., Whittle, P. W., Lascaris, E. & Finkelstein, A. (1997). Feeding innovations and forebrain size in birds. *Animal Behaviour*, **53**, 549–560.

Lefebvre, L., Nicolakakis, N. & Boire, D. (2002). Tools and brains in birds. *Behaviour*, **139**, 939–973.

Lefebvre, L., Reader, R. M. & Sol, D. (2004). Brains, innovations and evolution in birds and primates. *Brain, Behavior and Evolution*, **63**, 233–246.

Lonsdorf, E. V. (2006). What is the role of mothers in the acquisition of termite-fishing behaviors in wild chimpanzees (*Pan troglodytes schweinfurthii*)? *Animal Cognition*, **9**, 36–46.

Lycett, S. J., Collard, M. & McGrew, W. C. (2010). Are behavioural differences among wild chimpanzee communities genetic or cultural? An assessment using tool-use data and phylogenetic methods. *American Journal of Physical Anthropology*, **142**, 461–467.

Magat, M. & Brown, C. (2009). Laterality enhances cognition in Australian parrots. *Proceedings of the Royal Society of London B*, **276**, 4155–4162.

Maravita, A. & Iriki, A. (2004). Tools for the body (schema). *Trends in Cognitive Sciences*, **8**, 79–86.

Mather, J. A. (2008). Cephalopod consciousness: behavioural evidence. *Consciousness and Cognition*, **17**, 37–48.

Matsuzawa, T. (1994). Field experiments on use of stone tool by chimpanzees in the wild. In R. W. Wrangham, W. C. McGrew, F. B. M. de Waal & P. Heltne (eds.) *Chimpanzee Cultures* (pp. 351–370). Cambridge, MA: Harvard University Press.

Matsuzawa, T., Biro, D., Humle, T., *et al.* (2001). Emergence of culture in wild chimpanzees: education by master-apprenticeship. In T. Matsuzawa (ed.) *Primate Origins of Human Cognition and Behavior* (pp. 557–574). Tokyo: Springer.

McGrew, W. C. (1992). *Chimpanzee Material Culture: Implications for Human Evolution*. Cambridge: Cambridge University Press.

McGrew, W. C. (2010). Chimpanzee technology. *Science*, **328**, 579–580.

Mehlhorn, J., Hunt, G. R., Gray, R. D., Rehkämper, G. & Güntürkün, O. (2010). Tool-making New Caledonian crows have large associative brain areas. *Brain, Behavior and Evolution*, **75**, 63–70.

Morton, J. M., Plentovich, S. & Sharp, T. (1999). *Reproduction and Juvenile Dispersal of Mariana Crows (*Corvus kubaryi*) on Rota 1996–1999*. Honolulu: US Fish and Wildlife Service, Pacific Islands EcoRegion.

Mulcahy, N. J. & Call, J. (2006). How great apes perform on a modifed trap-tube task. *Animal Cognition*, **9**, 193–199.

Nicolakakis, N. & Lefebvre, L. (2000). Forebrain size and innovation rate in European birds: feeding, nesting and confounding variables. *Behaviour*, **137**, 1415–1427.

Packard, A. (1972). Cephalopods and fish: the limits of convergence. *Biological Reviews*, **47**, 241–307.

Paoletti, M. G., Buscardo, E., Vanderjagt, D. J., *et al.* (2003). Nutrient content of termites (Syntermes soldiers) consumed by Makiritare Amerindians of the Alto Orinoco of Venezuela. *Ecology of Food and Nutrition*, **42**, 177–191.

Pierce, J. D. (1986). A review of tool use in insects. *The Florida Entomologist*, **69**, 95–104.

Plentovich, S., Morton, J. M., Bart, J., *et al.* (2005). Population trends of Mariana crow *Corvus kubaryi* on Rota, Commonwealth of the Northern Mariana Islands. *Bird Conservation International*, **15**, 211–224.

Potter, E. F. (1970). Anting by wild birds, its frequency and probable purpose. *Auk*, **87**, 692–713.

Price, E. E., Lambeth, S. P., Schapiro, S. J. & Whiten, A. (2009). A potent effect of observational learning on chimpanzee tool construction. *Proceedings of the Royal Society of London B*, **276**, 3377–3383.

Rawlins, D. R. & Handasyde, K. A. (2002). The feeding ecology of the striped possum *Dactylopsila trivirgata* (Marsupialia: Petauridae) in far north Queensland, Australia. *Journal of Zoology*, **257**, 195–206.

Read, D. W. (2008). Working memory: a cognitive limit to non-human primate recursive thinking prior to hominid evolution. *Evolutionary Psychology*, **6**, 676–714.

Reader, S. M. & Laland, K. N. (2002). Social intelligence, innovation, and enhanced brain size in primates. *Proceedings of the National Academy of Sciences USA*, **99**, 4436–4441.

Reader, S. M. & Laland, K. N. (2003). Animal innovation: an introduction. In S. M. Reader & K. N. Laland (eds.) *Animal Innovation* (pp. 3–35). Oxford: Oxford University Press.

Rutledge, R. & Hunt, G. R. (2004). Lateralized tool use in wild New Caledonian crows. *Animal Behaviour*, **67**, 327–332.

Rutz, C., Bluff, L. A., Reed, N., et al. (2010). The ecological significance of tool use in New Caledonian crows. *Science*, **329**, 1523–1526.

Ruxton, G. D. & Hansell, M. H. (2011). Fishing with a bait or lure: a brief review of the cognitive issues. *Ethology*, **117**, 1–9.

Savage, C. (2005). *Crows: Encounters with the Wise Guys of the Avian World*. Vancouver: Greystone Books.

Schiller, P. H. (1952). Innate motor actions as a basis of learning: manipulative patterns in the chimpanzee. In C. H. Schiller (ed.) *The Development of a Modern Concept* (pp. 264–287). New York: International Universities Press.

Seed, A. M., Tebbich, S., Emery, N. J. & Clayton, N. S. (2006). Investigating physical cognition in rooks (*Corvus frugilegus*). *Current Biology*, **16**, 697–701.

Seed, A. M., Call, J., Emery, N. J. & Clayton, N. S. (2009). Chimpanzees solve the trap problem when the confound of tool-use is removed *Journal of Experimental Psychology*, **35**, 23–34.

Shettleworth, S. J. (2009). Animal cognition: deconstructing avian insight. *Current Biology*, **19**, R1039–R1040.

Shettleworth, S. J. (2010). *Cognition, Evolution and Behavior*. 2nd edn. New York: Oxford University Press.

Shumaker, R. W., Walkup, K. R. & Beck, B. B. (2011). *Animal Tool Behaviour: The Use and Manufacture of Tools by Animals*. Baltimore, MD: Johns Hopkins University Press.

St Amant, R. & Horton, T. E. (2008). Revisiting the definition of animal tool use. *Animal Behaviour*, **75**, 1199–1208.

Sterelny, K. (2006). The evolution and evolability of culture. *Mind and Language*, **21**, 137–165.

Striedter, G. F. (2005). *Principles of Brain Evolution*. Sunderland, MA: Sinauer Associates.

Takeshita, H., Fragaszy, D., Mizuno, Y., *et al.* (2005). Exploring by doing: how young chimpanzees discover surfaces through actions with objects. *Infant Behavior and Development*, **28**, 316–328.

Taylor, A. H. & Gray, R. D. (2009). Animal cognition: Aesop's fable flies from fiction to fact. *Current Biology*, **19**, R731–R732.

Taylor, A. H. & Hunt, G. R., Holzaider, J. C. & Gray, R. D. (2007). Spontaneous metatool use in New Caledonian crows. *Current Biology*, **17**, 1504–1507.

Taylor, A. H., Hunt, G. R., Medina, F. S. & Gray, R. D. (2009a). Do New Caledonian crows solve physical problems through causal reasoning? *Proceedings of the Royal Society of London B*, **276**, 247–254.

Taylor, A. H., Roberts, R., Hunt, G. R. & Gray, R. D. (2009b). Causal reasoning in New Caledonian crows: ruling out spatial analogies and sampling error. *Communicative and Integrative Biology*, **2**, 311–312.

Taylor, A. H., Elliffe, D., Hunt, G. R. & Gray, R. D. (2010). Complex cognition and behavioural innovation in New Caledonian crows. *Proceedings of the Royal Society of London B*, **277**, 2637–2643.

Taylor, A. H., Elliffe, D., Hunt, G. R., et al. (2011a). New Caledonian crows learn the functional properties of novel tool types. *PLoS ONE*, 6, ez6887.

Taylor, A. H., Hunt, G. R. & Gray, R. D. (2011b). Context-dependent tool use in New Caledonian crows. *Biology Letters*, **8**, 205–207.

Taylor, A. H., Miller, R. & Gray, R. D. (2012a). New Caledonian crows reason about hidden causal agents. *Proceedings of the National Academy of Sciences USA*, **109**, 16389–16391.

Taylor, A. H., Knaebe, B. & Grey, R. D. (2012b). An end to insight? New Caledonain crows can spontaneously solve problems without planning their actions. *Proceedings of the Royal Society of London B*, published online October 24, 2012.

Tebbich, S., Taborsky, M., Fessl, B. & Blomqvist, D. (2001). Do woodpecker finches acquire tool-use by social learning? *Proceedings of the Royal Society of London B*, **268**, 2189–2193.

Tebbich, S., Taborsky, M. & Fessl, B. (2002). The ecology of tool-use in the woodpecker finch (*Cactospiza pallida*). *Ecology Letters*, **5**, 656–664.

Tebbich, S., Seed, A. M., Emery, N. J. & Clayton, N. S. (2007). Non-tool-using rooks *Corvus frugilegus* solve the trap-tube task. *Animal Cognition*, **10**, 225–231.

Teschke, I., Cartmill, E. A., Stankewitz, S. & Tebbich, S. (2011). Sometimes tool use is not the key: no evidence for cognitive adaptive specializations in tool-using woodpecker finches. *Animal Behaviour*, **82**, 945–956.

Thouless, C. R., Fanshawe, J. H. & Bertram, B. C. R. (1989). Egyptian vultures *Neophron percnopterus* and ostrich *Struthio camelus* eggs: the origins of stone throwing behavior. *Ibis*, **131**, 9–15.

Timmermanns, S., Lefebvre, L., Boire, D. & Basu, P. (2000). Relative size of the hyperstriatum ventrale is the best predictor of feeding innovation rate in birds. *Brain, Behavior and Evolution*, **56**, 196–203.

Tomback, D. F. (1986). Observations on the behavior and ecology of the Mariana crow. *The Condor*, **88**, 398–401.

Troscianko, J., von Bayern, A. M. P., Chappell, J., Rutz, C. & Martin, G. R. (2012). Extreme binocular vision and straight bill facilitate tool use in New Caledonian crows. *Nature Communications*, **3**, 1110, doi: 10.1038/ncomms2111.

van Schaik, C. P. & Knott, C. D. (2001). Geographic variation in tool use on *Neesia* fruits in orangutans. *American Journal of Physical Anthropology*, **114**, 331–334.

van Schaik, C. P., Deaner, R. O. & Merrill, M. Y. (1999). The conditions for tool use in primates: implications for the evolution of material culture. *Journal of Human Evolution*, **36**, 719–741.

van Schaik, C. P., Ancrenaz, M., Borgen, G. & Galdikas, B. (2003). Orangutan cultures and the evolution of material culture. *Science*, **299**, 102–105.

von Bayern, A. M. P., Heathcote, R. J. P., Rutz, C. & Kacelnik, A. (2009). The role of experience in problem solving in innovative tool use in crows. *Current Biology*, **19**, 1965–1968.

Weir, A. A. S., Chappell, J. & Kacelnik, A. (2002). Shaping of hooks in New Caledonian crows. *Science*, **297**, 981.

Whiten, A., Goodall, J., McGrew, W. C., *et al.* (1999). Cultures in chimpanzees. *Nature*, **399**, 682–685.

Whiten, A., Goodall, J., McGrew, W. C., *et al.* (2001). Charting cultural variation in chimpanzees. *Behaviour*, **138**, 1481–1516.

Wolpert, D. M. & Flanagan, J. R. (2001). Motor prediction. *Current Biology*, **11**, R729–R732.

6 Understanding differences in the way human and non-human primates represent tools: The role of teleological-intentional information

April M. Ruiz
Yale University, Department of Psychology

Laurie R. Santos
Yale University, Department of Psychology

Introduction: redefining man or redefining tools?

On a morning in 1960, Jane Goodall made an observation that would forever change the way scientists think of our own species' place in the animal kingdom: she observed a non-human animal fashioning and using a tool. There, for the first time, Goodall witnessed the famous Gombe chimpanzee David Greybeard fishing for termites. She watched as, over and over, he grabbed a twig, stripped off its leaves, placed it inside a termite mound, and then retracted it to lick off a pile of termites. Even on that morning, Goodall recognized the significance of her observation (Goodall, 1986). At the time, scientists had assumed that humans were the only species capable of a cognitive feat like Greybeard's termite fishing. Indeed, sophisticated tool use had long been heralded as one of the key differences between humans and other animals. With a single observation, Goodall had challenged this understanding of non-human cognition. She excitedly detailed her findings in a telegram to her mentor, the anthropologist Louis Leakey, who replied with his now famous rejoinder: "Now we must redefine 'man,' redefine 'tool,' or accept chimpanzees as humans."

In the five decades that have followed Goodall's original observation, scientists are still struggling with the particulars of Leakey's interpretational challenge. On the one hand, researchers have learned much more about the impressive nature of non-human tool use, thereby redefining what it means to be a tool-using creature. We now know, of course, that humans and chimpanzees are not alone in their use and design of tools. Since Goodall's original observations, scientists have documented cases of tool use in nearly every taxa of the animal kingdom (see reviews in Beck, 1980; Hauser, 2000). We've observed capuchins using hammers (Ottoni & Izar, 2008; Chapter 10), orangutans using spears (van Schaik *et al.*, 2003), cephalopods using costumes (Finn *et al.*, 2009) and crows making fishing hooks (Weir *et al.*, 2002; see also Chapter 5). Indeed, non-human tool use is now known to be both

Tool Use in Animals: Cognition and Ecology, eds. Crickette M. Sanz, Josep Call and Christophe Boesch. Published by Cambridge University Press.

varied – involving a variety of different kinds and combinations of tools – and flexible – with many species employing tools to solve an array of different kinds of problems.

On the other hand, despite learning more about the impressive nature of non-human tool use, there is no denying that humans are special when it comes to the world of tools. Chimpanzees have termite probes and stone hammers, but humans have microscopes and bulldozers, not to mention airplanes, cellphones, microwaves and supercomputers. A quick look at any human living in the modern world reveals that human tool use can be far more varied, complex and specialized than that of any non-human studied to date. More so than any other creature, humans have developed an environment brimming with tools. Unlike non-human animals that tend to use tools for only a small subset of their daily tasks, humans use tools to alter nearly every facet of our daily lives. In doing so, our species has developed a material culture that is undoubtedly unique in both its complexity and its scope.

The puzzle for scientists, then, is the question of *why* human tool use is so different than that of other animals. Clearly, many non-human animals have the cognitive skills to make and use an impressive array of tools. Why, then, don't these species also experience a human-like explosion of tool use and material culture? What differences at the cognitive level account for the wide gap between human and non-human tool use? More specifically, what cognitive capacities are required not just for being a tool user, but also for being a *human-like* tool user?

In this chapter we present one hypothesis for the unique nature of human tool use. In contrast to some previous accounts (e.g., Povinelli, 2000), we argue that the human species' superiority does not emerge because of a human-unique prowess in physical or causal cognition. Indeed, we review recent work suggesting that many species possess a human-like degree of cognitive sophistication when it comes to reasoning about the physical and functional properties of a good tool. Instead, we argue that although humans and other primates share a sophisticated ability to recognize the functionally relevant aspects of tools, they differ greatly in how they reason about the *socially relevant* aspects of tool use and design (see Hernik & Csibra, 2009 for a similar argument). We review recent developmental data on intentional reasoning in humans and the consequences that this emerging understanding has for our understanding of tools. We then review recent social cognition work in non-human primates[1] to suggest that human and non-human primates may differ in their use of intentional information when representing tools. Specifically, we will argue that primates tend to weight physical information more than intentional information when representing the tools around them. We then discuss how this socio-cognitive difference could have led to the wide gap between the kinds of tool use we see in humans versus other species.

What non-human primates know about the physics of tools

As any good human carpenter can attest, one important aspect of being a good tool user is knowing which tool is needed for a specific job. At a physical level, this usually means

[1] Throughout our review of animal tool use, we've chosen to focus only on tool use in the primate order. That said, we believe our analysis of what makes tool use unique will apply equally to non-human animals in other taxa (for a review of work in this area, see Hunt *et al.*'s and Tebbich's chapters in this volume).

recognizing how a tool's physical properties will affect its ability to bring about the desired goal. If your goal is to drive a nail into a board, then you'll need a tool with certain kinds of physical properties: in this case, one with a hard surface rather than a soft one. In contrast, if you need to clean up a spill, then you'll need a tool made out of an absorbent material rather than one that's waterproof. The capacity to recognize which physical properties are pertinent to a tool's functionality has long been thought to be one of the important cognitive components of successful tool use (e.g., Tomasello & Call, 1997; Povinelli, 2000; Santos *et al.*, 2003; Hauser & Santos, 2007; Visalberghi *et al.* 2009)

Given limitations in the scope of non-human primate tool use relative to that of humans, one might initially assume that primates lack an awareness of the functional properties of objects. Indeed, only a decade ago many primate researchers shared this assumption, arguing that primates probably lacked the ability to recognize which physical affordances made an object a good tool (e.g., Tomasello & Call, 1997; Povinelli, 2000). More recent experimental work, however, has demonstrated that several primate species seem to recognize the functionally relevant aspects of potential tools, particularly when these features are readily observable (see reviews in Hauser & Santos, 2007; Penn & Povinelli, 2007). In a typical study, primates are presented with an out-of-reach food reward and given a choice of possible tools with which to obtain the food (capuchins: Fujita *et al.*, 2003; Cummins-Sebree & Fragaszy, 2005; chimpanzees: Furlong *et al.*, 2008; macaques: Ueno & Fujita, 1998; Maravita & Iriki, 2004; lemurs: Santos *et al.* 2005a; marmosets: Spaulding & Hauser 2005; tamarins: Hauser 1997; Hauser *et al.* 2002a; Hauser *et al.* 2002b; Santos *et al.* 2005b; Spaulding & Hauser 2005; vervet monkeys: Santos *et al.*, 2006b). Across a number of different kinds of manipulations, primates generally perform well on these tasks, reliably differentiating between tools that can and cannot be used to obtain the food. Furlong *et al.* (2008), for example, presented chimpanzees with a situation in which different kinds of tools could be used to take in an out-of-reach piece of food. They observed that chimpanzees spontaneously attend to the feature of rigidity when choosing a possible pulling tool, selectively choosing rakes with rigid tops over ones with flimsy tops (for similar results on an analogous task in other species, see: cotton-top tamarins: Santos *et al.*, 2006b; vervet monkeys: Santos *et al.*, 2006b). Primates also seem to recognize that some aspects of a tool's shape matter for its function; ring-tailed lemurs, for example, reliably choose tools with hook-like shapes at the top over tools with shapes that are less effective at hooking a piece of food (see Figure 6.1). In this way, lemurs seem to recognize how different kinds of shapes can affect a tool's functionality (for similar results on an analogous task in other species, see capuchins: Cummins-Sebree & Fragaszy, 2005; cotton-top tamarins: Hauser, 1997; Hauser *et al.*, 2002a; Hauser *et al.*, 2002b; Santos *et al.*, 2005b; Spaulding & Hauser, 2005; great apes: Marin-Manrique & Call, 2010). Finally, primates are successful at ignoring salient perceptual changes that don't affect a tool's function; having been trained to use a pulling tool of one color, vervet monkeys reliably ignore salient color changes even though they avoid tools of a new shape and rigidity (Santos *et al.*, 2006b; for similar results, see: capuchins: Cummins-Sebree & Fragaszy, 2005; cotton-top tamarins: Hauser, 1997; marmosets, lemurs: Santos *et al.*, 2005a).

Figure 6.1 The lemur tool task used in Santos *et al.* (2005a). Lemurs were given a choice of tools whose shape varied. Subjects reliably chose tools with shapes that were functionally relevant for the pulling task.

This attention to the functional features of tools has also been observed in recent experimental work with primates living outside of captivity (see review in Chapter 10). For example, Santos *et al.* (2003) examined whether free-ranging macaques understood the functional features of a tool by using an expectancy violation looking-time study. In this study, macaques were allowed to watch as a human experimenter used a tool to push a grape along a stage. After being familiarized to one kind of tool, monkeys saw several test trials in which the experimenter used a novel tool, one that had changed either in a functionally relevant feature (e.g., shape) or a functionally irrelevant feature (e.g., color). Santos and colleagues found that monkeys looked at the display significantly longer when a newly shaped tool appeared to push the grape, but showed no increase in looking when a newly colored tool performed the same action. In this way, macaques appear to recognize that shape properties are more relevant than color for a tool's function, even in cases when they themselves don't have an opportunity to directly act on the tools. In another set of studies, Visalberghi *et al.* (2009) presented stone-hammer-using wild-bearded capuchin monkeys with novel stone-hammer tools that varied in size and weight. They found that monkeys reliably attended to these features, selectively choosing hammers that were large and heavy enough to crack nuts. Monkeys attended to the correct dimensions even when these features were pitted against each other, selectively choosing a heavy stone that looked small over a light stone that looked larger. Taken together, this body of work suggests that many primates can use a potential tool's observable properties to determine whether a tool will be effective for a certain kind of function. As such, primates' limited tool use cannot be solely the result of a lack of functional understanding at the level of physical affordances.

Although primates do well on tool tasks when dealing with perceptually obvious physical affordances, there is some evidence that primates perform more poorly when a tool's causal properties are less perceptually obvious. For example, primates fail tool-choice tasks that require them to take into account causal forces such as gravity and support. Cotton-top tamarins and vervet monkeys, for example, perform at chance on a tool-choice task that requires them to attend to the substrate on which a tool acts, ignoring small traps that could impede a tool's trajectory (Santos *et al.*, 2006b). Even more competent tool users, such as chimpanzees and capuchins, do poorly on tool tasks that involve unobservable forces like gravity and support (Visalberghi & Trinca, 1989; Povinelli, 2000; Girndt *et al.*, 2008; Seed *et al.*, 2009, but see Hanus & Call, 2008 and Visalberghi *et al.* 2009 for some examples of primates successfully representing unobservable forces). In this way, non-human primates' functional understanding of the physics of tool use appears to be limited to cases in which the physical affordances are perceptually obvious.

While non-human primates' difficulties in reasoning about the unobservable properties of tools may seem like an obvious spot in which their cognitive limitations might hinder tool use, it's also important to note that limitations in non-obvious causal understanding are not only the case for non-human species. Indeed, a growing body of work suggests that even *human* primates possess an unusually limited understanding of the unobserved causal forces that affect a complex tool's function. Although humans in principle have the potential to understand unobservable causal forces as complex as quantum mechanics, work by Keil and colleagues has documented that most people actually know relatively little about the causal properties of even simple tools when probed directly (see review in Keil, 2006). For example, Rozenblit and Keil (2002) found that adult humans can't report how simple mechanical tools – such as zippers, can openers and cylinder locks – actually work. Although participants obviously can be made to understand how these objects work when explanations are provided, in the absence of specific training most people are unable to report how these very simple tools actually function. In an even more surprising failure, Lawson (2006) found that people from a variety of backgrounds (including bike experts) failed to understand even the simplest aspects of how a bicycle works. In practice, then, people also appear to have very limited causal knowledge of the unobservable features of the tools, even when they use such tools every day. Given that humans also appear to have a rather limited physical understanding of the simple tools, how does our species come to successfully deal with and effectively use the many causally complex tools that make up the modern world? If humans don't fully grasp the physics of zippers and screwdrivers, how do they come to successfully use complex tools like microwaves and computers?

Human tool use in a social context

The answer may be obvious to anyone who's ever watched a friend play with his new iPhone, seen a child explain how to use the new videogame controller or observed a television chef use a new kitchen gadget. As humans, much of our understanding of how to use novel tools comes from watching other individuals. Human tool use naturally occurs in a social context; from the time we grow up, we are surrounded by informed conspecifics using complex tools with a specific goal in mind. In this way, we often use social information to go beyond a tool's physical properties in order to figure out how to use a variety of complex tools, even ones whose causal properties we don't fully comprehend at a physical level.

Recent work in developmental psychology has shown that human children come to develop this sort of socially mediated understanding of tools even within the first few years of life. One of the earliest emerging aspects of this understanding is the tendency to think about tools *teleologically* – in terms of *what an intentional agent designed them for* (see Kelemen, 1999a; German & Johnson, 2002; Kelemen & Carey, 2007; Hernik & Csibra, 2009). This teleological stance – representing tools in terms of how they are typically used by others and what kinds of goal-directed actions they were designed to perform – seems to be incredibly important in children's early construal of tools, sometimes even trumping information about a tool's physical properties. By kindergarten, children think that a tool "is" whatever it was designed for, not what it could potentially be used for (Kelemen, 1999a). In one example, Kelemen (1999b) introduced children to a novel tool that was invented to stretch clothes that got shrunk in the wash, but also could be used by people to help stretch their backs. Even though this object was physically able to do both tasks, children reliably reported that this tool *was* a "clothes stretcher," namely the function for which the object was originally invented. Kelemen and colleagues have observed similar teleological biases in even young children. Casler and Kelemen (2007) introduced 24-month-old children to a novel tool performing a particular function (e.g., a "bell-ringer") and children were then allowed to generalize this tool's function to other tasks for which it was equally physically suited (e.g., "pasta crusher"). Even after a lengthy delay, they found that two-year-old children only used the object for its original function, using it to ring bells but not to crush pasta. These results suggest that by two years of age children think of tools not in terms of their physical affordances, but in terms of the goal for which other agents typically use the tools.

In addition to thinking about tools in terms of what they are used for, children also pay special attention to the intentions behind an agent's tool use. Children, for example, seem to care a lot about the intent of a tool's designer and the goal for which this individual intentionally created and used the tool (see Bloom, 1996). Kelemen (1996b), for example, found that children discredit objects that are accidentally used in a certain way, suggesting that an agent's intent has a big impact on how children think a tool should be employed. Bloom (1996) has argued that humans naturally construe tools in terms of their intended history, weighing intentional aspects of a tool's design even more heavily than a tool's physical properties when deciding what a tool is and how it should be used. Bloom points out that we readily label a tool based on what it was designed to do, even in cases where that object no longer has the physical properties needed to fulfill its original goal. In line with this claim, Kemler Nelson *et al.* (2004) showed children common household tools that were broken (e.g., a fork with its tines missing, a safety pin with its receiving end bent) and were thus no longer able to physically perform the actions they were designed to perform. In spite of this, children still labeled these objects as "forks" and "safety pins." In this way, children appear to categorize an object as a particular kind of thing based on its intentional history, not on its physical properties or affordances.

In all of the above cases, children's combined teleological-intentional stance toward tools allows them to go beyond physical affordances alone when representing how a tool works and what it can be used for. These socially mediated aspects of our human tool representations allow for several of the more impressive aspects of human tool use (see review in Hernik & Csibra, 2009). First, our teleological-intentional stance allows us to represent and use any tools demonstrated by a social agent, even when that tool's physical affordances are causally opaque. Humans are, of course, renowned for using artifacts whose physical affordances are both hidden and relatively complex (e.g., a computer with many complex inside parts), a skill that likely results from our ability to glean functional information from others' object-directed actions even in the absence of physically relevant information (see discussion in Hernik & Csibra, 2009).

Nevertheless, our teleological-intentional stance is not without its problems. Humans are sometimes so susceptible to social information about what a tool is for that such information prevents us from making use of a tool's obvious other physical affordances (see review in Hernik & Csibra, 2009). For example, psychologists have long known that adult humans are susceptible to *functional fixedness*, a bias in which we can only view a tool as capable of performing the function for which it was designed. In an original version of this task (Duncker, 1945), adult participants failed to realize that the physical properties of a tool designed to be used as a container (e.g., a tack box) also rendered it capable of being used as a support. Recently, German and colleagues have demonstrated that functional fixedness is a widespread phenomenon; it has been observed in children as young as seven (German & Defeyter, 2000; Defeyter & German, 2003), and also in populations that lack technologically rich environments, such as the Shuar of Ecuadorian Amazonia (German & Barrett, 2005). Humans' early reliance on social information when learning about artifacts can also lead children into a bias known as *overimitation*, in which children follow an adult's way of operating a tool so closely that they incorporate actions that they know are causally irrelevant (Horner & Whiten, 2005; Lyons *et al.*, 2007; McGuigan *et al.*, 2007; Kenward *et al.*, 2010; Nielsen & Tomaselli, 2010). In one overimitation study, Lyons *et al.* (2007) presented children with a novel artifact – a puzzle box that could be opened to reveal a toy. Children were then allowed to watch as an adult demonstrator opened the box. The demonstrator performed two kinds of actions when opening the puzzle box: a causally relevant action, that is, one that was physically necessary for opening the box (e.g., removing a door that blocked the toy); and a causally irrelevant action (e.g., tapping on the top of the box) that even the children knew was irrelevant to actually retrieving the toy. Lyons and colleagues found that children were much more likely to perform the irrelevant actions after watching the adult demonstrator perform them. Merely watching an adult intentionally act on an object therefore appears to change the way children will interact with that object (see also Horner & Whiten, 2005; McGuigan *et al.*, 2007). Indeed, Lyons and colleagues observed that children continued to make this error even when specifically trained to recognize "silly" unnecessary actions and told to not copy these, suggesting that this phenomenon may be more automatic than previously suspected. Indeed, children who grow up in technologically sparse cultures show similar levels of overimitation as children in Western cultures (Nielsen & Tomaselli, 2010).

Lyons and colleagues have interpreted these overimitation results as evidence that another agent's social actions can drastically change the way children represent the causal aspects of how a tool works. Specifically, they have argued that watching someone use a tool intentionally can alter the way a child thinks that tool works physically, especially in cases in which adults act on the tool in inefficient ways. As such, overimitation suggests that children may regularly override their knowledge of an object's functional properties when faced with conflicting social information. In this way, social information seems to deeply affect the way children come to learn about how a tool works.

Is there a social context to non-human primate tool use?

The work reviewed above suggests that social information has a critical impact on the way humans represent and learn to use tools. Do similar factors affect non-human primates' understanding of tools? Put differently, do primates also take a teleological-intentional stance when representing tools? Before launching into whether non-human primates bring an intentional understanding to the tools they use, it's worth noting that primates *do* possess the socio-cognitive capacities needed for a teleological-intentional construal. More specifically, there is a growing body of work in the domain of primate theory of mind suggesting that non-human primates can in fact represent other individuals' actions in terms of their underlying intentions. Chimpanzees, for example, have been shown to selectively imitate a human experimenter's intended actions rather than the accidental actions they actually demonstrate (see Myowa-Yamakoshi & Matsuzawa, 2000; Tomasello & Carpenter, 2005; Buttelmann *et al.*, 2007; see Meltzoff, 1995 for a version of this task in human infants). There is also evidence that chimpanzees (Call *et al.*, 2004) and capuchin monkeys (Phillips *et al.*, 2009) can distinguish between a human who is unwilling to share food and one who intends to share food but is unable to do so (see Behne *et al.*, 2005 for a similar study in human infants). Both chimpanzees and capuchins leave a testing area sooner when dealing with an unwilling experimenter than when dealing with an experimenter who intends to share but has become clumsy. These and other results (see review in Tomasello *et al.*, 2005) suggest that several non-human primate species can detect and use information about an agent's goals and intentions when reasoning about others' actions. These recent theory of mind studies suggest that non-human primates do possess the intentional representations needed for the kind of social construal humans take when representing tools.

The question, then, is whether non-human primates actually *use* the intentional information they represent in theory of mind tasks when learning about how a novel tool works. Unfortunately, only a limited body of work has addressed this issue directly. The work performed to date, however, suggests that primates may be limited in their use of teleological-intentional information. In an early study, Nagell *et al.* (1993) compared children's and chimpanzees' performance on a tool task in which a demonstrator modeled how to use a rake-shaped tool to obtain out-of-reach rewards. In one condition, the participants saw the demonstrator model an effective strategy: placing the rake with the tines facing up, such that the flat edge was able to rake in more rewards. In the other condition, participants saw the demonstrator use a less effective strategy: positioning the rake with the tines down, such that rewards were able to slip through. Children learned to use the tool differently depending on which demonstration they saw, performing worse (i.e., using the tool in such a way that they retrieved fewer rewards) when they saw the demonstrator use the less effective strategy. Chimpanzees, in contrast, were unaffected by the demonstrator's strategy; although they benefited from seeing a model perform the action, they performed equally well on the task no matter which of the two modeled strategies they saw. This early result established that chimpanzees are less affected by the social aspects of tool use they see than human children are.

In a more recent study, Horner and Whiten (2005) directly examined whether chimpanzees were as susceptible as human children to intentionally perform irrelevant actions. They gave chimpanzees and children puzzle boxes like those used by Lyons *et al.* (2007). The subjects were then allowed to watch as an adult human demonstrator illustrated how to open the box using both causally relevant and irrelevant actions. In one condition, the box was opaque, making it difficult for children and chimpanzees to understand which actions were causally relevant and which were irrelevant. Horner and Whiten found that both children and chimpanzees copied the demonstrator's actions in this condition. When chimpanzees were not given any information about the box's physical affordances, they could socially learn from the demonstrator's actions just as well as the human children did. Horner and Whiten also presented both subject groups with a second condition in which the puzzle box was transparent rather than opaque; as such, the physical affordances of the box were made more perceptually obvious, clearly showing which actions were causally relevant and which were irrelevant. Here, Horner and Whiten observed a striking difference in the performance of the two subject groups. As in the study by Lyons *et al.* (2007), children overimitated in the transparent condition, ignoring what they knew to be true about the box's physical affordances after watching an adult intentionally act on the box. Chimpanzees, in contrast, completely ignored the experimenter's irrelevant actions in the transparent condition; when performing the action themselves, chimpanzees eliminated obviously unnecessary steps, acting on the box in the most efficient way (Horner & Whiten, 2005).

Interpreting non-human primates' inability to use teleological-intentional information to represent tools

Although non-human primates represent intentions in a number of different theory of mind tasks, the studies performed to date on primate overimitation demonstrate that non-human primates may not take the kind of teleological-intentional stance that humans do when representing tools. Although other primates can learn socially how to use a tool, they are more swayed by physical information than by a model's intentional actions. While humans represent tools as objects that are *for* the purpose of achieving a specific intention or goal, non-human primates appear to focus more on the physical properties of tools; other primates tend to ignore information about why a tool was designed and how it has historically been used by others. It is possible, then, that this is one of the differences that makes human tool use unique – humans employ their intentional understanding to make sense of the tools around them, but non-human primates do not (or perhaps more accurately, cannot) employ the same understanding when representing tools (see Hernik & Csibra, 2009 for a similar conclusion).

The difficult question now facing researchers is *why* non-human primates don't use a teleological-intentional construal when thinking about tools. Hernik and Csibra (2009) offer an interesting analysis of this question, providing one possible way to interpret the unique nature of humans' teleological-intentional stance. They argue that non-human primates may fail to employ a teleological-intentional stance in part because other primates cannot attend to goals and intentions in the same way as humans; in their view, other primates "do not search for goals" (p. 36). Their review hints that non-human primates might lack more general representational structures for representing goal-directed actions, and thus do not have systems to make sense of goal-directed actions that involve tools. If non-human primates did lack intentional understanding in the way Hernik and Csibra suggest, then they would not be able to bring social information to bear when analyzing how tools work.

Although we agree that non-human primates may not find goals as salient as our own species does, we disagree with Hernik and Csibra's implication that non-human primates *cannot* attend to goals. As we reviewed above, there is strong evidence from studies of theory of mind that primates *can* attend to intentional information when they're not reasoning about others' tool use – non-human primates distinguish between accidents and intentions (Tomasello & Carpenter, 2005), seek out other agents with good intentions (Call *et al.*, 2004; Phillips *et al.*, 2009) and sometimes imitate others' intended actions (Buttelmann *et al.*, 2007). Moreover, chimpanzees do sometimes use information about a person's intentions when representing how a tool works (Horner & Whiten, 2005); the difference, though, is that chimpanzees *only* seem to employ intentional information when the tool in question lacks any obvious physical affordances that might provide clues as to how that particular tool works.

Faced with this pattern of results, we favor a slightly different interpretation than Hernik and Csibra (2009). We argue that non-human primates can attend to intentional information, even when representing tools, but often choose not to weight this information very heavily. Consider, for example, chimpanzees' performance in Horner and Whiten (2005). When the puzzle box tool had no obvious physical affordances (the opaque condition), chimpanzees attended to a demonstrator's intentionally directed actions on the box and copied these actions well in order to obtain the food. In contrast, when the puzzle box had obvious physical affordances, chimpanzees ignored the demonstrator's intended actions. Unlike children, chimpanzees appear to weight physical information more heavily than others' intentional cues when trying to learn how a new tool works. In this way, chimpanzees may be able to use intentional cues when representing tools, but fail to do so in cases where more obvious physical cues are available.

Our idea that chimpanzees and humans differ in their weighting of teleological-intentional cues makes a few relevant predictions when you consider how such a differential weighting might play out in the actual way that humans and non-human primates tend to learn about new tools. Consider perhaps the most prolific non-human tool user, the chimpanzee. Even though chimpanzees use a varied set of tools, they still grow up in a world where nearly all tools available have perceptually obvious functional properties. In a chimpanzee's natural habitat, the most complex tools (e.g., Chapter 8) involve very simple, observable physical affordances: large, heavy objects like hammers; skinny, probing objects like termite-fishing poles; etc. Any situation involving the use of these tools would be a case in which the perceptual aspects of the problem were obvious and, as such, chimpanzees would not be expected to attend to intentional information when watching others using tools. If chimpanzees weight physical information over social information, it would likely mean that they rarely, if ever, resort to thinking of their own real-world tools in terms of intentional information; all tool problems they face would be easily understood using physics alone. In this way, a slightly differential weighting of physical and social information of the kind we've argued for here could lead to large differences in the way chimpanzees see the tools around them. The same would be true for other tool-using non-human primates, who also lack tools with perceptually opaque features in their natural environment.

Our idea that non-human primates selectively weight physical over social information makes a few testable predictions about non-human primates' use of the teleological-intentional stance when watching others' tool use. First, non-human primates should not show the kinds of errors that humans show when given problem-solving tasks involving novel tools. Put differently, non-human primates should be *less* functionally fixed than humans (e.g., Duncker, 1945) – they should be just as successful on an insight problem-solving task when presented with a tool with a known function as when presented with a completely novel tool. This prediction is, in some sense, counter-intuitive. Recall that functional fixedness hinders human tool use, making human participants less good at coming up with solutions to novel tool problems. Our analysis, then, would predict that non-human primates could, in some situations, outperform adult humans on tool tasks, particularly in cases when successful tool use requires inhibiting teleological-intentional information. Primate researchers could therefore profit from developing this kind of functional-fixedness test for non-human primates, perhaps borrowing designs used in developmental studies (e.g., German & Defeyter, 2000).

Another prediction that follows from our analysis concerns ways to teach non-human primates to become better tool users. Our view is that non-human primates *can* attend to teleological-functional properties, but tend not to find this information salient. One possibility that follows from this view is that one might be able to induce non-human primates to use tools with more causally opaque affordances, thereby making intentional information more salient (see, for example, Hanus *et al.*, 2011). Theory of mind studies have found ways of making non-human primates attend more directly to intentional information by making goal-directed gestures more obvious or contextually salient (e.g., setting experiments in a competitive context; see Hare, 2001; Lyons and Santos, 2006; Santos *et al.*, 2006a). Future research may be able to use similar manipulations in tool-use studies in order to force non-human primates to attend to intentional information. We would predict that such manipulations could naturally nudge non-human primates toward more complex tool-use understanding.

In conclusion, then, we have argued that a small difference in attention toward different kinds of information may have led to a cascade of qualitative differences in the kind of tool use of which a species is capable. Humans' sophisticated modern tool use could be predicated on our (perhaps unique) tendency to naturally track social over physical information when watching goal-directed actions on objects. In contrast, we've argued that non-human primates tend to weight physical over social information when watching agents act on objects. This subtle difference in non-human primates' weighting of information may limit the kinds of things they can learn about tools that lack obvious perceptual affordances. In this way, a slight difference in a species' attention to physical versus intentional information may have tipped the balance for our own species, allowing us to ratchet up our material culture in ways no other species has experienced. In this way we've tried to offer an answer to Leakey's challenge. Rather than redefining "man" or "tools," we may need to take a more subtle approach – recognizing that how a species naturally attends to different kinds of information can have large cascading effects on the kinds of material culture they can naturally develop.

References

Beck, B. B. (1980). *Animal Tool Behavior*. New York: Garland Press.

Behne, T., Carpenter, M., Call, J. & Tomasello, M. (2005). Unwilling versus unable: infants' understanding of intentional action. *Developmental Psychology*, **41**, 328–337.

Bloom, P. (1996). Intention, history, and artifact concepts. *Cognition*, **60**, 1–29.

Buttelmann, D., Carpenter, M., Call, J. & Tomasello, M. (2007). Enculturated chimpanzees imitate rationally. *Developmental Science*, **10**, F31–F38.

Call, J., Hare, B. H., Carpenter, M. & Tomasello, M. (2004). "Unwilling" versus "unable": chimpanzees' understanding of human intentional action? *Developmental Science*, **7**, 488–498.

Casler, K. & Kelemen, D. (2007). Reasoning about artifacts at 24 months: the developing teleo-functional stance. *Cognition*, **103**, 120–130.

Cummins-Sebree, S. and Fragaszy, D. (2005). Choosing and using tools: capuchins use a different metric than tamarins. *Journal of Comparative Psychology*, **119**, 210–219.

Defeyter, M. & German, T. (2003). Acquiring an understanding of design: evidence from children's insight problem solving. *Cognition*, **89**, 133–155.

Duncker, K. (1945). On problem solving. *Psychological Monographs*, **58**, 5.

Finn, J. K., Tregenza, T. & Norman, M. D. (2009). Defensive tool use in a coconut-carrying octopus. *Current Biology*, **19**, R1069–R1070.

Fujita, K., Kuroshima, H. & Asai, S. (2003). How do tufted capuchin monkeys (*Cebus apella*) understand causality involved in tool use? *Journal of Experimental Psychology: Animal Behavior Processes*, **29**, 233–242.

Furlong, E. E., Boose, K. J. & Boysen, S. T. (2008). Raking it in: the impact of encultration on chimpanzees' tool use. *Animal Cognition*, **11**, 83–97.

German, T. & Barrett, H. (2005). Functional fixedness in a technologically sparse culture. *Psychological Science*, **10**, 1–5.

German, T. & Defeyter, M. (2000). Immunity to functional fixedness in young children. *Psychonomic Bulletin and Review*, **7**, 707–712.

German, T. & Johnson, S. A. (2002). Function and the origins of the design stance. *Journal of Cognition and Development*, **3**, 279–300.

Girndt, A., Meier, T. & Call, J. (2008). Task constraints mask great apes' ability to solve the trap-table task. *Journal of Experimental Psychology: Animal Behavior Processes*, **34**, 54–62.

Goodall, J. (1986). *The Chimpanzees of Gombe: Patterns of Behavior*. Cambridge, MA: Belknap Press of Harvard University Press.

Hanus, D. & Call, J. (2008). Chimpanzees infer the location of a reward based on the effect of its weight. *Current Biology*, **18**, R370–R372.

Hanus, D., Mendes, N., Tennie, C. & Call, J. (2011). Comparing the performances of apes (*Gorilla gorilla*, *Pan troglodytes*, *Pongo pygmaeus*) and human children (*Homo sapiens*) in the floating peanut task. *PLoS ONE*, **6**(6), e19555.

Hare, B. (2001). Can competitive paradigms increase the validity of social cognitive experiments on primates? *Animal Cognition*, **4**, 269–280.

Hauser, M. (1997). Artifactual kinds and functional design features: what a primate understands without language. *Cognition*, **64**, 285–308.

Hauser, M. (2000). *Wild Minds: What Animals Really Think*. New York: Henry Holt.

Hauser, M. D. & Santos, L. R. (2007). The evolutionary ancestry of our knowledge of tools: from percepts to concepts. In E. Margolis & S. Laurence (eds.) *Creations of the Mind: Theories of Artifacts and Their Representation* (pp. 267–288). Oxford: Oxford University Press.

Hauser, M., Pearson, H. & Seelig, D. (2002a). Ontogeny of tool use in cotton-top tamarins, *Saguinus oedipus*: innate recognition of functionally relevant features. *Animal Behaviour*, **64**, 299–311.

Hauser, M., Santos, L., Spaepen, G. & Pearson, H. (2002b). Problem solving, inhibition and domain-specific experience: experiments on cotton-top tamarins, *Saguinus oedipus*. *Animal Behaviour*, **64**, 387–396.

Hernik, M. & Csibra, G. (2009). Functional understanding facilitates learning about tools in human children. *Current Opinion in Neurobiology*, **19**, 34–38.

Horner, V. & Whiten, A. (2005). Causal knowledge and imitation/emulation switching in chimpanzees (*Pan troglodytes*) and children (*Homo sapiens*). *Animal Cognition*, **8**, 164–181.

Keil, F. C. (2006). Explanation and understanding. *Annual Review of Psychology*, **57**, 227–254.

Kelemen, D. (1999a). Functions, goals and intentions: children's teleological reasoning about objects. *Trends in Cognitive Sciences*, **12**, 461–468.

Kelemen, D. (1999b). The scope of teleological thinking in preschool children. *Cognition*, **70**, 241–272.

Keleman, D. & Carey, S. (2007). The essence of artifacts: developing the design stance. In E. Margolis & S. Lawrence (eds.) *Creation of the Mind: Essays on Artifacts and Their Representation* (pp. 394–403). Oxford: Oxford University Press.

Kemler Nelson, D., Holt, M. & Egan, L. (2004). Two- and three-year-olds infer and reason about design intentions in order to categorize broken objects. *Developmental Science*, **7**, 543–549.

Kenward, B., Karlsson, M. & Persson, J. (2010). Over-imitation is better explained by norm learning than by distorted causal learning. *Proceedings of the Royal Society of London B: Biological Sciences*, **278**(1709), 1239–1246.

Lawson, R. (2006). The science of cycology: failures to understand how everyday objects work. *Memory and Cognition*, **34**, 1667–1675.

Lyons, D. E. & Santos, L. R. (2006). Ecology, domain specificity, and the evolution of theory of mind: is competition the catalyst? *Philosophy Compass* (published online August 2006). doi: 10.1111/j.1747-9991.2006.00032.x.

Lyons, D. E., Young, A. G. & Keil, F. C. (2007). The hidden structure of overimitation. *Proceedings of the National Academy of Sciences USA*, **104**, 19751–19756.

Maravita, A. & Iriki, A. (2004). Tools for the body (schema). *Trends in Cognitive Science*, **18**, 79–86.

Marin-Manrique, H. & Call, J. (2010). Spontaneous use of tools as straws in great apes. *Animal Cognition*, **14**, 213–226.

McGuigan, N., Whiten, A., Flynn, E. & Horner, V. (2007). Imitation of causally opaque versus causally transparent tool use by 3- and 5-year-old children. *Cognitive Development*, **22**, 353–364.

Meltzoff, A. N. (1995). Understanding the intentions of others: re-enactment of intended acts by 18-month-old children. *Developmental Psychology*, **31**, 838–850.

Myowa-Yamakoshi, M. & Matsuzawa, T. (2000). Imitation of intentional manipulatory actions in chimpanzees (*Pan troglodytes*). *Journal of Comparative Psychology*, **114**, 381–391.

Nagell, K., Olguin, K. & Tomasello, M. (1993). Processes of social learning in the tool use of chimpanzees and human children. *Journal of Comparative Psychology*, **107**, 174–186.

Nielsen, M. & Tomaselli, K. (2010). Overimitation in Kalahari bushman children and the origins of human cultural cognition. *Psychological Science*, **21**, 729–736.

Ottoni, E. & Izar, P. (2008). Capuchin monkey tool use: overview and implications. *Evolutionary Anthropology*, **17**, 171–187.

Penn, D. C. & Povinelli, D. J. (2007). Causal cognition in human and nonhuman animals: a comparative, critical review. *Annual Review of Psychology*, **58**, 97–118.

Phillips, W., Barnes, J. L., Mahajan, N., Yamaguchi, M. & Santos, L. R. (2009). “Unwilling” versus “unable”: capuchins’ (*Cebus apella*) understanding of human intentional action? *Developmental Science*, **12**, 938–945.

Povinelli, D. (2000). *Folk Physics for Apes*. New York: Oxford University Press.

Rozenblit, L. R. and Keil, F. C. (2002). The misunderstood limits of folk science: an illusion of explanatory depth. *Cognitive Science*, **26**, 521–562.

Santos, L., Miller, C. & Hauser, M. (2003). Representing tools: how two nonhuman primate species distinguish between functionally relevant and irrelevant features of a tool. *Animal Cognition*, **6**, 269–281.

Santos, L. R., Mahajan, N. & Barnes, J. (2005a). How prosimian primates represent tools: experiments with two lemur species (*Eulemur fulvus* and *Lemur catta*). *Journal of Comparative Psychology*, **119**, 394–403.

Santos, L. R., Rosati, A., Sproul, C., Spaulding, B. & Hauser, M. D. (2005b). Means–means–end tool choice in cotton-top tamarins (*Saguinus oedipus*): finding the limits on primates’ knowledge of tools. *Animal Cognition*, **8**, 236–246.

Santos, L. R., Flombaum, J. I. & Phillips, W. (2006a). The evolution of human mind reading. In S. Platek, J. P. Keenan & T. K. Shackelford (eds.) *Evolutionary Cognitive Neuroscience* (pp. 433–456). Cambridge, MA: MIT Press.

Santos, L. R., Pearson, H. M., Spaepen, G. M., Tsao, F. & Hauser, M. D. (2006b). Probing the limits of tool competence: experiments with two non-tool-using species (*Cercopithecus aethiops* and *Saguinus oedipus*). *Animal Cognition*, **9**(2), 94–109.

Seed, A. M., Call, J., Emery, N. J. & Clayton, N. S. (2009). Chimpanzees solve the trap problem when the confound of tool use is removed. *Journal of Experimental Psychology: Animal Behavior Processes*, **35**, 23–34.

Spaulding, B. & Hauser, M. (2005). What experience is required for acquiring tool competence: experiments with two Callitrichids. *Animal Behaviour*, **70**, 517–526.

Tomasello, M. & Call, J. (1997). *Primate Cognition*. New York: Oxford University Press.

Tomasello, M. & Carpenter, M. (2005). The emergence of social cognition in three young chimpanzees. *Monographs of the Society for Research in Child Development*, **70**, vii–132.

Tomasello, M., Carpenter, M., Call, J., Behne, T. & Moll, H. (2005). Understanding and sharing intentions: the origins of cultural cognition. *Behavioral and Brain Sciences*, **28**, 675–691.

Ueno, Y. & Fujita, K. (1998). Spontaneous tool use by a tonkean macaque (*Macaca tonkeana*). *Folia Primatologica*, **69**, 318–324.

van Schaik, C. P., Ancrenaz, M., Borgen, G., *et al.* (2003). Orangutan cultures and the evolution of material culture. *Science*, **299**, 102–105.

Visalberghi, E. & Trinca, L. (1989). Tool use in capuchin monkeys: distinguishing between performing and understanding. *Primates*, **30**, 511–521.

Visalberghi, E., Addessi, E., Truppa, V., *et al.* (2009). Selection of effective stone tools by wild bearded capuchin monkeys. *Current Biology*, **19**, 213–217.

Weir, A., Chappell, J. & Kacelnik, A. (2002). Shaping of hooks in New Caledonian crows. *Science*, **297**(5583), 981.

7 Why do woodpecker finches use tools?

Sabine Tebbich
University of Vienna

Irmgard Teschke
Max Planck Institute for Ornithology

Introduction

Niko Tinbergen (1963) proposed four levels of analysis in seeking to explain why a given behavior exists: phylogenetic, functional, developmental and mechanistic. He postulated that only the integration of all four levels enables us to fully understand behavior. Animal tool use initially captivated the scientific world because of its resemblance to our own behavior, creating the impression that the origin of our own physical intelligence could be found in our close – and perhaps even distant – animal relatives. For a long time research on animal tool use has focused on the mechanistic and the ontogenetic level. The main question fueling this research was whether the cognitive abilities of humans and animals are on a continuum or whether one or several qualitative delimiting differences exist. Probably for this reason, most research has focused on primates and specifically on apes, our closest relatives. The anthropocentric approach has been helpful in drawing attention to the phenomenon of animal tool use. However, the empirical research on the cognitive abilities underlying this ability has revealed that even chimpanzees (*Pan troglodytes*), our closest relatives, do not possess a human-like understanding of the physical regularities governing tool use (Povinelli, 2000; Penn & Povinelli, 2007). A major contribution that this line of research has made to the field of comparative cognition is the growing awareness that a dichotomous distinction between high- and low-level processes may not be fruitful (Chappell, 2006). To date, the performance of New Caledonian crows (*Corvus moneduloides*) and chimpanzees in various tasks testing physical cognition indicates that their appreciation of these problems lies somewhere between a high-level understanding of the physical principles and low-level appreciation based on associative learning (Tomasello & Call, 1997; Bluff *et al.*, 2007; Emery & Clayton, 2009; Taylor *et al.*, 2009).

Research on tool-using birds is still in its early stages but has already revealed that they show complex tool-related behavior, despite having a brain that is much smaller than that of primates in absolute size. Habitual tool use in the wild has evolved in several bird species (Beck, 1980). For example, Egyptian vultures (*Neophron percnopterus*) throw stones on ostrich eggs (van Lawick-Goodall & van Lawick-Goodall, 1966), green-backed herons

Tool Use in Animals:Cognition and Ecology, eds. Crickette M. Sanz, Josep Call and Christophe Boesch.
Published by Cambridge University Press.

(*Butorides striatus*) use bait to catch fish (Walsh *et al.*, 1985), satin bower birds (*Ptilonorhynchus violaecus*) use bark-wads to paint their bower (Chaffer, 1945) and woodpecker finches use twigs or cactus spines to retrieve insects and spiders from tree holes (Eibl-Eibesfeldt, 1961; Curio & Kramer, 1964). Certainly the most diverse and complex avian tool use known is that of New Caledonian crows: these birds make and use at least three forms of tools to aid prey capture (Hunt, 1996; Hunt & Gray, 2002, 2004a, 2004b).

Despite the fact that this burgeoning new field of research represents a refreshing departure from the traditional orientation toward primates and has, in a brief period, contributed significantly to our understanding of tool use and its related mechanisms, little has changed with respect to the questions driving this work. The main focus in most studies of avian tool use to date has been the underlying mechanisms and ontogeny (reviewed in Emery & Clayton, 2009), while the selective forces that led to the evolution of tool use and ecological relevance of this behavior have received comparably little scientific attention. One reason for the scarcity of ecological studies might be the difficulties involved in gathering quantitative data from tool-using animals in the wild. For example, field biologists are generally privy to only a small proportion of their study subject's life, and their observations are often limited to a certain season. Additionally, these animals often live in areas that are difficult to access for topological, geographic or political reasons. Considerable headway has recently been made in overcoming this problem in New Caledonian crows, but with potential application to a wide variety of other animals through the clever application of miniature cameras (Rutz *et al.*, 2007) and stable isotope analysis of feathers, blood and putative food sources to uncover the nutritional significance of tool use in New Caledonian crows (Rutz *et al.*, 2010).

Woodpecker finches are one of the few tool-using species that can be observed with relative ease in the wild and can be easily habituated to aviaries, which allows controlled experimental investigation. An additional advantage is that woodpecker finches live sympatrically with a closely related, non-tool-using species, the small tree finch (*Camarhynchus parvulus*). Their close phylogenetic relationship makes this an ideal species group for inter-specific comparisons because confounding variables such as differences in ecology and morphology are minimized. Finally, the variation in the tool-using abilities of woodpecker finches between populations has the potential to yield new insights regarding the ecological relevance of tool use and facilitates the investigation of the effect of tool-using experience on cognitive abilities. These favorable conditions set the stage for our study in which we were able to compile a large data set on woodpecker finches and some of their relatives, providing us with the opportunity to integrate all four of Tinbergen's levels of investigation and to paint a comprehensive picture of tool use in this species.

The woodpecker finch

Woodpecker finches are endemic to the Galápagos Islands and belong to the group of tree finches, a subgroup of the Darwin's finch clade. They are mainly insectivorous but sometimes also feed on fruits and nectar (Tebbich *et al.*, 2004). They are most famous for their habitual use of either cactus spines or twigs as tools (Eibl-Eibesfeldt & Sielman, 1962, 1965). Although the tool use of woodpecker finches is context-specific, it is not inflexible. Woodpecker finches modify tools depending on the task at hand, shortening them if they are too long and breaking off transverse twigs or leaves that could hinder the insertion of a twig into tree holes. Anecdotal observations of tool-using behavior also exist for the warbler finch (*Certhidea olivacea*) (Hundley, 1963), the cactus finch (*Geospiza scandens*) (Millikan & Bowman, 1967) and the mangrove finch (*Cactospiza heliobatis*) (Curio & Kramer, 1964), but habitual tool use could not be confirmed (Tebbich *et al.*, 2004; Fessl *et al.*, 2011). In order to understand why woodpecker finches use tools at the functional level, more general information on Darwin's finches and on the ecological conditions under which this group has evolved is essential.

Darwin's finches and the ecological conditions on the Galápagos Islands

The Darwin's finch clade comprises a group of 14 closely related species, 13 of which live on the Galápagos Archipelago and one that lives on Cocos Island. The Darwin's finch group radiated only recently in evolutionary history and differs mainly in beak morphology, which neatly reflects the various foraging niches into which the species have radiated (Grant, 1986).

Although the Galápagos Islands are situated on the equator, environmental conditions are harsh and unpredictable (Grant, 1985). The Humboldt current brings cold water to the Ecuadorian coast, which causes an unusually dry and highly seasonal climate, with a short rainy season from January to April and a dry season for the remainder of the year. Particularly at elevations near sea level, annual rainfall is low (0–300 mm year^{-1}) (Hamann, 1981). In addition to the annual climatic fluctuation in rainfall, the onset of the dry and wet season varies from year to year and the climate is strongly influenced by the irregular El Niño phenomenon, which is characterized by unusually high sea temperatures and heavy rainfall. El Niño events are often followed by severe droughts that cause high mortality in finch populations (Grant & Boag, 1980). Probably as an adaptation to these inhospitable conditions Darwin's finches have developed a suite of unusual behaviors and use of food types that are highly unusual for passerines. These unusual behaviors include the aforementioned tool use of the woodpecker finch and the pecking of the sharp-beaked ground finch (*Geospiza difficilis*) at the developing feather shafts of boobies in order to draw blood (Bowman & Billeb, 1965). This latter species also breaks large booby eggs by bracing their beak against the ground and kicking the egg with both feet to dash it against rocks or push it over a ledge (Schluter, 1984). The large cactus finch (*Geospiza conirostris*), the medium ground finch (*Geospiza fortis*) and the small ground finch (*Geospiza fuliginosa*) use a similar method to dislodge stones so that they can search the exposed ground underneath for arthropods (DeBenedictis, 1966). The small and medium ground finches also glean ticks from tortoises and iguanas (MacFarland & Reeder, 1974). Other unusual food types eaten by Darwin's finches include the afterbirth of sea lions (Grant, 1999), decaying fish and the dead young and undigested fecal remains of seabirds (Bowman & Billeb, 1965).

Due to strict regulations for the protection of Galápagos wildlife and the remoteness of some of the Darwin's finch populations, hardly anything is known about the ecological relevance and the underlying mechanisms of these extraordinary behaviors. The tool use of the woodpecker finch is an exception because woodpecker finches occur on the inhabited island of Santa Cruz. Here, they are easily accessible for field studies and the necessary infrastructure for experimental testing in aviaries is available. Additionally, woodpecker finches on this island occupy very diverse habitats ranging from the harsh and unpredictable coastal areas to a lush cloud forest at higher altitude, where the availability of food resources is more stable. This provided us with the opportunity to test whether tool use provides a measurable advantage where and when it is used.

Concerning the adaptive value of tool use, several complementary possibilities have been suggested. Tool use could be important in collecting a substantial portion of the diet, could extend the feeding range to include otherwise inaccessible prey (Alcock, 1972; Parker & Gibson, 1977; Beck, 1980) and could provide prey of especially high nutritional value (Hladik, 1977; Nishida & Hiraiwa, 1982; Goodall, 1986; Yamakoshi, 1998). Tool use may even compensate for the lack of morphological adaptations and enable residence in otherwise uninhabitable areas (Alcock, 1972; Parker & Gibson, 1977; Beck, 1980). Selection to exploit concealed food resources should be particularly strong in the absence of extractive foraging competitors (Orenstein, 1972; Kenward *et al.*, 2006). In summary, all of these factors should affect the availabilities of concealed and non-concealed food resources, which in turn should influence the profitability of tool use relative to "conventional" foraging techniques and thus the strength of selection on tool use (Rutz & St Clair, in press).

Function: why do woodpecker finches use tools?

We studied the ecology of woodpecker finches on Santa Cruz Island between 1994 and 1998. On this island, as on all of the larger islands of the Galápagos Archipelago, distinct vegetation zones are situated along an altitudinal gradient stretching from deserts at or near sea level to lush cloud forest and moorland at high altitudes (Hamann, 1981). We concentrated our observations on two very distinct vegetation zones: the Arid Zone near the coast and the so-called Scalesia Zone, which ranges 300–600 m above sea level (asl). The Arid Zone begins just inland of the coast and extends up to an elevation of about 80–120 m asl. It is covered by a semi-desert forest consisting of deciduous trees, shrubs and cacti. In this vegetation zone the pronounced seasonality of the climate affects the abundance and distribution of arthropods and thus influences the feeding behavior of the woodpecker finches (Tebbich *et al.*, 2002). During the dry season, when bushes and trees have no leaves, woodpecker finches mostly search for food under the bark of dead trees; in the wet season they also forage on leaves and in lichens (Tebbich *et al.*, 2002, 2004). The Scalesia Zone is an evergreen cloud forest extending to the southern and eastern slopes of the island and is dominated by the tree-like composite *Scalesia pedunculata*. Moisture evaporated from the sea is concentrated in an inversion layer at 300–600 m asl, so it rains all year round in the

Scalesia Zone (Hamann, 1981). Tree trunks and branches in this zone are densely covered with epiphytes, mostly mosses. Moss and leaves are the main foraging substrate of woodpecker finches in the Scalesia Zone during both seasons. We measured the abundance of arthropods in the most important feeding substrates during the dry and the wet season in both study areas and found that arthropod abundance was significantly lower in the Arid Zone than in the Scalesia Zone, and also that arthropods under the bark were harder to access in the Arid Zone. Food abundance was stable throughout the year in the Scalesia Zone, but in the Arid Zone it was significantly lower during the dry season than in the wet season.

Our observations on the frequency of tool use support the hypothesis that it is influenced by the accessibility of concealed versus unconcealed prey: woodpecker finches hardly ever used tools in the stable Scalesia Zone where food is abundant and easy to access, but spent 50% of their foraging time and obtained nearly 50% of their prey using tools in the Arid Zone during the dry season. Thus, among the few tool-using species for which tool-use time budgets are available, woodpecker finches are certainly among the top tool users in terms of frequency of tool use (Mann *et al.*, 2008). However, during the wet season they only used tools for 12% of their foraging time. The reason that woodpecker finches use tools less frequently when food is abundant probably lies in the time costs associated with this foraging behavior: we found that median duration of prey extraction with tools was longer than when conventional foraging techniques were used (Tebbich *et al.*, 2002).

Apart from being of quantitative importance, tool use also provides qualitative advantages. We found that the use of tools extends the feeding range to encompass otherwise inaccessible prey and helps to access prey of high energetic value. Three prey types – spiders, spider egg sacks and Orthoptera – were only obtained with the help of tools, probably because they hide in crevices and tree holes which are not accessible by pecking (Tebbich *et al.*, 2002). Additionally, prey obtained with tool use was significantly larger and had a higher energy content. Thus, even though the foraging success measured as prey items per time spent foraging was similar with tool use and with other foraging techniques, tool use was more profitable (Tebbich *et al.*, 2002).

When we assess the advantage of tool use by comparing the woodpecker finch to its relatives, we find that the woodpecker finch is the only Darwin's finch species that can extract arthropods from tree holes. One other tree finch species, the large tree finch (*Camarhychus psittacula*) can access insects from under bark, but not from tree holes (Tebbich *et al.*, 2004). The powerful biting beak of this species allows it to crack the bark of twigs and to access mining arthropods. However, this morphological adaptation comes at a cost: the curvature of the large tree finch beak increases biting power but also impedes pecking, probing and gleaning. Thus, this morphological adaptation may have caused a narrowing of the niche and dietary specialization of the large tree finch. In contrast, the use of tools extends the function of the beak only temporarily, thus complementing an already versatile array of feeding techniques in the woodpecker finch (Tebbich *et al.*, 2004). The fact that only woodpecker finches among Darwin's tree finches regularly use tools also hints at potential differences in the ontogenetic pathways of this species and other Darwin's finches.

Ontogeny: how do woodpecker finches acquire tool use?

In an experimental study we investigated whether tool use in woodpecker finches is acquired socially. This seemed plausible since previous studies have shown that several forms of tool use in primates develop via social learning (Tomasello *et al.*, 1987; Nagell *et al.*, 1993; Whiten *et al.*, 1996). We took whole broods from the Scalesia Zone and the upper Agricultural Zone, which is an area that was formerly part of the Scalesia Zone but has been transformed into farmland. We split each brood into two groups: half of the chicks were reared with a tool-using model, and the other half were reared with a non-tool-using model. We found that young woodpecker finches that never had the opportunity to watch tool use develop this ability with similar aptitude and reached distinct developmental steps that marked the appearance of new tool-oriented behavior at a similar age as their siblings that were given the chance to observe tool use in adult woodpecker finches (Tebbich *et al.*, 2001). We concluded that, in contrast to chimpanzees, social learning is not necessary for the acquisition of this behavior in woodpecker finches. Instead, the developmental process seems to be strongly dependent on genetically fixed components. Interestingly, New Caledonian crows also appear to have a specific genetic predisposition for tool use, as demonstrated by the finding that they develop basic use of stick tools without a tool-using model (Kenward *et al.*, 2005). However, in contrast to our study, a tool-using demonstrator (a human in the study on New Caledonian crows) stimulated faster development of tool use in juvenile New Caledonian crows (Kenward *et al.*, 2006). Field observations also show that New Caledonian crow parents scaffold the development of wide *Pandanus* tool manufacture and use in juveniles for up to one year (Holzhaider *et al.*, 2010a). Juveniles stay close to their parents and are provided with discarded tools. The early exposure to this discarded tool might help juveniles to form a mental template of functional tool design (Holzhaider *et al.*, 2010b). Data on whether juveniles need social input to learn to make hooked twig tools and stepped *Pandanus* tools are missing.

Our data on the ecological relevance of tool use in woodpecker finches and information about their social system can shed some light on the reasons for the strong genetic predetermination of tool use in this species. For one thing, in contrast to socially living primates, woodpecker finches are solitary and thus parents are likely to be the only available tool-using models. In such a system, reliance on social transmission from parents to offspring during an apparently very short period of association would be a highly risky endeavor. Where the likelihood of encountering important social information is uncertain, selection for a development process based on genetically fixed components could be advantageous, especially given that tool use provides an important part of the woodpecker finch's diet and seems crucial to survival during the dry season in the Arid Zone.

Although our experiment showed that the development of tool use is based on a very specific genetic predisposition, we were able to demonstrate that non-social, individual learning does play an important role during the ontogeny of tool use in serving to improve the efficiency of this behavior (Tebbich *et al.*, 2001). Five individuals developed tool-using techniques that deviated from the tool use performed by birds in the wild, most likely because our artificial crevices differed from natural crevices and tree holes. At some point during the study, each of these birds dropped their tool into the artificial crevice and pulled it out with an upward motion of their beak, thereby levering the prey to within reach at the front of the crevice. After initial success with this technique, the five birds significantly increased their use of this method. These and other observations on learning in tool-using woodpecker finches have altered our conception of how this behavior develops. The ontogenetic unfolding of this complex behavior is determined by a very specific genetic component, but is enhanced through individual learning (Tebbich *et al.*, 2001).

Despite the apparently strong genetic foundation of this behavior, we found a surprising disparity in the ability to use tools among adult woodpecker finches that we caught in the wild and tested in captivity. All woodpecker finches from the Arid Zone used tools in captivity, while fewer then 50% of the birds from the Agricultural Zone and the Scalesia Zone were able to extract prey using tools, and also did not learn to do so after repeated exposure to a tool-using model. Nevertheless, all of our naive juveniles developed the ability to use tools despite the fact that they originated from the Agricultural Zone (Tebbich *et al.*, 2001). This puzzling contradiction suggests that woodpecker finches might have a sensitive ontogenetic phase during which tool use must be learned. Furthermore, it is possible that juveniles need specific environmental conditions during the sensitive period in order to develop tool use. This is the case in black rats (*Rattus rattus*), which learn to extract pine seeds from the cones by systematically stripping the cone's scale in a spiral pattern. As in our study, adult rats were unable to acquire the behavior through social experience (Aisner & Terkel, 1992) and experimental studies suggest that rat pups must learn to strip the cones within a restricted age period (Zohar & Terkel, 1996). In this period, exposure to cones from which the first rows of scales have already been removed seems to be a key factor in the learning process (Zohar & Terkel, 1991). In our study, the adults which were caught in the highlands and the young that we raised in captivity were exposed to different environments as juveniles. In captivity, the young birds were exposed to conditions that probably encouraged the development of tool use (close proximity of tools and prey available in open crevices), whereas the adult birds came from an area where tool use was not necessary and probably also constrained by a low number of tree holes (Tebbich *et al.*, 2002). However, to conclusively confirm the hypothesis that the development of tool use in woodpecker finches takes place during a sensitive phase, more controlled experimentation would be needed. Besides the question of whether tool-using abilities must be learned during a developmental window, another important issue to address is the type and level of cognition involved in learning tool use.

Mechanism: what causes woodpecker finches to use tools?

Enhanced cognitive abilities that are related to a propensity to use tools could range from fast trial-and-error learning to an improved ability to process physically relevant stimuli, or even qualitatively different cognitive abilities such as causal reasoning about physical problems. The ability to abstract the underlying causal structure of a physical problem provides the greatest flexibility in that it does not only allow the formation of task-specific solutions, but also the transfer of acquired knowledge to completely novel situations.

If tool use is indeed associated with enhanced cognitive abilities, two evolutionary scenarios are possible (Kacelnik, 2009). In the first scenario, cognitive abilities that evolved in other contexts preceded and facilitated the evolution of tool use. For example, valuable food that is difficult to access and an unpredictable environment might have pushed some animals to evolve cognitive abilities for coping with these challenges. Such adaptations might then have acted as the cognitive foundation from which tool use developed. If this is the case, the enhancement of cognitive competence should not be strictly limited to tool-using species, but should also extend, for example, to closely related, non-tool-using extractive foragers.

Alternatively, tool use itself may have driven the evolution of cognition in the form of a positive feedback loop: the initial evolution of tool use might have exposed tool-using individuals to new, context-specific selection pressures, which in turn stimulated the improvement of learning abilities. Both general learning abilities, such as trial-and-error learning, and more specific cognitive adaptations that are solely tailored to solve the physical problems underlying tool use, could be selected for via this feedback loop. In woodpecker finches it is plausible that tool-using individuals had a selective advantage because they had access to more high-quality food during the dry season. Once the technique spread through the population, traits that improved the efficiency of this advantageous behavior may have been favored by selection.

These enhanced cognitive abilities of tool users could be qualitative – i.e., only tool users appreciate the function of hooks – but also could be of a quantitative nature – i.e., tool users learn spatial relationships that are relevant when using tools faster than non-tool users.

One way to identify an adaptive specialization is the comparative approach, which may take two forms. One is to compare distantly related tool-using species and test whether the use of tools correlates with a convergent evolution of enhanced cognitive abilities. The other is to compare tool-using species with closely related non-tool-using relatives and see whether their cognitive abilities diverge. The strongest support for cognitive adaptations associated with tool use is found when comparisons of two or more distantly related groups yield similar patterns of divergence within the group and convergence between the groups (Harvey & Pagel, 1991). Unfortunately, tool use in animals is rare and therefore the study of correlated evolution within groups that contain several tool-using species is hardly possible. Thus, we have to rely on multiple pair-wise comparisons from different taxa. Most comparative research on tool use to date has focused on the comparison of distantly related tool-using species to identify patterns of convergence, while comparisons between closely related tool-using and non-tool-using species are rare. However, a full understanding of the true relationship between tool use and physical cognition is only possible through the synthesis of both approaches. Additionally, few truly comparative data sets exist. Data are often gathered using different paradigms, training procedures or variations on the same paradigm which, though often are well-intentioned improvements on the original, greatly complicate interpretation from a comparative perspective (Santos *et al.*, 2006).

In our attempt to unravel whether tool use in woodpecker finches evolved in conjunction with enhanced cognitive abilities, we pursued both approaches by comparing the performance of woodpecker finches in various learning tasks to other distantly related tool-using species and to a closely related non-tool-using one.

For this purpose we caught adult woodpecker finches and small tree finches and held them in aviaries at the Charles Darwin Research Station on Santa Cruz Island for the duration of the experiments. In the physical tasks, we first tested whether the animals could solve an initial problem, and in some tasks whether they could assess the problem in advance or only when given many learning trials. In the latter case the birds had to reach a learning criterion in a certain number of trials, which was dependent on the task difficulty (ranging from 8 to 160 trials). The success criterion required a bird to make 14 or more correct choices within two consecutive blocks of ten trials (for more details, see Tebbich & Bshary, 2004; Tebbich *et al.*, 2010; Teschke *et al.*, 2011). For some tasks, following success in the initial problem we subsequently tested whether and how quickly birds were able to transfer the acquired knowledge to a conceptually similar but perceptually novel configuration of each problem. We deem this to be an ecologically relevant ability for most tool-using species even if they only use tools in one context. In the transfer tasks birds only received up to 30 trials, since the emphasis was on what rule they could immediately apply to the problem at hand.

Comparison between distantly related tool-using species

In order to test whether the cognitive abilities of distantly related tool users show convergent patterns, we tested woodpecker finches in three experimental designs: the trap-tube task, a tool-modification task and a tool-length task (Tebbich & Bshary, 2004). The trap-tube task, the two-trap-tube task and modification tasks have previously been used to test primate species (Visalberghi *et al.*, 1995), and New Caledonian crows have been tested in the tool-length task (Chappell & Kacelnik, 2002; Tebbich & Bshary, 2004).

The trap-tube task

In this task an animal must use a tool to extract a food reward from a horizontal tube. The tube has a "trap" along its length into which the food will drop if pulled or pushed over it (Figure 7.1a). In order to solve this task, an animal has to use the tool to push or pull the food from the correct end of the tube so that the food can be extracted without falling into the trap. They must learn to do this at above chance level in a series of trials. The traditional control used to assess how an animal has solved this task is the inverted trap-tube in which the trap is now on top of the tube and therefore non-functional (Figure 7.1b). The reasoning underlying this control is that if animals have understood the causal nature of the task, they should no longer avoid an ineffective trap (Visalberghi & Limongelli,

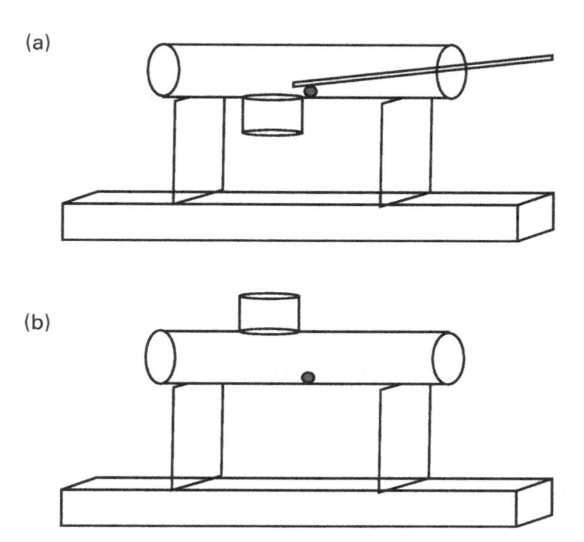

Figure 7.1 (a) The trap-tube task; (b) the inverted trap-tube.

1994). However, this control task was later criticized by Silva *et al.* (2005), who argued that there is no incentive to stop avoiding the trap and found that even some humans continued to avoid the inverted trap. In series of early studies, 3 out of 12 chimpanzees (Limongelli *et al.*, 1995; Povinelli, 2000) and one out of four capuchin monkeys (*Cebus apella*) (Visalberghi & Limongelli, 1994) learned to avoid the trap and gain access to the food by pushing the food away from themselves. However, the successful primates subsequently failed the control test in which the trap was non-functional by continuing to avoid the trap. This clearly demonstrated that they had solved the initial task via a simple procedural rule, namely "insert the tool into the opening farthest from the reward," and therefore showed no understanding of the physical problem (Visalberghi & Limongelli, 1994). Some recent studies have demonstrated that the manner of task presentation plays an important role in learning and performance in this task. For example, several ape species perform better and are even capable of passing the transfer task when the tube is wide enough to allow them to pull the food toward themselves (Mulcahy & Call, 2006).

In our study, one of six woodpecker finches solved the initial trap-tube task. In the transfer task with the inverted tube, this bird reverted to random choice of sides and was therefore the first individual of all animals tested in this task to pass the control test – an indication of causal understanding. However, more detailed analysis revealed that the bird achieved this result by making careful observations of the effect of his action upon the reward: In most trials he inserted the tool a few times from both sides, carefully moving the reward (Tebbich & Bshary, 2004) and switching sides whenever the food moved closer to the trap. While this does not necessarily exclude causal understanding as an explanation for this subject's behavior, there is a simple and more parsimonious associative explanation of his actions – namely use of the spatial relationship between food and trap as a cue.

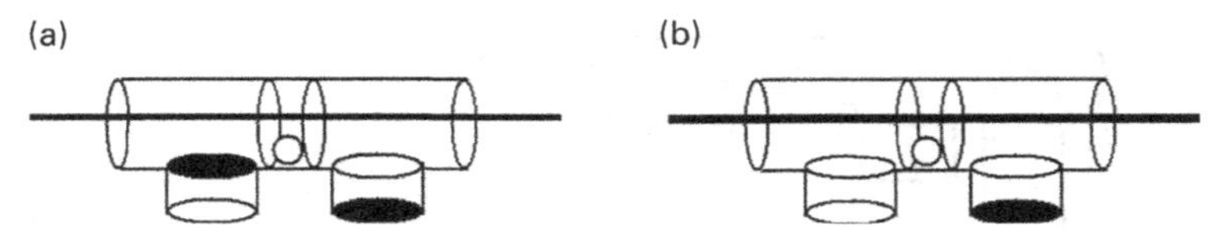

Figure 7.2 The two-trap-tube task: (a) Tube A; (b) Tube B. In both cases, the correct decision is to pull the left-hand side of the stick.

Two-trap-tube task

The two-trap-tube task that we used in our more recent experiments (Teschke & Tebbich, 2011) is a modification of the original trap-tube. It was developed to test non-tool-using rooks (*Corvus frugilegus*) (Seed *et al.*, 2006; Tebbich *et al.*, 2007) and provides multiple transfer tasks in which the arbitrary visual task features are systematically varied while the underlying causal task properties are conserved. The task features two "traps" along a horizontal tube with a reward positioned between them (Figure 7.2). One trap is functional – i.e., the food can fall in and be trapped. The other is non-functional: the food must be moved toward it to be obtained. A stick with two clear discs attached near the middle is pre-inserted into the apparatus. Thus, the food can be moved by pulling the stick from either trap opening, which enables the training of non-tool-using species in this task. In Tube A the food can be extracted by pulling it across the top of the non-functional trap; in Tube B it has to be pulled so that it falls through the bottomless, non-functional "trap," where it can be recovered from below. Half of the birds were initially presented with Tube A; the other half were tested with Tube B (Figure 7.2). In the transfer task they always received the version of the tube that they had not seen before.

Only 3 out of 14 tested woodpecker finches were able to solve the initial task, but none of these birds were able to solve the transfer task. These results confirmed our conjecture from the single-trap-tube experiment, namely that woodpecker finches apparently only use simple visual cues in combination with a simple procedural rule in solving trap-tube problems. This is in contrast to the performance of rooks, New Caledonian crows and chimpanzees that were also tested in analogous two-trap paradigms. Seven of the eight rooks solved the initial problem and also succeeded in the first transfer task. One bird was even successful in two additional transfer tasks, a problem most likely solved by forming concepts about surface continuity (Seed *et al.*, 2006). Likewise, three out of six New Caledonian crows solved a modified version of the initial task and were subsequently successful on two transfer tasks but failed in the third (Taylor *et al.*, 2009). Another recent study demonstrated that chimpanzees could solve the two-trap tasks, including two transfer tasks, but their performance was better when they could move the food directly with their fingers rather than when they were required to use a tool (Seed *et al.*, 2009).

The tool-length task

In the tool-length task, food was presented at four different distances in a clear Perspex tube, and subjects were presented with tools of different lengths with which

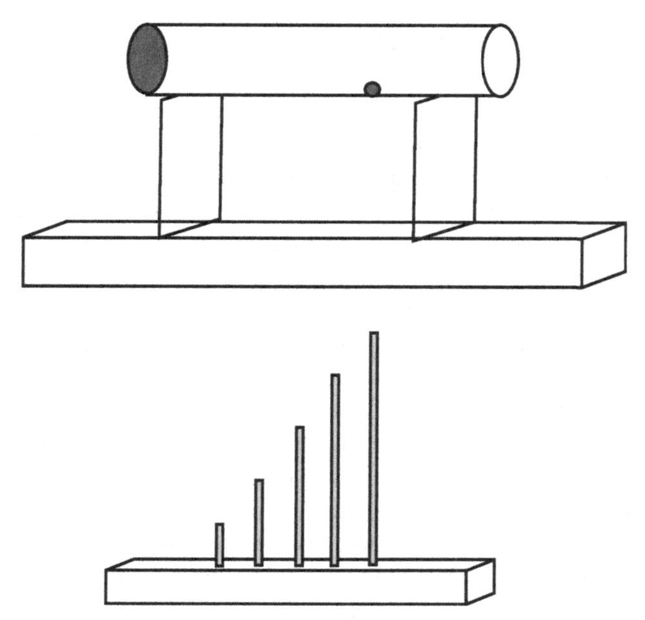

Figure 7.3 The tool-length task.

they could retrieve the food (Figure 7.3). The goal was to test whether woodpecker finches are capable of choosing a tool of proper length for a given task on a first-choice basis. We found that, similar to New Caledonian crows (Chappell & Kacelnik, 2002), three out of five finches chose a tool of sufficient length at a level significantly above chance. However, they did not match the tools to a given distance, but rather they chose tools that were longer than necessary for the task at hand.

The modification task

In the modification task, finches were presented with H- and S-shaped tools which had to be modified by removing a transverse piece to make them suitable for the retrieval of food from a tube. Like several primate species (Visalberghi *et al.*, 1995; Mulcahy *et al.*, 2005), four out of six woodpecker finches were able to solve the tasks; also like primates, they were unable to assess the problem in advance: they first inserted the unmodified tool before actually removing the transverse pieces and retrieving the food.

In sum, apart from the tool-length task, we found no evidence that woodpecker finches were able to assess the problem in advance or reliably transfer their knowledge to new tasks. Overall, we found no indication that this species was able to appreciate the underlying physical problem. Rather, it seemed that tool use in this species is based on a general learning mechanism such as fast trial-and-error learning and stimulus generalization.

Comparison between closely related species

To test whether tool use may have led to quantitative adaptive specializations (i.e., faster learning about the necessary physical relationships between tools and modified objects) we compared the performance of woodpecker finches and a closely related non-tool-using species, the small tree finch, in a series of physical tasks. However, it is also possible that tool use may have evolved in conjunction with general cognitive abilities, leading to a domain-general enhancement of learning, or that neither general nor specialized cognitive adaptations evolved along with tool use. Therefore, in addition to the physical tasks, we also tested birds in two general learning tasks: one tested the ability to unlearn a previously learned association, while the other tested performance in a novel operant task. These two tasks provided us with clues as to what extent general learning abilities might differ between species, and how this might fit into an explanation of the inter- and intraspecific patterns found in the specialized physical tasks.

The physical tasks fell into two subcategories: one which involved the passive use of tools in which the birds had to pull a stick; and one task that investigated physical cognition without incorporating the use of tools. This distinction allows a more precise identification of the level of adaptation, namely whether tool use evolved in conjunction with a general increase in physical cognition or with adaptations specific to the physical problems underlying the use of tools. We predicted that if tool use is associated with enhanced domain-specific cognition in the physical domain of tool use, woodpecker finches should outperform small tree finches in the physical tasks, but not in the general learning tasks (Tebbich *et al.*, 2010; Teschke, 2011). If this adaptive specialization is specific to the domain of tool use, woodpecker finches should only excel in tasks that involve the use of sticks (the cane task, below) but not in tasks where this is not required (the seesaw task, below). Additionally, the natural variation in tool-use abilities among woodpecker finches allowed us to test whether the experience with tool use hones specific or general cognitive abilities.

Cane task

This experiment is based on an experiment conducted by Hauser (1997) and was designed to test the birds' sensitivity to the functional configuration between a food reward and the tool used to attain it, a physical cognitive ability that should be important in tool use. In the initial task, birds were repeatedly given a choice between two canes. In each trial, a food reward was placed on the inside of the hooked portion of one of the canes and on the outside of the other. The canes and food rewards were contained in a clear encasement so that only pulling the cane with the food reward contained within its hooked portion would move the food reward to within reach of the bird in a given trial (Figure 7.4a). In the transfer tasks (Figure 7.4b–e) birds were confronted with slight variations of the spatial relationships between canes and rewards. The fourth transfer task was an exception to the other tasks because in each trial of this task, both rewards were placed inside the hooked portion of the canes. However, since one of the two canes was

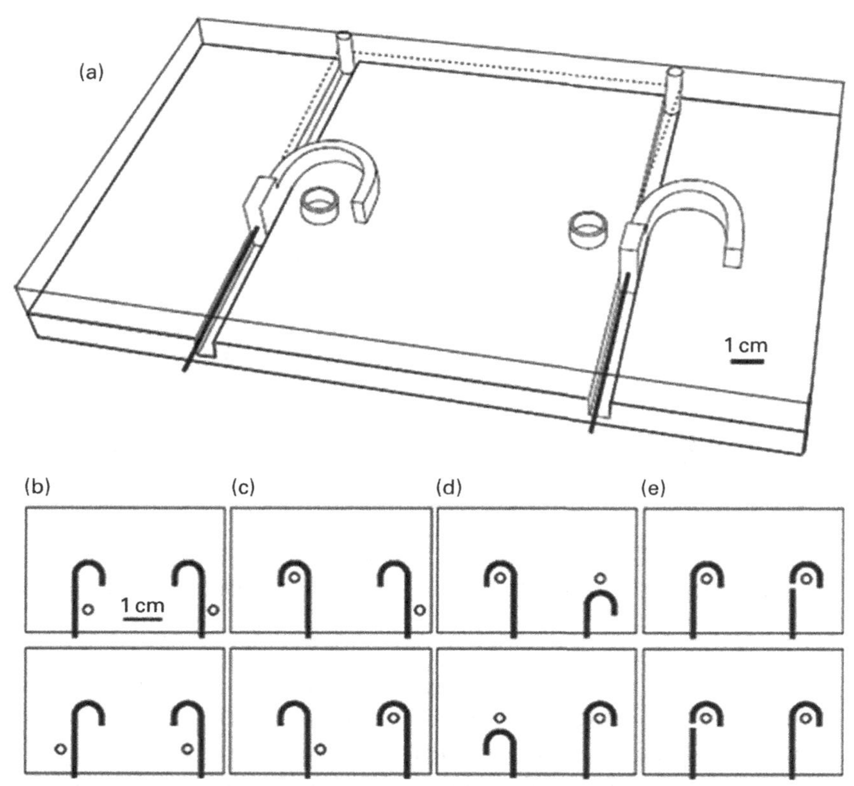

Figure 7.4 (a) The cane task apparatus with canes and rewards arranged as in the initial test condition. (b)–(e) Transfer tasks 1–4.

always bisected and therefore non-functional it could not be used to retrieve the reward in this task (Figure 7.4e). We deliberately designed the transfer tasks such that they could be solved via simple mechanisms such as stimulus generalization because we were interested in quantitative species differences in these abilities.

We found that all six small tree finches tested chose the correct canes in the initial task above chance, while 8 of 12 woodpecker finches were successful. Furthermore, we found no differences in the speed of acquisition. None of the birds were able to solve all transfer tasks and we found no differences in transfer ability between species. We expected woodpecker finches with tool-using experience in particular to outperform inexperienced individuals in the cane task, because this task most closely mimics natural tool use. However, though fewer non-tool-using woodpecker finches solved the task than tool-using woodpecker finches, the difference was not significant.

Seesaw task

This task was designed to test sensitivity to surface continuity without necessitating the use of tools. The apparatus consisted of a seesaw platform with either a central gap

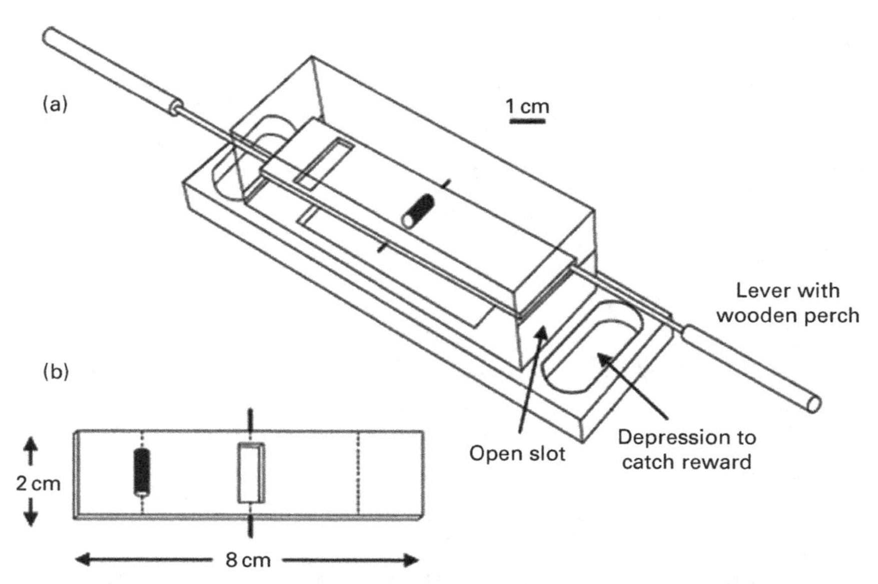

Figure 7.5 (a) The seesaw task apparatus depicted in the food-central condition. (b) The gap-central condition.

or a gap which was off-center (Figure 7.5a, b). The platform was encased in clear Perspex and was attached to two wooden dowels. The apparatus could be operated by perching on one of the dowels which tilted the seesaw platform and caused the food reward to roll down the platform. Perching on the correct dowel caused the food to roll down the continuous portion of the platform and out of the apparatus, while perching on the wrong one caused the food to fall into the gap, from where it was inaccessible. The transfer task was simply the condition of the task they had not seen before (Figure 7.5a, b). Surprisingly, small tree finches outperformed woodpecker finches in this task: five out of six small tree finches reached the success criterion in the initial task compared to only two of six woodpecker finches (one tool user and one non-tool user). However, none of the successful subjects were successful in the subsequent transfer task.

Novel box-opening task

This task was designed to test for differences in the ability to solve a novel operant task. The task apparatus was a small box made of opaque Perspex with a transparent lid (Figure 7.6). The lid was attached to the back edge of the box with hinges and could be opened by pushing the overlapping lip of the lid upwards. Birds were given up to eight sessions of 25 minutes to successfully open the box once. Woodpecker finches outperformed small tree finches in this task: 8 of the 18 tested woodpecker finches opened the box, but none of the 8 small tree finches did so. The most likely reason that woodpecker finches were better at opening the box was that they contacted the box more frequently. This is in line with their

Figure 7.6 Woodpecker finch pictured looking into the box-opening apparatus.

extractive foraging ecology: In foraging for food embedded in substrate, they must be highly persistent in their attempts to extract the food – this mostly entails long bouts of pecking.

Reversal task

The reversal task is a classic paradigm designed to estimate flexibility in terms of the capacity to inhibit a previously learned rule. This experiment consisted of an initial acquisition phase and a reversal phase. In the acquisition phase, subjects had to choose between two colors, one of which was the rewarded (S+) stimulus. Once a subject learned this initial color discrimination, in the reversal phase the color-reward contingency was reversed. The two species did not differ in the number of trials they needed to reach the criterion in the acquisition phase and the reversal phase. However, small tree finches made significantly fewer errors in the reversal phase (Tebbich *et al.*, 2010; Teschke *et al.*, 2011).

In summary, in contrast to our prediction, small tree finches outperformed woodpecker finches in the reversal task and the seesaw task, and performed equally well in the cane task. Woodpecker finches only performed significantly better than their non-tool-using relatives in the box-opening task. Additionally, we found no effect of tool experience on test performance between tool-using and non-tool-using woodpecker finches. The relevance of these results is that they provide no evidence that tool use in woodpecker finches evolved in association with cognitive adaptive specializations in the domain of tool use, nor do they contribute any evidence suggesting that experience with tools hones general or specialized cognitive abilities. Nevertheless, we are aware that it is possible that future work with different tasks will demonstrate a connection between tool use and other cognitive abilities. For instance, one possibility is that woodpecker finches possess enhanced cognitive abilities that are specific to active tool use or tool modification, but for obvious reasons we were unable to assess this in a comparison with a non-tool-using species (Teschke *et al.*, 2011).

Evolution: how did tool use evolve?

Our comparative study yielded no evidence for an adaptive specialization in the physical domain in woodpecker finches. This has implications for understanding the evolutionary roots of tool use in woodpecker finches and possibly even the evolution of Darwin's finches more generally. Indeed, the fact that the small tree finches performed equally well or better than woodpecker finches in the physical tasks indicates that the founding members of the Darwin's finch clade probably possessed certain cognitive characteristics, such as unusual flexibility, that laid the groundwork for the development of a number of unusual feeding strategies, among these tool use (Tebbich *et al.*, 2010). Recently, an experimental study on non-tool-using corvids suggested that the necessary cognitive foundation for tool use also preceded this ability in corvids (Bird & Emery, 2009). The authors found that the capacity of rooks, a non-tool-using corvid species, to solve sophisticated problems related to tool use, such as spontaneous modification and implementation of a variety of tools to suit a given task, rivals that of New Caledonian crows. However, a proper comparison of both species with exactly the same experiments would be needed to confirm this idea. Apart from this, comparison of transfer performance in the various tasks in corvids and Darwin's finches suggests that the level of cognition underlying tool use differs in these two groups. Whereas the performance of rooks and New Caledonian crows shows that they are capable of higher-level cognition, allowing transfer of knowledge to novel situations (Seed *et al.*, 2006; Bluff *et al.*, 2007; Taylor *et al.*, 2009), the performance of woodpecker finches suggests that they apply nothing more than enhanced trial-and-error learning and stimulus generalization to the problems at hand. The seesaw transfer tasks and the two-trap-tube tasks yielded no indication of an ability to extract general rules in solving physical problems. This difference in underlying mechanisms might form the basis for the higher diversity of tool use in New Caledonian crows compared with woodpecker finches.

Nevertheless, the behavioral flexibility underlying tool use in woodpecker finches and probably other unusual foraging techniques in Darwin's finches is exceptional. One indication of this is that Darwin's finch species learn the reversal phase of the reversal task significantly faster than pigeons and three species of corvids (Tebbich *et al.*, 2010). Furthermore, this clade exhibits a significantly higher number of innovative foraging techniques than New World jays, which were found to have the most diverse repertoire of innovative behavior among passerines (Overington *et al.*, 2009). This might be the consequence of enhanced flexibility in the founding population of Darwin's finches and might have brought about their comparatively rapid and broad radiation, with respect to other Galápagos clades (Tebbich *et al.*, 2010). It is also possible that the special ecological conditions on the Galápagos Islands spurred the acquisition of flexibility in foraging.

Conclusion

We have spent the bulk of this chapter illustrating how the adoption of a comprehensive approach in attempting to understand animal tool use encompassing all of Tinbergen's four questions is necessary in drawing closer to a broad understanding of animal tool use. We can now offer a more comprehensive answer to the question of why woodpecker finches use tools, though some elements of our answer remain speculative. We propose that woodpecker finches use tools because they are fast and flexible learners; perhaps because they evolved from a flexible stem species and because the harsh ecological conditions selected for behavioral flexibility. Presumably tool use, like other unusual foraging techniques, began as a behavioral innovation. At its inception, this behavior was probably transmitted socially and was also possibly supported by fast trial-and-error learning. As time passed, the innovative behavior gradually became anchored into the behavioral repertoire, with some components modified in an iterative process of learning and subsequent genetic assimilation (Hardy, 1965). This scenario fits in with the ontogeny of tool use that we observe today: It is a complex behavior comprising genetically fixed components that are augmented by trial-and-error learning. In its current form the behavior does not require social transmission from parents to offspring for its subsistence.

When we turn to the question of why this behavior has genetically fixed components at all, we find that ecology is important, but so are other factors. Woodpecker finches use tools because tools allow them to reach otherwise inaccessible prey that is particularly rich in nutrients, an ability that is probably essential to survival in the arid parts of the archipelago, where the abundance of easily accessible prey is highly variable. The ecological importance of this behavior may be one of the keys to understanding why this behavior has genetically fixed components. If a new behavior provides a selective advantage, phenotypes that require less environmental input and learning to develop the behavior might be favored by selection. This is especially so when the cost of learning is high or the likelihood of encountering the necessary information is low. The latter condition is certainly true for woodpecker finches: as a solitary species with a short juvenile period, they have few opportunities to encounter social information (in contrast to most primates and some corvids).

Genetic fixation of behavioral components need not go hand in hand with decreased behavioral flexibility and depressed learning abilities (Bateson, 2004). Indeed, if the use of tools enables access to food resources that are rich in energy but otherwise difficult to access, it seems plausible that selection might act to enhance cognitive abilities that make the use of tools more effective and applicable to a wider range of contexts. However, we did not find any evidence for enhanced cognitive abilities in the physical domain in woodpecker finches. Indeed, their non-tool-using relatives performed equally well or even exceeded their performance in physical and general learning tasks. This indicates that the evolution of cognitive abilities necessary for tool use in woodpecker finches may have preceded the evolution of this unusual behavioral adaptation. In this scenario, capacities such as curiosity and fast trial-and-error learning that were shared by all

Darwin's finches were co-opted for the specific woodpecker finch niche and also led to the evolution of unusual, complex foraging techniques in other Darwin's finches. Our data provide evidence that flexible tool use, even that involving task-specific modification and tool selectivity, can be based on relatively simple cognitive processes. From a functional perspective, all that matters is that tool use provides a selective advantage which in turn depends on whether the cost–benefit ratio is favorable. For example, if an advantageous behavior is based on simple cognitive mechanisms that do not require the evolution of a costly neural apparatus, then this solution should be favored by selection when the costs of evolving the necessary neural substrate for complex cognition outweigh the benefits of using tools in a wider range of contexts.

Future directions

One substantive issue arises if tool use requires no special cognitive capacity: In this case, we would expect tool use to be very common, and yet it is not. Hansell and Ruxton (2008) argue that the use of tools may seldom be useful compared with anatomical adaptations, but this does not seem to be a very comprehensive answer (see also Chapter 5). Instead, we consider it more likely that the simultaneous occurrence of factors that facilitate the evolution of tool use may be rare. Like Hansell and Ruxton (2008) and Kacelnik (2009), we suggest that ecological pull factors include unexploited niches that are more accessible with the help of tools, but also ecological push factors such as seasonal food limitation. However, for tool use that is based on an individual behavioral innovation and is transmitted socially, the necessary preconditions may also include behavioral promoters such as the tendency to approach novel situations and to manipulate objects, as well as quick operant learning. Support for the idea of specific behavioral promoters comes from a study that compared the ontogeny of object-oriented behaviors in captive-bred New Caledonian crows and ravens (*Corvus corax*). The authors observed an increase in object combinations which can be seen as precursors for tool use in New Caledonian crows, but a decrease in ravens over the same time period (Kenward *et al.*, 2011). This study also suggests food caching as a precursor for tool use in corvids since it entails inserting of objects into small cavities. The ontogenetic comparison between New Caledonian crows and ravens showed that the developmental patterns of caching and tool use are strikingly similar.

Additionally, the dissemination of newly acquired tool-use abilities is more likely in social than in non-social species and depends on the ability to learn socially. The importance of some of these for the evolutionary and ontogenetic development and deployment of tool use could be tested with a comparative analysis.

One factor that could limit the more frequent innovation of tool use is its initial breakthrough: at least in stick-tool use, the incomplete behavioral sequence (i.e., probing into a crevice with a stick) is not rewarded and a certain proficiency has to be achieved before it is successful, as our observations of the ontogeny of this behavior indicate. A more detailed analysis of data on the acquisition of tool use by non-tool-using rooks (Bird & Emery, 2009) could shed light on this question. Other useful results bearing on this problem could be gleaned from literature and research on nest building, since this could be a precursor for tool-using abilities (Hansell & Ruxton, 2008).

Certainly a particularly fascinating aspect of tool use and of behavioral innovations more generally is their relationship with flexibility and the interplay of these factors in the evolutionary process. The appearance of new and unusual behavior can catalyze the construction of a novel niche, thereby exposing proficient individuals or a proficient population to novel selective pressures, which in turn can lead to evolutionary diversification and adaptive radiation (Odling-Smee, 2003; Laland & Sterelny, 2006). The idea that there is a positive association between behavioral flexibility and species richness is supported by several studies (Nicolakakis & Lefebvre, 2000; Sol, 2003; Sol *et al.*, 2005). Such a process might explain the radiation of Darwin's finches, where high flexibility appears to have led to the emergence of novel foraging techniques which may in turn have been an important force in the adaptive radiation of this clade (West-Eberhard, 2003; Grant & Grant, 2008; Price, 2008; Tebbich *et al.*, 2010). Exploring the relationship between novel behaviors, behavioral flexibility and evolutionary diversification is an exciting new avenue of research. However, in pursuing this field, we re-emphasize our conviction of the importance of the integration of all four of Tinbergen's questions.

References

Aisner, R. & Terkel, J. (1992). Ontogeny of pine cone opening behaviour in black rats, *Rattus rattus*. *Animal Behaviour*, **214**, 327–336.

Alcock, J. (1972). The evolution of the use of tools by feeding animals. *Evolution*, **26**, 464–773.

Bateson, P. (2004). The active role of behaviour in evolution. *Biology and Philosophy*, **19**, 283–298.

Beck, B. (1980). *Animal Tool Behavior: The Use and Manufacture of Tools by Animals*. New York: Garland STPM Press.

Bird, C. & Emery, N. (2009). Insightful problem solving and creative tool modification by captive rooks. *Proceedings of the National Academy of Sciences USA*, **106**, 10370–10375.

Bluff, L. A., Weir, A. A. S., Rutz, C., Wimpenny, J. H. & Kacelnik, A. (2007). Tool-related cognition in New Caledonian crows. *Comparative Cognition and Behavior Reviews*, **2**, 1–25.

Bowman, R. I. & Billeb, S. L. (1965). Blood-eating in a Galápagos finch. *The Living Bird*, **4**, 29–44.

Chaffer, N. (1945). The spotted and satin bower-birds: a comparison. *Emu*, **44**, 161–181.

Chappell, J. (2006). Avian cognition: understanding tool use. *Current Biology*, **16**, R244.

Chappell, J. & Kacelnik, A. (2002). Tool selectivity in a non-primate, the New Caledonian crow (*Corvus moneduloides*). *Animal Cognition*, **5**, 71–78.

Curio, E. & Kramer, P. (1964). Vom Mangrovenfinken (*Cactospiza heliobates* Snodgrass und Heller). *Zeitschrift f. Tierpsychologie*, **21**, 223–234.

DeBenedictis, P. A. (1966). The bill-brace feeding behavior of the Galapagos finch *Geospiza conirostris*. *Condor*, **68**, 206–208.

Eibl-Eibesfeldt, I. (1961). Über den Werkzeuggebrauch des Spechtfinken *Camarhynchus pallidus* (Slater und Slavin). *Zeitschrift f. Tierpsychologie*, **18**, 343–346.

Eibl-Eibesfeldt, I. & Sielman, H. (1962). Beobachtungen am Spechtfinken *Cactospiza pallida* (Sclater und Salvin). *Journal of Ornithology*, **103**, 92–101.

Eibl-Eibesfeldt, I. & Sielman, H. (1965). Werkzeuggebrauch beim Nahrungserwerb. *Encyclopedia Cinematographica, E597 Publikationen zum wissenschaftlichen Film*, **1**(A), 385–390.

Emery, N. J. & Clayton, N. S. (2009). Tool use and physical cognition in birds and mammals. *Current Opinion in Neurobiology*, **19**, 27–33.

Fessl, B., Loaiza, A. D., Tebbich, S. & Glyn Young, H. (2011). Feeding and nesting requirements of the critically endangered mangrove finch (*Camarhynchus heliobates*). *Journal of Ornithology*, **152**(2), 453–460.

Goodall, J. (1986). *The Chimpanzees of Gombe*. Cambridge, MA: Harvard University Press.

Grant, P. R. (1985). Climatic fluctuations on the Galápagos Islands and their influence on Darwin's finches. *Ornithological Monographs*, **36**, 471–483.

Grant, P. R. (1986). *Ecology and Evolution of Darwin's Finches*. Princeton, NJ: Princeton University Press.

Grant, P. R. (1999). *Ecology and Evolution of Darwin's Finches*. 2nd edn. Princeton, NJ: Princeton University Press.

Grant, P. R. & Boag, P. T. (1980). Rainfall on the Galápagos and the demography of Darwin's finches. *Auk*, **97**, 227–244.

Grant, P. R. & Grant, R. B. (2008). *How and Why Species Multiply: The Radiation of Darwin's Finches*. Princeton, NJ: Princeton University Press.

Hamann, O. (1981). Plant communities of the Galápagos Islands. *Dansk Botanisk Archiv*, **34**, 1–63.

Hansell, M. & Ruxton, G. D. (2008). Setting tool use within the context of animal construction behaviour. *Trends in Ecology and Evolution*, **23**, 73–78.

Hardy, A. (1965). *The Living Stream*. London: Collins.

Harvey, P. H. & Pagel, M. D. (1991). *The Comparative Method in Evolutionary Biology*. Oxford: Oxford University Press.

Hauser, M. D. (1997). Artifactual kinds and functional design features: what a primate understands without language. *Cognition*, **64**, 285–308.

Hladik, C. M. (1977). Chimpanzees of Gabon and chimpanzees of Gombe: some comparative data on the diet. In T. H. Clutton-Brock (ed.) *Primate Ecology* (pp. 481–501). New York: Academia Press.

Holzhaider, J. C., Hunt, G. R. & Gray, R. D. (2010a). The development of pandanus tool manufacture in wild New Caledonian crows. *Behaviour*, **147**, 553–586.

Holzhaider, J. C., Hunt, G. R. & Gray, R. D. (2010b). Social learning in New Caledonian crows. *Learning and Behavior*, **38**, 206–219.

Hundley, M. H. (1963). Notes on methods of feeding and the use of tools in the Geospizinae. *Auk*, **80**, 372–373.

Hunt, G. R. (1996). Manufacture and use of hook-tools by New Caledonian crows. *Nature*, **379**, 249–251.

Hunt, G. R. & Gray, R. D. (2002). Species-wide manufacture of stick-type tools by New Caledonian crows. *Emu*, **102**, 349–353.

Hunt, G. R. & Gray, R. D. (2004a). The crafting of hook tools by wild New Caledonian crows. *Proceedings of the Royal Society of London B: Biological Sciences*, **271**, S88–S90.

Hunt, G. R. & Gray, R. D. (2004b). Direct observations of pandanus-tool manufacture and use by a New Caledonian crow (*Corvus moneduloides*). *Animal Cognition*, **7**, 114–120.

Kacelnik, A. (2009). Tools for thought or thoughts for tools?. *Proceedings of the National Academy of Sciences USA*, **106**, 10071–10072.

Kenward, B., Weir, A. A. S., Rutz, C. & Kacelnik, A. (2005). Tool manufacture by naive juvenile crows. *Nature*, **433**, 121.

Kenward, B., Rutz, C., Weir, A. A. S. & Kacelnik, A. (2006). Development of tool use in New Caledonian crows: inherited action patterns and social influence. *Animal Behaviour*, **72**, 1329–1343.
Kenward, B., Schloegl, C. & Rutz, C. (2011). On the evolutionary and ontogenetic origins of tool-oriented behaviour in New Caledonian crows (*Corvus moneduloides*). *Biological Journal of the Linnean Society*, **102**, 870–877.
Laland, K. & Sterelny, K. (2006). Perspective: seven reasons (not) to neglect niche construction. *Evolution*, **60**, 1751–1762.
Limongelli, L., Boysen, S. T. & Visalberghi, E. (1995). Comprehension of cause–effect relations in a tool-using task by chimpanzees (*Pan troglodytes*). *Journal of Comparative Psychology*, **109**, 18–26.
MacFarland, C. & Reeder, W. (1974). Cleaning symbiosis involving Galápagos tortoise and two species of Darwin's finches. *Zeitschrift für Tierpsyhologogie*, **34**, 464–483.
Mann, J. S. B., Watson-Capps, J. J., Gibson, Q. A., et al. (2008). Why do dolphins carry sponges? *Plos ONE*, **3**, 3868.
Millikan, G. C. & Bowman, R. I. (1967). Observations on Galápagos tool-using finches in captivity. *Living Bird*, **6**, 23–41.
Mulcahy, N. J. & Call, J. (2006). How great apes perform on a modified trap-tube task. *Animal Cognition*, **9**, 193–199.
Mulcahy, N. J., Call, J. & Dunbar, R. (2005). Gorillas (*Gorilla gorilla*) and orangutans (*Pongo pygmaeus*) encode relevant problem features in a tool-using task. *Journal of Comparative Psychology*, **119**, 23–32.
Nagell, K., Olguin, R. S. & Tomasello, M. (1993). Processes of social learning in the tool use of chimpanzees (*Pan troglodytes*) and human children (*Homo sapiens*). *Journal of Comparative Psychology*, **107**, 174–186.
Nicolakakis, N. & Lefebvre, L. (2000). Forebrain size and innovation rate in European birds: feeding, nesting and confounding variables. *Behaviour*, **137**, 1415–1429.
Nishida, T. & Hiraiwa, M. (1982). Natural history of a tool-using behaviour by wild chimpanzees in feeding upon wood-boring ants. *Journal of Human Evolution*, **11**, 73–99.
Odling-Smee, F. J., Laland, K. N. & Feldman, M. W. (2003). *Niche Construction: The Neglected Process in Evolution*. Princeton, NJ: Princeton University Press.
Orenstein, R. I. (1972). Tool-use by the New Caledonian crow (*Corvus moneduloides*). *Auk*, **89**, 674–676.
Overington, S. E., Morand-Ferron, J., Boogert, N. J. & Lefebvre, L. (2009). Technical innovations drive the relationship between innovativeness and residual brain size in birds. *Animal Beahviour*, **78**, 1001–1010.
Parker, S. T. & Gibson, K. R. (1977). Object manipulation, tool use and sensomotor intelligence as feeding adaptations in cebus monkeys and great apes. *Journal of Human Evolution*, **6**, 623–641.
Penn, D. C. & Povinelli, J. P. (2007). Causal cognition in humans and nonhuman animals: a comparative critical review. *Annual Review of Psychology*, **58**, 97–118.
Povinelli, D. J. (2000). *Folk Physics for Apes: A Chimpanzee's Theory of How the World Works*. Oxford: Oxford University Press.
Price, T. (2008). *Speciation in Birds*. Greenwood Village, CO: Roberts & Company Publishers.
Rutz, C. & St Clair, J. (in press). The evolutionary origins and ecological context of tool use in New Caledonian crows. *Behavioral Processes*.
Rutz, C., Bluff, L. A., Weir, A. A. S. & Kacelnik, A. (2007). Video cameras on wild birds. *Science*, **318**, 765.
Rutz, C., Bluff, L. A., Reed, N., *et al.* (2010). The ecological significance of tool use in New Caledonian crows. *Science*, **329**, 1523–1526.

Santos, L. R., Pearson, H., Spaepen, G. M., Tsao, F. & Hauser, M. D. (2006). Probing the limits of tool competence: experiments with two non-tool-using species (*Cercopithecus aethiops* and *Saguinus oedipus*). *Animal Cognition*, **9**, 94–109.

Schluter, D. (1984). Feeding correlates of breeding and social organization in two Galápagos finches. *Auk*, **101**, 59–68.

Seed, A. M., Tebbich, S., Emery, N. & Clayton, N. S. (2006). Investigating physical cognition in rooks (*Corvus frugilegus*). *Current Biology*, **16**, 697–701.

Seed, A. M., Call, J., Emery, N. J. & Clayton, N. S. (2009). Chimpanzees solve the trap problem when the confound of tool-use is removed. *Journal of Experimental Psychology: Animal Behaviour Processes*, **35**, 23–34.

Silva, F. J., Page, D. M. & Silva, K. M. (2005). Methodological-conceptual problems on the study of chimpanzees' folk physics: how studies with adult humans can help. *Learning and Behaviour*, **32**, 47–58.

Sol, D. (2003). *Behavioural Flexibility: A Neglected Issue in the Ecological and Evolutionary Literature?* Oxford: Oxford University Press.

Sol, D., Stirling, D. G. & Lefebvre, L. (2005). Behavioral drive or behavioral inhibition in evolution: subspecific diversification in Holarctic passerines. *Evolution*, **59**, 2669–2677.

Taylor, A., Hunt, G., Medina, F. & Gray, R. (2009). Do New Caledonian crows solve physical problems through causal reasoning? *Proceedings of the Royal Society of London B: Biological Sciences*, **276**, 247–254.

Tebbich, S. & Bshary, R. (2004). Cognitive abilities related to tool use in the woodpecker finch, *Cactospiza pallida*. *Animal Behaviour*, **67**, 689–697.

Tebbich, S., Taborsky, M., Fessl, B. & Blomqvist, D. (2001). Do woodpecker finches acquire tool-use by social learning? *Proceedings of the Royal Society of London B: Biological Sciences*, **268**, 2189–2193.

Tebbich, S., Taborsky, M., Fessl, B. & Dvorak, M. (2002). The ecology of tool-use in the woodpecker finch (*Cactospiza pallida*). *Ecology Letters*, **5**, 656–664.

Tebbich, S., Taborsky, M., Fessl, B., Dvorak, M. & Winkler, H. (2004). Feeding behavior of four arboreal Darwin's finches: adaptations to spatial and seasonal variability. *Condor*, **106**, 95–105.

Tebbich, S., Seed, A. M., Emery, N. & Clayton, N. S. (2007). Non-tool-using rooks (*Corvus frugilegus*) solve the trap-tube task. *Animal Cognition*, **10**, 225–231.

Tebbich, S., Sterelny, K. & Teschke, I. (2010). The finches' tale: adaptive radiation and behavioural flexibility. *Philosophical Transactions of the Royal Scociety of London B*, **365**, 1099–1109.

Teschke, I. & Tebbich, S. (2011). Physical cognition and tool-use: performance of Darwin's finches in the two-trap tube task. *Animal Cognition*, **14**, 555–563.

Teschke, I., Cartmill, E., Stankewitz, S. & Tebbich, S. (2011). Sometimes tool-use is not the key: no evidence for cognitive adaptive specializations in tool-using woodpecker finches. *Animal Behaviour*, **82**, 945–956.

Tinbergen, N. (1963). On aims and methods of ethology. *Zeitschrift für Tierpsychologie*, **20**, 410–433.

Tomasello, M. & Call, J. (1997). Tools and causality. In J. Call (ed.) *Primate Cognition* (pp. 57–99). New York: Oxford University Press.

Tomasello, M:., Davis-Dasilva, M. & Camak, L. (1987). Observational learning of tool-use by young chimpanzees. *Human Evolution*, **2**, 175–183.

van Lawick-Goodall, J. & van Lawick-Goodall, H. (1966). Use of tools by the Egyptian vulture, *Neophron percnopterus*. *Nature*, **212**, 1468–1469.

Visalberghi, E. & Limongelli, L. (1994). Lack of comprehension of cause–effect relations in tool-using capuchin monkeys (*Cebus apella*). *Journal of Comparative Psychology*, **108**, 15–22.

Visalberghi, E., Fragaszy, D. M. & Savage-Rumbaugh, S. (1995). Performance in a tool-using task by common chimpanzees (*Pan troglodytes*), bonobos (*Pan paniscus*), an orangutan (*Pongo pygmaeus*), and capuchin monkeys (*Cebus apella*). *Journal of Comparative Psychology*, **109**, 52–60.

Walsh, J. F., Grunewald, J. & Grunewald, B. (1985). Green-backed herons (*Butorides striatus*) possibly using a lure and using apparent bait. *Journal of Ornithology*, **126**, 439–442.

West-Eberhard, M. (2003). *Developmental Plasticity and Evolution*. Oxford: Oxford University Press.

Whiten, A., Custance, D. M., Gomez, J.-C., Teixidor, P. & Bard, K. A. (1996). Imitative learning of artificial fruit processing in children (*Homo sapiens*) and chimpanzees (*Pan troglodytes*). *Journal of Comparative Psychology*, **110**, 3–14.

Yamakoshi, G. (1998). Dietary responses to fruit scarcity of wild chimpanzees at Bossou, Guinea: possible implications for ecological importance of tool use. *American Journal of Physical Anthropology*, **106**, 283–295.

Zohar, O. & Terkel, Y. (1991). Acquisition of pine cone stripping behaviour in black rats. *International Journal of Comparative Psychology*, **5**, 1–5.

Zohar, O. & Terkel, J. (1996). Social and environmental factors modulate the learning of pine-cone stripping techniques by black rats, *Rattus rattus*. *Animal Behaviour*, **51**, 611–618.

Part III

Ecology and culture

8 The social context of chimpanzee tool use

Crickette M. Sanz
Washington University, Department of Anthropology

David B. Morgan
Lester E. Fisher Center for the Study and Conservation of Apes, Lincoln Park Zoo
Wildlife Conservation Society, Congo

Although several animal species exhibit some form of tool use (Shumaker *et al.*, 2011), there are relatively few animals which flexibly use a diverse repertoire of implements on a regular basis within their natural environments. As shown in this volume, hominins, chimpanzees, orangutans, some capuchins and corvids are the exceptions. Van Schaik *et al.* (1999) have suggested that the evolution of material culture in primates is dependent upon the intersection of four primary factors: manipulative skills, cognitive abilities, suitable ecological niches and social tolerance. Although one cannot entirely dismiss the possibility of differences in manipulative skills and cognitive abilities within species, there are intriguing differences in diversity and types of tool use among populations of wild chimpanzees (*Pan troglodytes*) which are not entirely explicable by environmental circumstances, and so have been attributed to social influences (Whiten *et al.*, 1999, 2001; Möbius *et al.*, 2008). With the exception of a few developmental studies, the social context of tool use remains largely unexplored in these apes.

Primates show differing degrees of social cohesion, and varying responses to fluctuations in the availability of resources in their environment. The abundance and distribution of important food resources dictates not only population density, but also social tolerance. Social tolerance varies across species, but may also differ within species (among populations, groups and even individuals). Resource scarcity may incite feeding competition among conspecifics, which could cause primates to avoid spending time in close proximity, whereas bountiful resources may attract conspecifics to forage at the same site. Social tolerance in gregarious foraging could enhance social learning by allowing primates to forage in close proximity to each other, providing a relaxed social atmosphere in which attention may be focused on a task, and enabling subordinate individuals to participate in close proximity foraging without risk of theft or aggression by conspecifics (van Schaik, 2003). The "opportunities for social learning hypothesis" predicts that higher degrees of social tolerance should result in a larger number of customary technical skills exhibited by primates (van Schaik, 2003). An extension of the opportunities for social learning hypothesis is that higher degrees of

Tool Use in Animals: Cognition and Ecology, eds. Crickette M. Sanz, Josep Call and Christophe Boesch.
Published by Cambridge University Press.

social tolerance could be associated with the transmission of more complex tool behaviors. Pradhan *et al.* (2012) have proposed that variation in sociability accounts for intraspecific and interspecific differences in the simple and cumulative technology of chimpanzees and orangutans. In this chapter we examine the social context of tool use in the chimpanzee population residing in the Goualougo Triangle, which is known to exhibit relatively complex technology. We report on the degree of gregariousness within this wild chimpanzee population, and attempt to elucidate pathways of information transmission within and between communities in this region.

Coussi-Korbel and Fragaszy (1995) have proposed that the degree of social tolerance exhibited by a species and coordination in time and space between conspecifics will not only affect the likelihood of individuals attaining information, but also the type of information that is attained. Species with egalitarian relationships are predicted to have more frequent and extensive behavioral coordination between individuals in both time and space than those with despotic relationships, in which certain individuals may actively avoid others. Individuals may be coordinated in space when an individual approaches a place where another individual previously engaged in a tool task, but is no longer present. In this setting, the naive individual has the opportunity to gain information about the task and potential quality of the site based on physical alterations that persist in the environment. Information may also be gained by interacting with discarded tools. For example, at sites where chimpanzees fish for termites, discarded fishing probes may be found on or around the periphery of the termite nest. Termite-nest puncturing tools may also be found littered around a termite nest or even embedded within the matrix of a nest. These tools persist for several months, and may be reused by subsequent visitors. Another example is the wooden and stone hammers used for nut cracking that are often transported to trees producing nuts, and remain in close proximity to anvils, which may be covered with tell-tale nut shells from previous tool-using bouts (Boesch & Boesch, 1984; Chapter 11). This is also the case for nut-cracking sites of capuchin monkeys (Visalberghi *et al.*, 2007; Chapter 10).

Behavioral coordination in time and space requires that two individuals are at the same locality at the same time, and engaging in similar activities. In this setting the naive individual has the possibility to gather information about the substrate, tool action and outcome from an experienced individual. Lonsdorf (2006) and Humle *et al.* (2009) have shown the importance of social learning opportunities in the acquisition of tool-using behaviors of young chimpanzees. Youngsters whose mothers provided more opportunities for tool use in ant dipping showed more advanced skills than those with mothers who provided little opportunity (Humle *et al.*, 2009). As no active facilitation was observed in ant dipping, it seems that coordination in space was effective in promoting acquisition of tool-using skills. Further, infants of avid termite-fishing mothers at Gombe were more likely to be proficient tool users themselves (Lonsdorf, 2006). Chimpanzee mothers at Gombe were highly tolerant of their offspring's behaviors, such as reaching toward her tool or termites, stealing tools and investigating the termite mound even when these behaviors seemed to interfere with her food gathering. At the same site, McGrew (1977) documented coaction, which is when a chimpanzee allows another to touch either her hand or part of her tool during use. Coaction has also been identified as a potentially important aspect of social transmission of tool traditions in captive chimpanzees (Horner, 2010).

Social facilitation is an important aspect of the acquisition of tool-using skills (Lonsdorf, 2006; Hopper *et al.*, 2007; Humle *et al.*, 2009). However, the specific mechanisms of social information transfer and their relative contributions are difficult to determine, particularly in naturally occurring environments. Boesch (1991) compiled observations of social facilitation by chimpanzee mothers in their infant's acquisition of nut-cracking skills in the Taï forest in West Africa. The majority of interactions were characterized as stimulation and facilitation. Stimulation involved the mother leaving the hammer near the anvil, and facilitation occurred when a mother provided tools and/or intact nuts to the infant. The only two instances of active teaching observed among wild chimpanzees were documented in Boesch's (1991) study when a mother intervened to show her infant how to correctly orient an irregularly shaped hammer and in another case the mother demonstrated correct positioning of the nut to her offspring. Mothers were frequently observed to delay their nut cracking while infants ate nuts from her anvil or manipulated her nut-cracking tools, but it is also interesting to note that this behavior changed in relation to the infant's age and skill (Boesch & Boesch-Achermann, 2000). It is advantageous for a mother to allocate only the minimal level of assistance necessary for her offspring to succeed in acquiring the skill. Although the mother eventually benefits through the increasing independence of her offspring, she experiences an immediate cost of relinquishing food resources and/or functional tools to her offspring. In contrast, the recipient of assistance experiences both short- and long-term benefits from this exchange. For young chimpanzees, access to the appropriate materials and opportunity to exhibit a tool behavior seem to be important factors which affect the speed of acquisition and proficiency in using tools.

Active teaching was long considered a derived trait of hominins which emerged after the split from the last common ancestor with chimpanzees (see review in Hoppitt *et al.*, 2008). However, recent evidence of teaching from a wide range of taxa, including meerkats and ants, has challenged commonly held notions of social learning and teaching mechanisms (Franks & Richardson, 2006; Thornton & McAuliffe, 2006). Active teaching can be differentiated from other types of socially biased learning by active participation of the instructor. Caro and Hauser (1992) identified teaching in situations in which the instructor modified his/her behavior in the presence of a naive observer, with the specifications that the modification in behavior was at the cost of the instructor and resulted in the observer acquiring information that previously was less accessible. Knowledgeable practitioners may provide opportunities for the learner to practice skills, or they may also coach the individual with encouraging or punishing feedback. To identify the specific mechanisms which underpin the transmission of technological skills and the factors shaping the evolutionary origins of social learning, it is important to carefully examine and compare the social contexts of tool-use acquisition and maintenance within and across species.

Chimpanzee society and tool use

Wild chimpanzees have one of the most expansive and varied distributions of any living primate. Their range extends across equatorial Africa and encompasses habitats ranging from woodland savannahs to dense lowland rainforests. Unlike animal species who live in stable groups, chimpanzees (and other species exhibiting fission-fusion sociality) regularly adjust the size of their foraging parties to available resources. An entire chimpanzee group (referred to as a community) may comprise 20–148 individuals (Mitani, 2006). However, most foraging subgroups consist of fewer than five individuals (Taï: Boesch, 1996; Boesch & Boesch-Achermann, 2000; Gombe: Halperin, 1979; Bossou: Sugiyama, 1984). It has been widely accepted that a combination of ecological, demographic and social factors interact to determine subgroup size and composition (Goodall, 1986; Chapman *et al.*, 1995; Boesch & Boesch-Achermann, 2000; Anderson *et al.*, 2002; Mitani *et al.*, 2002).

Every population of wild chimpanzees has also been documented to exhibit at least one type of tool-using behavior, and several groups have a diverse repertoire of tool types (McGrew, 1992; Sanz & Morgan, 2007). Technological repertoires shown by a particular community range from 6 to 22 different tool types (Sanz & Morgan, 2007). The composition of these tool kits differs between populations, and sometimes adjacent groups (McGrew & Collins, 1985; McGrew, 1992; Boesch, 2003; Sanz & Morgan, 2007). Although ecological factors shape some of the differences between groups, other behaviors have been identified as putative cultural variants which presumably rely on social transmission to be maintained at a habitual or customary level within chimpanzee society (Whiten *et al.*, 1999, 2001).

The structural dynamics of fission-fusion societies pose specific challenges to the maintenance of behavioral variants through social transmission. The entire membership of a chimpanzee community is rarely, if ever, assembled in the same locality, but rather comprises several subgroups which vary in size and composition throughout the day. Van Schaik (2003) used time spent in a social subgroup (or party) as a proxy for opportunities for social learning, and found that time spent in foraging parties was positively related to the number of feeding tools and putative cultural variants exhibited within a chimpanzee community. Although vertical transmission (between mother and her offspring) is the most frequent means of skill transmission, van Schaik (2003) claims that horizontal transmission (between mature individuals) provides the only plausible explanation for variation across populations in the size of a population's tool kit or cultural repertoire. Social intersections at some types of tool-using sites may be rare and particularly uncommon among certain individuals, such as adult females residing in the peripheral areas of the group's range. However, there are tool-harvested resources such as nuts which may attract large parties of mixed age and sex composition to the same locality within a community range (Boesch & Boesch-Achermann, 2000).

Within fission-fusion societies, individuals may exhibit preferences for associating with particular conspecifics more than others. Individual identity or pre-existing relationships may also have an effect on whether or how information is transmitted within a group (Coussi-Korbel & Fragaszy, 1995). Frequent association and close proximity among certain individuals is more likely to promote information transfer than among loose associates. The specific type of interaction between individuals is also likely to affect the type and amount of information transferred. Some types of tool-using skills are relatively simple and therefore more likely to be invented by individuals with little or no social input. Leaf sponging is a behavior with relatively few components (Sanz &

Morgan, 2010) which has been invented by captive chimpanzee populations provided with the appropriate materials and setting (Kitahara-Frisch & Norikoshi, 1982). In contrast, more complex technical skills may require more specific input about raw materials, tool manufacture, tool actions and results. Acquisition of more complex skills may require exposure or proximity to a model, as has been shown in captive studies (Hopper *et al.*, 2007).

Until recently it was widely held that chimpanzee tool use was characterized by a direct relationship between a single tool and its goal. Several study populations have now been reported to exhibit hierarchically structured use of multiple tools and flexibility in using a tool for multiple goals (Sanz & Morgan, 2007, 2010; Boesch *et al.*, 2009). Further, some tools have specific design features, such as material selectivity or pre-modification of tool form, which increase the complexity of the task (Sanz *et al.*, 2009; Sanz & Morgan, 2010). The chimpanzee population in the Goualougo Triangle located in northern Republic of Congo exhibits a wide range of tool-using behaviors, several of which involve the regular use of tool sets – multiple tools used in sequence to accomplish a task. In contrast to previous reports of the rarity of tool sets, these behaviors are habitual in termite (Sanz *et al.*, 2004), driver ant (Sanz *et al.*, 2010) and honey gathering (Sanz & Morgan, 2009) contexts within this chimpanzee population, and possibly across the range of the central subspecies (*Pan troglodytes troglodytes*).

In this chapter we review the social settings in which tool use occurs within the Goualougo Triangle chimpanzee population, whose members regularly exhibit complex tool-using behaviors. We compare different tool-using contexts to determine if coordination in space, time and social interactions provides varied opportunities for social transmission of tool-using skills. Social networks at tool sites are also examined to determine if the observed degree of social contact could facilitate horizontal transmission within and between groups. In particular, we are interested in identifying the potential means by which the relatively complex tool-using behaviors of these chimpanzees have been so effectively maintained over both time and space. Insights from these wild apes will enable us to better understand the mechanisms underlying the large-scale similarity of technology among our hominin ancestors.

Methods

Study site

The Goualougo Triangle is located within the southern portion of the Nouabalé-Ndoki National Park (16°51′–16°56′ N; 2°05′–3°03′ E) in northern Republic of Congo. The study area is composed of evergreen and semi-deciduous lowland forest, with altitudes ranging between 330 m and 600 m. The climate can be described as transitional between the Congo-equatorial and sub-equatorial climatic zones. Rainfall and temperature were recorded daily at the Goualougo Triangle base camp. Rainfall was 1650 mm in 2007 and 1676 mm in 2008. The average minimum and maximum temperatures were 21.5°C and 24.2°C in 2007, and 21.5°C and 24.1°C in 2008.

Data collection

Direct observations of the chimpanzees in the Goualougo Triangle have been ongoing since February 1999. Chimpanzee tool-using behaviors have been documented by direct observation throughout this time and by remote video monitoring since 2003 (for description of these methods, see Sanz *et al.*, 2004). For all observations of tool-using behavior, we recorded the identification of the chimpanzee, type of object used, target of object, actions, context and/or goal of the tool-using behavior and the outcome. In addition, chimpanzee tool assemblages were also collected at termite and ant nests by several field teams conducting daily reconnaissance walks, and during monthly phenology circuits. We recorded the location, materials used to make the tool, length, width and any modifications to each tool.

We also examined the size and composition of subgroups engaging in feeding on particular types of food resources (termites, leaves, fruits). These measures serve as a proxy for social tolerance, or opportunities for social learning (van Schaik, 2003).

Dyadic association indices

Matrices of dyadic associations were calculated based on direct and remote video observations of sexually mature males and females. Associations were scored as presence of an individual in the same party. Association was quantified as:

$$\frac{x}{x + y_a + y_b},$$

where x is the number of sightings that include both chimpanzees, y_a is the number of sightings that included chimpanzee a but not chimpanzee b, and y_b is the number of sightings that included chimpanzee b but not chimpanzee a (Cairns & Schwager, 1987; Ginsberg & Young, 1992).

Network analysis

Matrices of association indices were treated as weighted networks (Newman, 2004; Barthélemy *et al.*, 2005; Boccaletti *et al.*, 2006). Binary networks consider ties between individuals (edges) as present or not, whereas weighted networks show heterogeneity in the connections between nodes. Typical analyses of social networks include measures of which individuals are best connected and how individuals are connected within a network (Newman, 2003). We adopted the standard measure of clustering coefficient to assess the degree to which individual nodes (in this case chimpanzees) tend to cluster together within a social network. More specifically, the clustering coefficient calculates the probability that adjacent nodes of a particular node are connected. This is based on the premise that two associates of a particular individual have a greater probability of knowing one another than two individuals chosen at random from a population. The weighted clustering coefficient takes into account that associations with some neighbors are more important than others (Holme *et al.*, 2007). Network analyses were performed using SOCPROG 2.3 (Whitehead, 2006).

Results

Coordination in space

Tool traces were frequently recovered from tool-using activities that occurred on the ground, such as termite and ant gathering. Tools were also occasionally recovered from sites where chimpanzees were observed to gather honey, which most often occurs in the forest canopy. However, individuals often placed their tools on branches in the high canopy, which prohibited tool collection by the authors. Because of this, the remainder of our investigations of coordination in space focus only on ant dipping and termite fishing. During circuits of active termite nests, we recovered an average of 3.0 fishing probes ($n=685$ tool assemblages, range = 1, 30) and 4.1 puncturing sticks ($n=94$ tool assemblages, range = 1, 32) per site. These assemblages often consisted of tools of different ages, indicating repeat visitation to the tool-using site. We also collected 284 tool assemblages (totaling 1060 tools) at ant nests in the Goualougo Triangle (Sanz *et al.*, 2010). The average number of ant-gathering tools recovered at each site was 3.7 tools ($n=284$, range = 1, 18). Thirty-six percent of these assemblages recovered at ant nests contained two types of tools: nest-perforating tools and ant-dipping probes. Although we have documented seasonal peaks in some types of tool use (Chapter 3; Sanz *et al.*, 2010), there were ample possibilities for chimpanzees to encounter tool sets at termite nests throughout the year. Due to the high degree of material selectivity and modification of tools exhibited in tool manufacture by this chimpanzee population, their tools were easily distinguishable from other detached vegetation around insect nests (see Figure 8.1).

Coordination in time and space

The average number of individuals in chimpanzee parties at termite nests was 2.23 ± 1.57 individuals ($n=388$, range = 1, 14), smaller than parties observed in other contexts (4.98 ± 4.19, $n=606$). This difference was significant for both males and females compared across the two settings ($\text{Wilcoxon}_{\text{Males}}$ $Z=-3.011$, $p=0.002$; $\text{Wilcoxon}_{\text{Females}}$ $Z=-4.107$, $p<0.001$). However, party sizes in termite gathering were actually of intermediate size in comparison to party sizes in other foraging contexts (Figure 8.2). Party sizes were larger when feeding on leaves, fruits or flowers, and smaller when feeding on meat, bark or pith.

Our results indicate ample opportunity for vertical transmission among these apes, as parties visiting termite nests were most often composed of mothers with their dependent offspring (37% of parties). Lone individuals were the next most frequent visitors to termite nests, accounting for 32% of visitation. Adult parties and mixed parties of different age and sex classes accounted for only 12% of parties at termite nests. Small parties consisting of only mothers or a lone individual raises the question about the possibility of horizontal transmission in tool-using contexts. However, graphical analysis of dyadic association indices at termite nests showed a high degree of social connectivity among mature individuals within the Moto community, despite the rarity of adult-only and mixed-party associations (Figure 8.3). The association network of adult parties

Figure 8.1 A subterranean termite nest with discarded puncturing tools. These stick tools persist for several months and may be used by different individuals who visit the nest after the tools have been deposited.

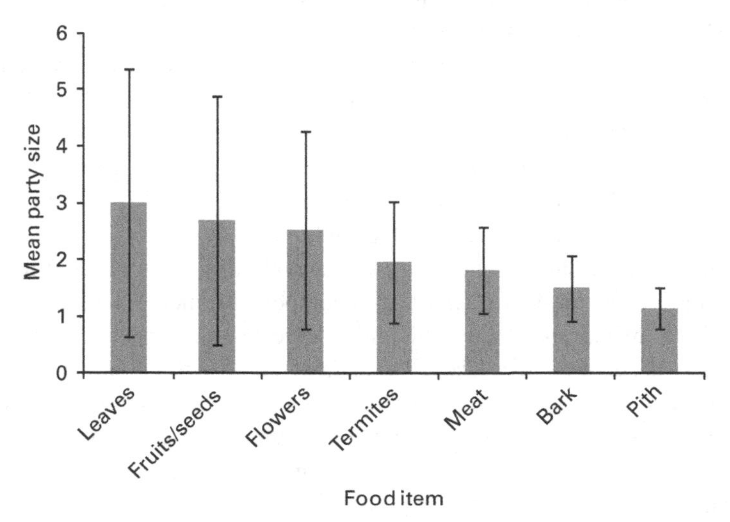

Figure 8.2 Chimpanzee party sizes across different foraging contexts. The number of individuals visiting termite nests is often smaller than those foraging on fruits or leaves.

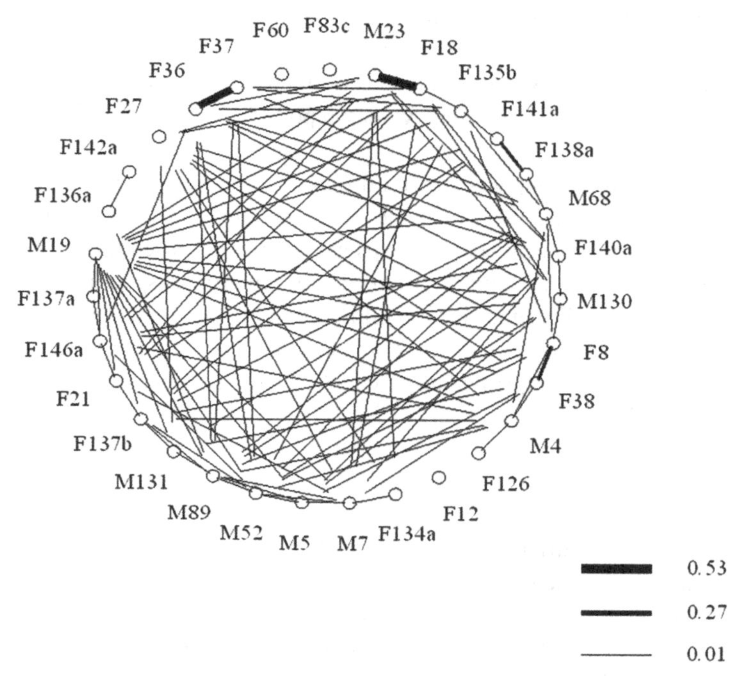

Figure 8.3 Associations of mature individuals in the Moto community at termite nests. Line weights depict strength of association between two individuals.

Table 8.1 Social clustering of chimpanzees in termite fishing and in other contexts, compared with social networks of dolphins and humans.

Taxa	Network	Clustering coefficient	Reference
Human	High-energy physics coauthors	0.73	Newman, 2001
Human	Company directors	0.59	Newman *et al.*, 2002
Human	Computer science coauthors	0.50	Newman, 2001
Human	Theoretical physics coauthors	0.43	Newman, 2001
Dolphin	Moray Firth dolphin associations	0.41, 0.58	Lusseau *et al.*, 2006
Dolphin	Doubtful Sound dolphin associations	0.30	Lusseau, 2003
Chimpanzee	Chimpanzee associations at termite nests	0.26	This study
Chimpanzee	Chimpanzee associations – contexts other than termite nests	0.19	This study
Human	Movie actors associations	0.08	Newman *et al.*, 2002
Human	Biomedical research coauthors	0.07	Newman, 2001

and mixed parties visiting termite nests was relatively dense, with most individuals showing several connections with other mature individuals within the community.

As shown in Table 8.1, clustering coefficients were higher in termite gathering than other feeding contexts, which indicates that although party sizes were relatively small in termite gathering, there was a higher degree of social connectedness and more potential pathways of horizontal transmission than in other feeding contexts. Network analytic measures also provide a means of comparing the degree of connectivity across species and different types of social networks; for example, the clustering coefficients of this chimpanzee population were within values reported from human and dolphin social networks.

We also found that chimpanzees from different communities spatially overlapped in their use of termite nests located in the boundary areas of their community ranges. However, apes from adjacent social groups were never observed to use the same site at the same time. In addition to female dispersal between groups, spatial overlap in tool site use extends the possibility of horizontal information transfer between groups via local or stimulus enhancement. Males were observed to inspect fresh tools found at termite nests within boundary areas (Supplemental Video #8.1), which is a behavior rarely observed in the core area of the range.

Social facilitation

We observed several different types of social interactions during tool-using episodes which could facilitate the transfer of information between individuals. First, individuals tolerate others in close proximity. In addition, immature chimpanzees often approached their mothers to observe their behavior. We rarely observed mothers allowing youngsters to gather termites directly from their tools. However, several females transferred tools and productive work sites to their offspring. During 101 hours of video-recorded tool use at termite nests, a total of 33 tool transfers were observed (Supplementary Video #8.2). Three types of tools (fishing probes, perforating twigs and puncturing tools) are used by the chimpanzee population to gather termite prey (Sanz *et al.*, 2004). Fishing probes or materials to manufacture fishing probes accounted for 97% of observed tool transfers, with only one exchange involving a puncturing tool and no observations of social exchanges involving perforating tools. More than half of the observed tool transfers (18 of 33 transfers) also involved gaining access to a termite tunnel for fishing. Coaction was also observed in this chimpanzee population.

Discussion

In this study we found that the social context of chimpanzee tool use is rich in opportunities for social learning which may serve to stimulate and maintain complex technologies within and between social groups across space and over time. Chimpanzees are not only gregarious, but highly tolerant of conspecifics at tool-using sites. While vertical transmission is likely to be the most common means of social information transfer within social groups, we also documented ample opportunities for indirect and direct transfer of information among mature conspecifics within and between chimpanzee communities in the Goualougo Triangle. While contact with tool traces or noticeable changes in the tool site substrate may suffice to maintain simple forms of tool use, the association of individuals at tool sites and transfer of tools are also likely contributors to the maintenance of the complex tool traditions exhibited by chimpanzees in this region. Future comparative studies could be conducted to determine whether the degree of spatial and temporal overlap at tool sites differs among chimpanzee populations, and if this is related to the diversity or complexity of their tool traditions.

Mother–offspring parties (mother with her dependent offspring) were the most frequently observed party composition at termite nests in the Goualougo Triangle. In addition to dependent offspring, mothers were often accompanied by their mature (subadult, adult) offspring. In addition to association within subgroups, mothers and offspring engaged in social interactions which sometimes involved transfer of tools or sharing of work sites. Parties containing multiple mothers and their offspring were also observed at tool sites. Although the kin relationships among adult females in the Goualougo Triangle population are not yet known, all mature subadult females in the main study communities have dispersed from their natal groups. This situation obviously differs from Gombe, where female chimpanzees often visit termite nests with their mature maternal kin (Lonsdorf, 2006). Within that chimpanzee population, it is not uncommon to have a mother visiting a termite nest with her daughters' families. The demographics of chimpanzee communities and patterns of female dispersal have implications for the transmission and prevalence of information within and between communities, which could dramatically affect patterns of tool-using skills. Opportunities for immature chimpanzees in the Goualougo Triangle to interact with competent tool users other than their mother arose in mixed parties, female parties and consortships. Similar to mothers, we found that mature individuals were highly tolerant of younger chimpanzees who occasionally approached to watch or even interfered with other individuals' tool-using activities. However, the number of individuals converging at termite nests was relatively low compared to other feeding contexts, such as leaf and fruit food sources.

Social facilitation was observed in the form of tool transfers between individual chimpanzees. These were often a result of requests from younger individuals toward their mothers or older siblings. The active transfer of raw materials and manufactured tools from one individual to another is seemingly unique among primates. Such transfers indicate sensitivity to another individual's request or need for a tool, and that the giver is willing to assume the immediate costs of reduced foraging and increased energy expended to locate another tool. Leaving the foraging party at a termite nest to procure tool materials may also place an individual at greater risk of leopard predation. Studies of captive populations have shed light on tool transfers by chimpanzees in controlled circumstances (Yamamoto *et al.*, 2009), but studies of natural populations are necessary to inform us of the ecological and social constraints within which tool use has evolved. Further systematic research is needed on the type of tool transfers (reactive versus proactive) and their prevalence with relation to group demographics in wild populations.

Transfer of information between social groups is an important and often overlooked aspect of the transmission of advantageous foraging strategies, particularly those involving tools. Foraging in boundary areas of a community range is often difficult to detect because it is done discreetly, so as not to draw the attention of neighboring conspecifics. However, our remote video recordings of chimpanzee visitation to termite nests revealed that individuals from different communities visited the same termite nests. Visits of different community members to a particular termite nest did not occur simultaneously, but rather were separated in time (sometimes less than an hour apart). Tools left at tool-use sites and changes to the tool-using substrate provided clues that chimpanzees have previously visited the site. We observed individuals arriving at a termite nest inspecting the fresh and recent tools left behind by conspecifics. This indirect interaction could be one potential means of facilitating transmission of technological information between groups. Further research on the degree of overlap at tool sites between chimpanzee communities, and its impact on promoting conformity in tool traditions over large spatial scales, is needed.

Fission-fusion sociality is an adaptation which enables organisms to flexibly respond to variable ecological and social circumstances. Chimpanzees form subgroups within their community range that are of optimal size and composition for the task at hand, such as foraging at a fruiting tree or conducting a boundary patrol along their territory's frontier. However, the fluidity of this social structure poses a challenge for information transfer within a social group. All of the group members of the community never assemble at a particular location, and so information must be transmitted within smaller party associations. The naturalistic context of tool use in wild chimpanzee communities provides an opportunity to study the means by which individuals maintain technological traditions. The majority of human hunter-gatherers also live in patrilocal fission-fusion societies, and so we can infer that the last common ancestor may have lived in a similar social setting.

Potts (2004) has suggested the fluctuations in the spatial distribution and temporal availability of particular food resources (fruits) and habitats (forests) since the Miocene are reflected in the evolutionary trajectory of great ape ecology, sociality and cognition. It is further suggested that the flexibility and adaptability of great apes may have provided a selective advantage in fluctuating paleoenvironments. Environmental changes can result in shifting ecological pressures (such as interspecific feeding competition) and/or emergence of opportunities which favor the invention of new technology among wild apes. Population size and distribution are also related to the likelihood of the invention and accumulation of socially transmitted information among hominins. The clade of living hominoids provides us with an opportunity to identify shared and derived traits related to the acquisition and maintenance of technological skills.

Identifying the specific social and environmental circumstances that either promote or suppress specific learning mechanisms and outcomes in natural populations will provide a deeper understanding of the evolutionary forces which have shaped the technological sophistication of our own species. Research is currently underway to document the development of complex tool-using skills in the chimpanzees of the Goualougo Triangle and compare their acquisition of tool-using skills to other wild chimpanzee populations. We also advocate research efforts to document the tool traditions of additional populations, as our understanding of the breadth of behavioral diversity has expanded with the study of additional chimpanzee communities. Climate change and anthropogenic disturbances are two factors which are likely to affect the ecological and social contexts of wild ape tool use in the future (van Schaik, 2001). As a result, the natural cultures of wild apes may be even more endangered than the individuals who harbor this knowledge.

Acknowledgments

We thank the Ministry of Forest Economy of the Republic of Congo and the Wildlife Conservation Society for their support and collaboration. We would also like to recognize the tireless dedication of J. R. Onononga, C. Eyana-Ayina, S. Ndolo, A. Nzeheke, W. Mayoukou, M. Mguessa, I. Singono and the Goualougo tracking team. Special thanks are due to J. M. Fay, B. Curran, P. Elkan, S. Elkan, P. Telfer, E. Stokes, M. Gately, T. Breuer, T. Nishihara, B. Djoni and D. Dos Santos. R. Mundry provided assistance with statistical analysis. This manuscript was greatly improved by discussions with C. Boesch, J. Call and E. Lonsdorf. Grateful acknowledgment of funding is due to the US Fish and Wildlife Service, National Geographic Society, Wildlife Conservation Society, Columbus Zoological Park, Brevard Zoological Park and Lincoln Park Zoo.

References

Anderson, D. P., Nordheim, E. V., Boesch, C. & Moermond, T. C. (2002). Factors influencing fission-fusion grouping in chimpanzees in the Taï National Park, Cote d'Ivoire. In C. Boesch, G. Hohmann & L. F. Marchant (eds.) *Behavioral Diversity in Chimpanzees and Bonobos* (pp. 90–101). Cambridge: Cambridge University Press.

Barthélemy, M., Barrat, A., Pastor-Satorras, R. & Vespignani, A. (2005). Characterization and modeling of weighted networks. *Physica A*, **346**, 34–43.

Boccaletti, S., Latora, V., Moreno, Y., Chavez, M. & Hwang, D. U. (2006). Complex networks: structure and dynamics. *Physics Reports*, **424**, 175–308.

Boesch, C. (1991). Teaching among wild chimpanzees. *Animal Behaviour*, **41**, 530–532.

Boesch, C. (1996). Social grouping in Taï chimpanzees. In W. C. McGrew, L. F. Marchant & T. Nishida (eds.) *Great Ape Societies* (pp.101–113). Cambridge: Cambridge University Press.

Boesch, C. (2003). Is culture a golden barrier between human and chimpanzee? *Evolutionary Anthropology*, **12**, 82–91.

Boesch, C. and Boesch, H. (1984). Mental map in wild chimpanzees: an analysis of hammer transports for nut cracking. *Primates*, **25**, 160–170.

Boesch, C. and Boesch-Achermann, H. (2000). *The Chimpanzees of the Taï Forest: Behavioural Ecology and Evolution*. Oxford: Oxford University Press.

Boesch, C., Head, J. & Robbins, M. M. (2009). Complex tool sets for honey extraction among chimpanzees in Loango National Park, Gabon. *Journal of Human Evolution*, **56**(6), 560–569.

Cairns, S. J. & Schwager, S. J. (1987). A comparison of association indices. *Animal Behaviour*, **35**, 1454–1469.

Caro, T. M. & Hauser, M. D. (1992). Is there teaching in nonhuman animals? *Quarterly Review of Biology*, **67**(2), 151–174.

Chapman, C. A., Chapman, L. J. & Wrangham, R. W. (1995). Ecological constraints on group size: an analysis of spider monkey and chimpanzee subgroups. *Behavioral Ecology and Sociobiology*, **36**, 59–70.

Coussi-Korbel, S. & Fragaszy, D. M. (1995). On the relation between social dynamics and social learning. *Animal Behaviour*, **50**, 1441–1453.

Franks, N. R. & Richardson, T. (2006). Teaching in tandem-running in ants. *Nature*, **439**, 153.

Ginsberg, J. R. & Young, T. P. (1992). Measuring association between individuals or groups in behavioural studies. *Animal Behaviour*, **44**, 377–379.

Goodall, J. (1986). *The Chimpanzees of Gombe: Patterns of Behavior.* Cambridge, MA: Belknap Press.

Halperin, S. (1979). Temporary association patterns in free ranging chimpanzees: an assessment of individual grouping preferences. In D. A. Hamburg & E. McCown (eds.) *The Great Apes* (pp. 491–499). Menlo Park, CA: Benjamin/Cummings.

Holme, P., Park, S. M., Kim, B. J. & Edling, C. R. (2007). Korean university life in a network perspective: dynamics of a large affiliation network. *Physica a-Statistical Mechanics and Its Applications*, **373**, 821–830.

Hopper, L. M., Spiteri, A., Lambeth, S. P., *et al.* (2007). Experimental studies of traditions and underlying transmission processes in chimpanzees. *Animal Behaviour*, **73**, 1021–1032.

Hoppitt, W. J. E., Brown, G. R., Kendal, R., *et al.* (2008). Lessons from animal teaching. *Trends in Ecology and Evolution*, **23**(9), 486–493.

Horner, V. (2010). The cultural mind of chimpanzees: how social tolerance can shape the transmission of culture. In E. V. Lonsdorf, S. R. Ross & T. Matsuzawa (eds.) *The Mind of the Chimpanzee: Ecological and Experimental Perspectives* (pp. 116–126). Chicago, IL: University of Chicago Press.

Humle, T., Snowdon, C. T. & Matsuzawa, T. (2009). Social influences on ant-dipping acquisition in the wild chimpanzees (*Pan troglodytes verus*) of Bossou, Guinea, West Africa. *Animal Cognition*, **12**, S37–S48.

Kitahara-Frisch, J. & Norikoshi, K. (1982). Spontaneous sponge-making in captive chimpanzees. *Journal of Human Evolution*, **11**(1), 41–47.

Lonsdorf, E. V. (2006). What is the role of mothers in the acquisition of termite-fishing behaviors in wild chimpanzees (*Pan troglodytes schweinfurthii*)? *Animal Cognition*, **9**, 36–46.

Lusseau, D. (2003). The emergent properties of a dolphin social network. *Proceedings of the Royal Society of London B: Biological Sciences*, **270**, S186–S188.

Lusseau, D., Wilson, B., Hammond, P. S., *et al.* (2006). Quantifying the influence of sociality on population structure in bottlenose dolphins. *Journal of Animal Ecology*, **75**(1), 14–24.

McGrew, W. C. (1977). Socialization and object-manipulation of wild chimpanzees. In S. Chevalier-Skolnikoff & F. E. Poirier (eds.) *Primate Bio-social Development: Biological, Social, and Ecological Determinants* (pp. 261–288). New York: Garland Publishing.

McGrew, W. C. (1992). *Chimpanzee Material Culture: Implications for Human Evolution.* Cambridge: Cambridge University Press.

McGrew, W. C. & Collins, D. A. (1985). Tool use by wild chimpanzees (*Pan troglodytes*) to obtain termites (*Macrotermes herus*) in the Mahale Mountains, Tanzania. *American Journal of Primatology*, **9**, 47–62.

Mitani, J. C. (2006). Demographic influences on the behavior of chimpanzees. *Primates*, **47**(1), 6–13.

Mitani, J., Watts, D. & Lwanga, J. (2002). Ecological and social correlates of chimpanzee party size and composition. In C. Boesch, G. Hohmann & L. F. Marchant (eds.) *Behavioural Diversity in Chimpanzees and Bonobos* (pp. 102–111). Cambridge: Cambridge University Press.

Möbius, Y., Boesch, C., Koops, K., Matsuzawa, T. & Humle, T. (2008). Cultural differences in army ant predation by West African chimpanzees? A comparative study of microecological variables. *Animal Behaviour*, **76**, 37–45.

Newman, M. E. J. (2001). The structure of scientific collaboration networks. *Proceedings of the National Academy of Sciences USA*, **98**, 404–409.

Newman, M. E. J. (2003). The structure and function of complex networks. *Society for Industrial and Applied Mathematics*, **45**, 167–256.

Newman, M. E. J. (2004). Analysis of weighted networks. *Physical Review E*, **70**, 056131.
Newman, M. E. J., Watts, D. J. & Strogatz, S. H. (2002). Random graph models of social networks. *Proceedings of the National Academy of Sciences USA*, **99**, 2566–2572.
Potts, R. (2004). Paleoenvironmental basis of cognitive evolution in great apes. *American Journal of Primatology*, **62**(3), 209–228.
Pradhan, G. R., Tennie, C. & van Schaik, C. P. (2012). Social organization and the evolution of cumulative technology of apes and hominins. *Journal of Human Evolution*, **63**(1), 180–190.
Sanz, C. M. & Morgan, D. B. (2007). Chimpanzee tool technology in the Goualougo Triangle, Republic of Congo. *Journal of Human Evolution*, **52**(4), 420–433.
Sanz, C. M. & Morgan, D. B. (2009). Flexible and persistent tool-using strategies in honey-gathering by wild chimpanzees. *International Journal of Primatology*, **30**(3), 411–427.
Sanz, C. & Morgan, D. (2010). Complexity of chimpanzee tool using behaviors. In E. V. Lonsdorf, S. R. Ross & T. Matsuzawa (eds.) *The Mind of the Chimpanzee: Ecological and Experimental Perspectives* (pp. 127–140). Chicago, IL: University of Chicago Press.
Sanz, C., Morgan, D. & Gulick, S. (2004). New insights into chimpanzees, tools, and termites from the Congo basin. *American Naturalist*, **164**(5), 567–581.
Sanz, C., Call, J. & Morgan, D. (2009). Design complexity in termite-fishing tools of chimpanzees (*Pan troglodytes*). *Biology Letters*, **5**(3), 293–296.
Sanz, C. M., Schöning, C. & Morgan, D. B. (2010). Chimpanzees prey on army ants with specialized tool set. *American Journal of Primatology*, **72**(1), 17–24.
Shumaker, R. W., Walkup, K. R. & Beck, B. B. (2011). *Animal Tool Behavior: The Use and Manufacture of Tools by Animals*. Baltimore, MD: Johns Hopkins University Press.
Sugiyama, J. (1984). Population dynamics of wild chimpanzees at Bossou, Guinea, between 1976–1983. *Primates*, **25**, 391–400.
Thornton, A. & McAuliffe, K. (2006). Teaching in wild meerkats. *Science*, **313**, 227–229.
van Schaik, C. P. (2001). Fragility of traditions: the disturbance hypothesis for the loss of local traditions in orangutans. *International Journal of Primatology*, **23**(3), 527–538.
van Schaik, C. P. (2003). Local traditions in orangutans and chimpanzees: social learning and social tolerance. In D. M. Fragaszy & S. Perry (eds.) *The Biology of Traditions: Models and Evidence* (pp. 297–328). Cambridge: Cambridge University Press.
van Schaik, C. P., Deaner, R. O. & Merrill, M. Y. (1999). The conditions for tool use in primates: implications for the evolution of material culture. *Journal of Human Evolution*, **36**, 719–741.
Visalberghi, E., Fragaszy, D., Ottoni, E., *et al.* (2007). Characteristics of hammer stones and anvils used by wild bearded capuchin monkeys (*Cebus libidinosus*) to crack open palm nuts. *American Journal of Physical Anthropology*, **132**(3), 426–444.
Whitehead, H. (2006). SOCPROG 2.3.
Whiten, A., Goodall, J., McGrew, W. C., *et al.* (1999). Cultures in chimpanzees. *Nature*, **399**, 682–685.
Whiten, A., Goodall, J., McGrew, W. C., et al. (2001). Charting cultural variation in chimpanzees. *Behaviour*, **138**, 1481–1516.
Yamamoto, S., Humle, T. & Tanaka, M. (2009). Chimpanzees help each other upon request. *PLoS ONE*, 4(10).

9 Orangutan tool use and the evolution of technology

Ellen J. M. Meulman
Anthropological Institute and Museum, University of Zürich

Carel P. van Schaik
Anthropological Institute and Museum, University of Zürich

Introduction

Commonly referred to as a hallmark of human evolution, tool use is often considered a complex skill. Paradoxically, however, tool use seems to be widespread in the animal kingdom and may consist of fairly simple behavioral actions. In this chapter we try to relate these somewhat contradictory views to the relatively rare occurrence of habitual and complex tool use in wild orangutans, especially when compared to wild chimpanzees. We propose that, in addition to the previously suggested factors (i.e., extractive foraging, social tolerance and intelligence), terrestriality may have been instrumental in the evolution of especially habitual (*sensu* McGrew & Marchant, 1997) and complex tool use, thus explaining the "orangutan tool paradox." Our preliminary comparison of eight orangutan and ten chimpanzee study populations (descriptively, via a principal component analysis [PCA], and by testing predictions related to the four factors) does indeed point in this direction.

Defining tool use

Although tool use has been defined in various ways (see Shumaker *et al.*, 2011 for a detailed discussion), we choose to follow the definition of Parker and Gibson (1977):

> Tool use is the manipulation of an object (the tool), not part of the actor's anatomical equipment and not attached to a substrate, to change the position, action, or condition of another object, either directly through the action of the tool on the object or of the object on the tool, or through action at a distance as in aimed throwing. (Modified from Parker & Gibson, 1977; Sanz & Morgan, 2007)

We did not adopt the new definition proposed by Shumaker *et al.* (2011) because we believe that the criterion that objects are "not attached to a substrate" is very important. This condition may be particularly relevant for the evolution of complex tool use, because detached objects can be more easily modified and can be incorporated more

Tool Use in Animals: Cognition and Ecology, eds. Crickette M. Sanz, Josep Call and Christophe Boesch. Published by Cambridge University Press.

flexibly into tool combinations/sequences. We also avoided making inferences about an animal's intentions when using tools (e.g., to alter more efficiently the form, position or condition of another object, organism or the user itself: Beck, 1980). Therefore, Parker and Gibson's (1977) definition seems the most appropriate for this chapter.

Tool use as a reflection of a cognitive gradient

The significance of tool use lies in what it reveals about the cognitive abilities of its users. Although cognitive abilities may be reflected in many tasks, tool use provides us with the clearest window into the cognitive abilities underlying animal behavior (Byrne, 1995). This is not because tool use requires advanced cognition *per se*, but rather because of the cognitive gradient that can be recognized when animals use objects. This ranges from the fairly simple manipulation of fixed substrates or borderline tool use to *true* tool use in which objects are detached from their substrate (although use may still be stereotypic and inflexible); additional steps of manufacture and modification (Beck, 1980; Boesch & Boesch, 1990; McGrew, 1992; Bentley-Condit & Smith, 2010); flexible tool use, in which the tools are adjusted to the task at hand (van Schaik *et al.*, 1996); and finally accumulated tool use (also: cumulative or associated tool use), in which multiple innovations (cf. Reader & Laland, 2002) may be combined for a single purpose (Parker & Gibson, 1977; Beck, 1980; Byrne, 1995; Bentley-Condit & Smith, 2010; Shumaker *et al.*, 2011). Flexible and cumulative tool use in particular reveal the operation of intentions or mental simulation and planning, rather than direct responses to stimuli (Byrne, 1995), and therefore can be considered intelligent (Parker & Gibson, 1977).

Defining "complex" tool use

For this chapter we focus especially on the complex end of the tool-use gradient because of our interest in "the orangutan tool paradox": i.e., the rare occurrence of complex tool use in wild orangutans, although such complex tool use is fairly common for rehabilitant or captive orangutans (cf. van Schaik, 2004). Complex tool use has been defined in varying ways (Sanz & Morgan, 2010; Shumaker *et al.*, 2011). Here we define complex tool use as tool use that includes more than one element (accumulated), because the number of constituent elements will generally be correlated with the difficulty of learning and because hominin technology is characterized by increasing accumulation (cf. Haidle, 2010; Pradhan *et al.*, 2012). Where known, accumulated techniques are also generally accompanied by flexibility (adjustment to the task at hand) and acquisition through social learning (any kind of learning that is triggered or influenced by other group members or conspecifics [cf. Fragaszy & Perry, 2003] and thus including also socially facilitated individual learning via, for example, stimulus enhancement). However, because these latter two aspects are less consistently reported in the literature, we focus on the accumulation criterion. This may not be perfect, in that some non-accumulated tool-use techniques may be cognitively challenging as well, but this is the most practical division of the complexity gradient.

Evolution of primate tool use

Apart from the insight it provides into cognitive abilities, tool use is also interesting from the perspective of human evolution. Among all tool-using taxa, primates are unique in the variation they show in tool-using contexts (Bentley-Condit & Smith, 2010). Nonetheless, habitual and complex tool use have often been considered a hallmark of hominins. Habitual tool use here refers to those tool-use variants that have been seen repeatedly in several individuals, consistent with some degree of social transmission (*sensu* McGrew & Marchant, 1997), excluding branch throwing in agonistic contexts, which is universal among primates. Within the primate order, only chimpanzees (McGrew, 2004b; Sanz & Morgan, 2007, 2009), orangutans (van Schaik *et al.*, 1996), some capuchins (Ottoni & Izar, 2008; Visalberghi *et al.*, 2009) and possibly some long-tailed macaques (Gumert *et al.*, 2009) are known to be capable of habitual tool use in natural conditions.

To explain the evolution of tool use in primates, van Schaik *et al.* (1999) proposed a socioecological model that includes a nested series of conditions. Tool use will be performed in broader contexts only when the primates engage in extractive foraging and are capable of dexterous manipulation (first two conditions). Species with more advanced innovative ability (intelligence) can also manufacture tools in both captivity and the wild (third condition). Subsequently, social tolerance allows for the spread of tool innovations within a population, allowing for habitual tool use and material culture (fourth condition). Finally, the ability for teaching in humans further allows for cumulative culture (fifth condition).

The potential role of terrestriality

Although the socioecological model explains the broad distribution of aspects of primate tool use, it cannot explain the rarity of complex tool use in all wild orangutans (van Schaik *et al.*, 1996) relative to chimpanzees (Whiten *et al.*, 2009). We would therefore like to propose to add terrestriality as a factor to the model. A terrestriality effect on tool innovations (especially for extractive foraging) and complex manipulations has already been suggested in various previous studies (e.g., McGrew, 2004a; Visalberghi *et al.*, 2005; Humle & Matsuzawa, 2009; Spagnoletti *et al.*, 2009; but see Boesch-Achermann & Boesch, 1994). However, here we propose that terrestriality may not only affect opportunities for (complex) tool innovations, but may also affect opportunities for socially facilitated tool-affordance learning (*sensu* Huang & Charman, 2005), because previously used tools are more easily encountered in a terrestrial setting (see also Meulman *et al.*, 2012). Terrestriality may especially promote the occurrence and transmission of complex tool use, because accumulated technology is less likely to be invented independently and therefore relies more critically on propitious learning conditions. Orangutans are arboreal and appear to lack complex tool use. Thus, they provide us with an excellent opportunity to study the conditions favoring the origins of complex tool use, and hence the foundation of hominin cumulative technology.

Orangutans

Among the great apes, orangutans are the least related to humans. The current consensus among paleoanthropologists is that the orangutan lineage and that of the other great apes separated around 14 mya (Kelley, 2002; Raaum *et al.*, 2005). Today, orangutans are only found on the islands of Sumatra and Borneo, in Southeast Asia. They are commonly subdivided into two species, the Sumatran *Pongo abelii* and the Bornean *Pongo pygmaeus* (Xu & Arnason, 1996; Warren *et al.*, 2001). The existing taxonomic subdivision of the three Bornean subspecies (*P. p. pygmaeus*, *P. p. wurmbii* and *P. p. morio*), described on the basis of morphological characteristics (Groves, 2001), however, does not adequately capture the genetic variation within this species (Arora *et al.*, 2010).

Orangutans are large-bodied great apes that live in habitats varying from coastal peat swamp forest to montane dryland rainforest. They mainly differ from the African great apes in that females are almost exclusively arboreal, and, despite variation in gregariousness across populations, are generally semi-solitary. Bornean males are more terrestrial, but almost exclusively solitary apart from brief consortships with females (Utami-Atmoko *et al.*, 2009).

Ecologically, orangutans are much like chimpanzees, being frugivorous and omnivorous foragers with a large dietary repertoire. This includes extractive foraging, which means that they extract food items from the matrices in which these items are embedded. Orangutans feed, for example, on seeds of *Polyalthia glauca* after first discarding the foul-tasting pulp, and remove the seeds of *Neesia* sp. without even touching the prickly matrix embedding them. Insects or their products (e.g., honey, larvae) are extracted from nests that are often located in tree holes, or picked up after pulling bole climbers off the trunk. Pith is extracted from hearts or stems of palm trees or the young twigs of *Dyera costulata*, and tree cambium is scraped off inner bark after first removing the outer bark of tree trunks. Because all these items are embedded in a matrix that is hard, or even dangerous, animals must learn to identify them as food and overcome their defenses. This strong reliance on extractive foraging leads us to expect abundant tool use in orangutans.

This study

Updating the orangutan tool catalog

In this chapter we have compiled all available information on wild orangutan tool use to create an updated overview of the orangutan tool repertoire and to compare this to the chimpanzee tool repertoire. To allow for fair comparison, and to exclude effects such as enculturation that are less directly relevant for understanding the occurrence and evolution of tool use in primates, only wild populations were considered.

Describing the variation in tool repertoires

To establish the main components distinguishing tool repertoires across sites, and to gain insight into the level of interdependency between outcome variables, one can conduct a

PCA. Eight outcome variables (or nine when we included nest variables – see discussion) were included as a potential source of variation in the tool repertoires. These included three context-related variables to help us to better discriminate between different aspects of the tool repertoire and innovation biases that may exist. Based on the socioecological model of van Schaik *et al.* (1999) we expect the following outcome variables to cluster together: (1) intelligence – physical comfort tool variants (non-extractive), total number of tool variants, complex tool variants; (2) extractive foraging – extractive foraging tool variants, subsistence tool variants, total number of tool variants; (3) social tolerance – cultural tool variants, communication tool variants, total number of tool variants. However, if we include the potential terrestriality effect, we expect terrestrial, extractive, cultural, complex, subsistence, communication and the total number of tool variants to cluster together because of the potential positive effect of terrestriality on extractive foraging, social learning and the acquisition of complex skills. The second component should then include physical comfort tool variants and the total number of tool variants.

Predictions of the new model

To gain some insight into the independent effects of each of the four predictor variables on the variation in tool repertoires across sites and the importance of interactions between predictor variables, a multiple regression analysis would have been ideal. Sample size, however, did not allow for a multiple regression analysis to predict the best explanatory model for each component extracted from the PCA. We decided to use bivariate analysis as an alternative method for testing the predictions regarding the effects of the four factors on the tool repertoires of the various orangutan and chimpanzee populations. Although these analyses ignore possible interaction effects among the factors, we believe they do help us understand the extent of the direct effects of the four factors on the tool repertoire. Hence, although preliminary, these results should give us an idea of the best explanatory model for the variation in tool repertoires.

We will now discuss the proxies used for each factor and develop predictions for the expected differences among orangutan populations and between orangutans and chimpanzees.

Testing the role of intelligence

Intelligence can be viewed as general cognitive ability (Deaner *et al.*, 2007; Reader *et al.*, 2011). Although the best proxy measure of such general cognitive abilities or intelligence is still highly debated, these studies showed that absolute measures of brain size provided a far better fit than body-size-corrected measures such as the encephalization quotient. We therefore considered absolute correlates of cranial capacity to be a valid proxy for intelligence, especially given that female great apes are quite similar in body size. An additional advantage of taking this measure is that it allows us to compare the different orangutan species and subspecies.

For females, cranial capacity is almost identical between chimpanzees and orangutans (Isler *et al.*, 2008). Among orangutans, however, *P. p. morio* (northeast Borneo) have significantly smaller cranial capacities than the Sumatran orangutans (*P. abelii*),

Table 9.1 Predicted differences in tool repertoire between orangutan populations and between orangutan and chimpanzee populations, depending on the potential terrestriality effect.

Role of	Measure	Kind of tools	OU-B vs. S	OU vs. CH (−Terrestriality)	OU vs. CH (+Terrestriality)
Intelligence	Cranial capacity	Non-extractive	B (*m*) < B, S (n*m*)	OU(n*m*) = CH	OU(n*m*) ≤ CH
		complex	B (*m*) < B, S (n*m*)	OU(n*m*) = CH	OU(n*m*) < CH
Extraction	Insectivory	Extractive	B (n*m*) < S	OU(n*m*) = CH	OU(n*m*) < CH
Opportunities for social learning	Social tolerance	Cultural	B (n*m*) < S	OU(n*m*) < CH	OU(n*m*) << CH

Notes

B = Bornean orangutans, S = Sumatran orangutans, *m* = *morio*, n*m* = non-*morio*, OU(n*m*) = non-*morio* orangutans, CH = chimpanzees, −/+Terrestriality: ex-/including potential terrestriality effect, respectively. The first three columns describe what has been tested, the last three columns describe the predictions for each comparison and test. For orangutans the ex- or inclusion of the potential terrestriality effect has no effect on the predictions.

with *P. p. wurmbii* (central Kalimantan/southern Borneo) being intermediate but closer to *P. abelii* (Taylor & van Schaik, 2007). We therefore considered *P. p. morio* somewhat less intelligent than the other orangutan subspecies (although this has not been formally tested). Nevertheless, we also report the results when *P. p. morio* were included in the analyses. We expected no differences between chimpanzees and (non-*morio*) orangutans in general (see also Deaner *et al.*, 2007; Reader *et al.*, 2011).

The prediction is that innovative ability, as proxied by intelligence, predicts total tool repertoire size (see also Reader & Laland, 2002). However, the latter may be confounded by other variables. First, given the known variation in reliance on extractive foraging, a cleaner estimate of the role of intelligence would be to examine the repertoire size of tool variants *not* used for extractive foraging. Second, variation among populations and species in opportunities for social learning may affect the likelihood that innovations persist. Thus, it is possible that the total repertoire is greater in species or populations with better opportunities for social learning (van Schaik, 2006). Hence, tool complexity may be a better measure of intelligence, although it in turn may be affected by terrestriality and opportunities for social learning (see below). We will therefore use the total repertoire of non-extractive tool variants and tool complexity as preliminary estimates of the effect of intelligence (Table 9.1).

Testing the role of extractive foraging

Currently few quantitative estimates for extractive foraging frequency exist. The effect of extractive foraging opportunities is best estimated by comparing the total repertoire of extractive tool variants. As almost all insectivory is extractive, insectivory may be the best proxy for estimating tendencies toward extractive foraging (van Schaik *et al.*, 1999). This is especially likely since other extractive activities are not amenable to support tool use (e.g., the extraction of cambium or bark by Bornean orangutans). Nevertheless, to validate this, we need to establish that most insect foraging is indeed extractive. Data confirm this (Tuanan: >95%, M. A. van Noordwijk, 2010, unpublished data; Suaq

Balimbing: >75%, Sitompul, 1995). Overall, Sumatran orangutans are more insectivorous than Bornean orangutans (11% of the total feeding time in Sumatran populations, about 5.7% for *P. p. wurmbii* and ca. 1.4% for *P. p. morio*) (Morrogh-Bernard *et al.*, 2009). It is commonly thought that chimpanzees rely more on extractive foraging than orangutans. However, the mean percentage of insectivory across chimpanzee populations is around 4% (Stumpf, 2007), similar to Bornean orangutans. Thus, until future work provides better estimates of the incidence of extractive foraging, it is parsimonious to expect that Sumatran orangutans have more extractive foraging tool variants than Bornean orangutans, but that there are no systematic species differences between the two great ape species (see Table 9.1).

Testing the role of social tolerance

Opportunities for social learning will depend on the degree of tolerant proximity (Coussi-Korbel & Fragaszy, 1995). Among orangutans, Sumatran populations are much more gregarious than the Bornean populations (van Schaik, 1999; van Noordwijk *et al.*, 2009), largely due to differences in forest productivity and food availability (van Schaik, 1999; Marshall *et al.*, 2009). Most chimpanzee populations are more gregarious than orangutans (van Schaik *et al.*, 2003c). Thus, chimpanzees have more opportunities for social learning than orangutans, and Sumatran orangutans have more than Bornean orangutans. The size of the cultural tool repertoire is usually considered to be a good estimate for the effect of opportunities for social learning (see Table 9.1).

Testing the role of terrestriality

The increased innovation tendencies are expected to primarily affect the number of (terrestrial) extractive tool variants. Moreover, as explained above, enhanced social learning opportunities should increase the number of complex tool variants within the repertoire more than the simple forms. We therefore predict that the socially tolerant terrestrial chimpanzees have more extractive (cultural) and complex (cultural) tool variants in their repertoire than the semi-solitary arboreal orangutans. This contrasts with the predictions of the socioecological model (see Table 9.1). Moreover, we expect that tools used on the ground are more complex than tool variants used in arboreal settings. Because orangutans are rarely terrestrial and usually solitary (aside from consortships), we expect no differences in tool complexity between the various orangutan populations due to terrestriality (Table 9.1).

Methods

Orangutan tool catalog

Despite the recent wave of interest in innovation and culture in orangutans, so far no complete tool catalogs have been compiled for wild orangutans (but see Fox & Bin'Muhammad, 2002). We therefore reviewed the literature on tool use, innovations and culture in wild orangutans (Russon *et al.*, 2009; van Schaik *et al.*, 2009), and added

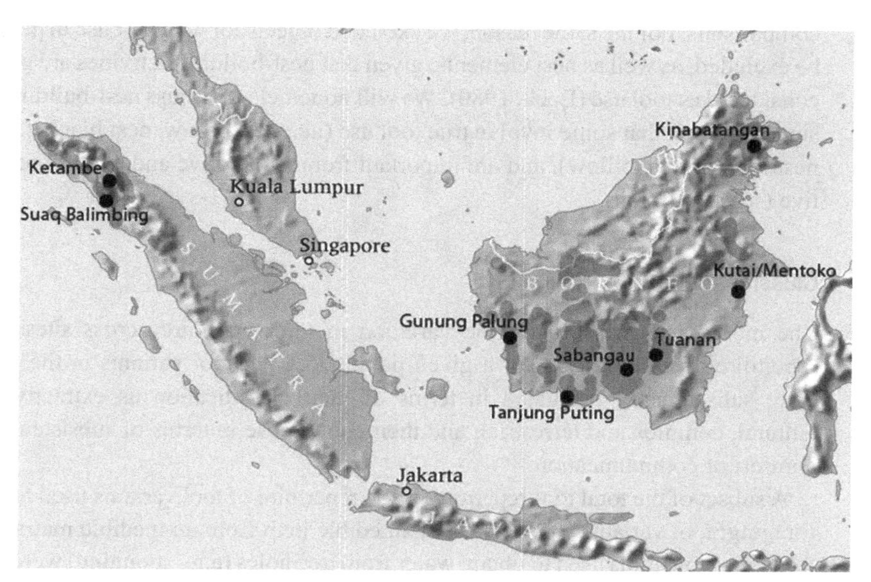

Figure 9.1 Map showing the locations of the eight orangutan study sites (black dots) that have been included in the tool catalog. Dark-shaded areas indicate orangutan distribution.

some unpublished observations from Suaq Balimbing, to construct a tool-use inventory for eight wild orangutan populations in Sumatra and Borneo (Figure 9.1). We trust that this provides us with the complete tool repertoire for orangutans at existing study sites (especially for the habitual tool variants), because tool-use behaviors are striking to observers and have been a focus of attention for at least three decades (van Schaik *et al.*, 1996; Fox & Bin'Muhammad, 2002).

Chimpanzee tool catalog

For the species comparison to chimpanzees we used the chimpanzee tool repertoire as reported by Sanz and Morgan (2007). We additionally included one new tool variant reported for Goualougo – "ant nest perforation" (Sanz *et al.*, 2010). For information about cultural status, tool complexity and terrestrial use, see Meulman *et al.* (2012).

Criteria for the inclusion of tool variants in the tool catalog

To be able to compare orangutan tool repertoires with those of chimpanzees, we applied the same criteria for the inclusion of tool variants as reported by Sanz and Morgan (2007) (see also Whiten *et al.*, 2001). Hence, dependent on the similarity of the action patterns, tool variants were split or lumped. Similar criteria have been reported in the literature on cultural behavioral variants in wild orangutans (van Schaik *et al.*, 2003a; Wich *et al.*, 2009).

"Accidental" innovations reflect the potential for innovation and flexible and complex tool use, and have also been included in the chimpanzee tool catalog described by Sanz and Morgan (2007). We therefore included them here as well to facilitate unbiased comparisons. For the same reason, we excluded objects for which or use in play could not be excluded, as well as nest elements, given that nest-building activities are generally not considered as tool use (Beck, 1980). We will nonetheless discuss nest-building elements, since we think that some involve true tool use (i.e., nest pillow, nest blanket, nest lining, nest roof, artistic pillow), and are important from a cognitive and evolutionary perspective (see discussion).

Classification of tool variants

The most basic measure for the variation in tool repertoire across sites is the total repertoire of tool variants for a given population. All tool variants in the tool catalog were subsequently evaluated in terms of their classification as extractive foraging, cultural, complex and terrestrial; and their context use in terms of subsistence, physical comfort or communication.

A subset of the total tool repertoire is the repertoire of tool variants used for extractive foraging: tool variants used to extract an edible item from an inedible matrix. Based on this definition, tools used to obtain water from tree holes (e.g., sponging) were considered extractive foraging tool variants, whereas tools used to obtain water from ponds or streams were not (e.g., algae scoop).

Another subset is the cultural tool repertoire of a population. We identified putative cultural tool variants as those behavioral patterns that are absent without ecological explanation in at least one community, yet achieve at least habitual status in at least one other community, excluding those that are species universals (i.e., at least habitual prevalence observed at each site and therefore thought to have more canalized development) (Whiten *et al.*, 1999). This approach has recently been validated for orangutans (Krützen *et al.*, 2011).

A fourth measure is the complexity of the tool variants. We classified tool variants as complex when the accumulation of tools, including the particular tool variant (e.g., in tool sets or combined tool use *sensu* McGrew, 2010), has been reported in the literature.

Tool variants may be used exclusively in arboreal conditions or in terrestrial conditions as well (a fifth measure or outcome variable). They can furthermore be classified according to the context in which they were used (*sensu* van Schaik *et al.*, 2006): subsistence, physical comfort and communication (outcome variables 6–8). In instances of doubt, tool variants were classified according to their direct purpose. Hence, tooth-cleaning tools, for example, were classified as physical-comfort tool variants because they were not used to assist feeding but used after feeding to enhance physical comfort or hygiene. Where multiple contexts were possible, we chose the predominant one (e.g., a branch swatter is mainly used to protect against insects while resting, but can also assist in feeding on bees' nests).

Statistical analysis

Given the small sample sizes, we used (if possible) non-parametric statistical tests with exact p-values (two-tailed), and also reported trends. The repertoires of the three orangutan subspecies were compared with the Kruskal–Wallis test. Mann–Whitney U tests were used to compare orangutan with chimpanzee repertoires. The distribution of tool variants over the three behavioral contexts was tested with a Chi-square over the 81 tool variants (chimpanzees: $n=43$, orangutans: $n=38$).

We conducted a PCA with orthogonal rotation (varimax) to extract the factors relevant for distinguishing tool repertoires of different study sites, and to look at the clustering of the various subsets of tool variants. Bivariate correlation analyses were conducted to test the effect of study duration on the eight outcome variables included in the PCA.

To test the effect of the four factors proposed to be relevant for the evolution of tool use, populations and (sub)species were compared with the Mann–Whitney U test by taking the mean number of tool variants per long-term study site, to control for sampling intensity (chimpanzees: $n=10$; *P. p. wurmbii*: $n=4$; *P. p. morio*: $n=2$; and *P. abelii*: $n=2$ study sites). Furthermore, we performed Mann–Whitney U tests to evaluate whether complex tool variants differed from simple tool variants recorded for orangutans and chimpanzees (total $n=81$) in being used more terrestrially and/or extractively.

Results

Tool catalogs

For wild orangutans a total of 38 (true) tool variants (excluding five nest-building variants that could also be considered true tool use) have been reported (see Table 9.2). This number includes a hitherto unpublished variant, the "straw tool": using a tool to drink water from a tree hole or hole in the liana bark ($n=3$ observations). The entire catalog included seven tool variants used for extractive foraging (amounting to 18% of the total repertoire); 16 tool variants that were potentially cultural (42%); two (5%) were used in terrestrial contexts; and zero variants contained multiple elements and hence were considered complex. All five nest elements (not included in the above-mentioned totals) were classified as non-cultural, complex, physical-comfort tool variants. Regarding the context, 13 (34%) of the orangutan tool variants were used in the subsistence context, 18 (47%) for physical comfort and 7 (18%) for communication.

Figure 9.2 shows the distribution of tool variants over the eight wild orangutan study sites included in the analysis. The figure indicates that differences regarding the number of tool variants between the subspecies are all in favor of *P. abelii*, against *P. p. morio*, with *P. p. wurmbii* being intermediate. Only the total number of tools differed significantly among the three subspecies ($\chi^2(2)=6.1$; $p=0.014$), whereas a trend was observed for the number of extractive tool variants: $\chi^2(2)=5.1$; $p=0.057$. The number of complex, cultural, subsistence, physical-comfort and communication tools did not differ significantly among the subspecies. A bivariate correlation analysis revealed no significant correlation of study duration with any of the eight outcome variables.

The chimpanzee catalog included 43 variants, including 23 (53%) extractive, 17 (40%) complex, 23 (53%) cultural and 32 (74%) terrestrial tool variants. With regard to the context, 26 (60%) of the chimpanzee tool variants were used for subsistence, 12 (28%) for physical comfort and 5 (12%) for communication. Sanz and Morgan (2007) also

Table 9.2 Orangutan tool catalog. An overview of all the tool variants (including nest elements) reported for wild orangutans; comparing their prevalence across eight long-term orangutan study populations.

				Study sites									
Tool variant	Cont.	EF	Cult.	GP	TP	SA	TU	KU	KI	KE	SB	Ref	Source
Leaf bundle ("doll")	2	0	0	R	R	A	R	A	A	R	A	1	T 21.2, 20.1
Auto-erotic tool	2	0	1	A	A	A	A	P	A	C	A	1	T 21.1, 20.1
Bee cover	2	0	0	A	A	A	R	A	P	P	R	1–3	T 21.2, 20.1
Branch cushion	2	0	1	R	H	H	C	?	A	H	C	1,3	T 21.1, 20.1
Branch hide	3	0	0	A	R	A	Ȧ	P	R	R	R	1–3	T 21.2, 20.1
Branch hook	2	0	0	A	A	R	R	?	A	?	R	1,3	T 21.2, 20.1
Branch reach fruit	1	0	0	A	R	A	A	A	A	A	A	2	
Branch scoop	1	1	1	A	A	H	A	A	A	A	H	1	T 21.1, 20.1
Branch as swatter	1	0	1	R	R	R	A	H	H	H	H	1	T 21.1, 20.1
Branch fan	2	0	0	?	?	?	?	?	?	?	?	2	
Branch dragging display	3	0	0	A	A	A	A	?	R	E	E	1	T 21.2, 20.1
Foam leaf body	2	0	1	A	A	H	A	?	A	A	A	1	T 21.1, 20.1
Club	3	0	0	R	A	A	A	A	A	A	A	2	
Leaf wipe	3	0	1	A	C	A	A	A	A	A	R	1,3	T 21.1, 20.1
Kiss-squeak leaves	3	0	1	C	A	R	H	H	A	R	R	1,3	T 21.1, 20.1
Leaf cushion	2	0	1	E	R	E	E	A	R	C	A	1,3	T 21.1, 20.1
Leaf glove (bite)	1	0	0	A	A	R	A	?	?	R	A	1,2	T 21.2, 20.1
Leaf glove (spine)	1	0	1	E	R	E	E	A	R	H	A	1,3	T 21.1, 20.1
Leaf napkin	2	0	1	A	A	A	A	C	A	R	R	1,3	T 21.1, 20.1
Poultice use	2	0	0	A	A	R	A	?	A	A	A	1	T 21.2, 20.1
Leaf scoop	1	0	0	R	A	A	A	A	A	A	A	1	T 21.2, 20.1
Sponging	1	1	0	A	A	R	A	A	A	R	A	1	T 21.2, 20.1
Moss cleaning	2	0	1	A	A	H	A	?	A	A	A	1	T 21.1, 20.1
Leaf wiper	2	0	0	R	A	R	A	A	A	A	A	1	T 21.2, 20.1
Aimed missile	3	0	0	C	C	C	C	C	C	C	C	2	
Nail cleaning	2	0	0	?	A	A	A	?	?	P	A	1	T 21.2, 20.1
Hat cover	2	0	0	C	C	C	C	C	C	C	C	1,2	T 21.2
Scratch with stick	2	0	1	A	R	R	R	H	A	A	A	1	T 21.1, 20.1
Snag crash	3	0	0	C	C	C	C	C	C	C	C	1	P 21.3.3

Snag riding	2	0	1	A	C	R	H	A	A	A	R	1,3	T 21.1, 20.1
Stick as chisel (1-Nest)	1	1	0	A	R	A	A	A	A	R	A	1	T 21.2, 20.1
Stick as chisel (2-Durian)	1	1	0	A	A	A	E	A	A	R	A	1	T 21.2, 20.1
Stick push spine	1	0	0	A	A	A	E	?	A	R	A	2	
Seed-extraction tool use	1	1	1	A	A	E	E	E	A	E	C	1	T 21.1, 20.1
Tree-hole tool use	1	1	1	A	A	A	A	A	A	A	C	1	T 21.1, 20.1
Straw tool	1	1	0	A	A	A	A	A	A	A	R	2,3	
Tooth cleaning (leaf)	2	0	1	H	A	A	H	?	C	A	A	1	T 21.1, 20.1
Tooth pick (stick)	2	0	0	?	A	A	A	?	?	P	R	1–3	T 21.2, 20.1
Artistic pillows	2	0	0	A	P	?	?	?	A	?	?	4	
Nest blanket	2	0	0	A	R	H	H	?	?	?	C	1	P 21.3.3
Nest lining	2	0	0	H/C	H/C	H/C	H/C	H/C	H/C	H/C	H/C	1	P 21.3.3
Nest pillow	2	0	0	H/C	H/C	H/C	H/C	H/C	H/C	H/C	H/C	1	P 21.3.3
Nest roof	2	0	0	A	?	H	C	C	C	C	C	1	T 20.1

Notes

Study sites include for Borneo: Gunung Palung (GP), Tanjung Putting (TP), Sabangau (SA) and Tuanan (TU), all *P. p. wurmbii*; and Kutai/Mentoko (KU) and Kinabatangan (KI), both *P. p. morio*; and for Sumatra: Suaq Balimbing (SB) and Ketambe (KE), both *P. abelii*. Prevalence of variants is referred to as: Absent (A), Present (P), Absent for ecological reasons (E), Habitual (H), Customary (C) or not known (?). Contexts are as in van Schaik *et al.* (2006a); subsistence (1), physical comfort (2) and communication (3). "Cultural" refers to the cultural status of the variant as described by Russon *et al.* (2009); Van Schaik *et al.* (2009). The classification of terrestrial versus exclusively arboreal tool variants, extractive foraging (column name "EF") and complex tool variants was based on the definitions described in the text in the method section. Nest variants are printed in italic and could all be classified as complex, whereas none of the other tool variants could be classified as complex. The "Ref." column mentions the most recent and complete references describing the particular tool variants and their presence data for the eight study populations (1 = Wich *et al.*, 2009; 2 = Shumaker *et al.*, 2011; 3 = Meulman, unpublished data, 4 = van Schaik *et al.*, 2003). The "Source" column additionally indicates which tables (T) or paragraphs (P) from Wich *et al.* (2009) were used to retrieve the data. Tool variants for which short names were not reported before, or for which the definition or description has been modified, are: "Bee cover" (cover hat/body with leafy branches or leaves against stinging bees – not swatting); "Branch hide" (combination of "hat hide humans" and "sneaky hat approach"); "Branch reach fruit" (use detached branch to reach incentive); "Branch fan" (fan themselves with branches for cooling); "Club" (tool for hitting conspecifics with a piece of bark during agonistic interactions); "Aimed missile" (throwing or aimed dropping of branches, large fruits or other objects toward terrestrial predators (or humans), apparently to drive them away); "Stick push spine" (use a long stick to push a spiny Durian fruit into a crevice and thus protect hands); and "Hat cover" (use of leaves/leafy branches as head cover to protect against rain/strong sun, etc. – different from "Bee cover" in that usually fewer leaves or branches are used, which are moreover held less closely to the body). Only the variants "Branch dragging display" and "Stick as chisel (1-Nest)" were (sometimes) used in terrestrial contexts.

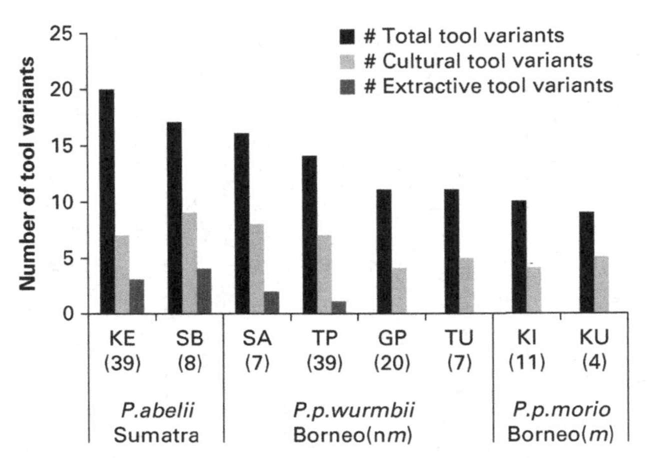

Figure 9.2 Overview of the number of tool variants, cultural tool variants and extractive-foraging tool variants per orangutan site. See the legend of Table 9.2 for study site abbreviations. Thirty-eight tool variants were recorded in total over all orangutan populations, of which 16 were cultural and seven were extractive-foraging tools. Both Sumatran populations (KE, SB) are on the higher end of the gradient with respect to the total number of tool variants, the number of cultural tool variants and the number of extractive-foraging tool variants. In parentheses is the approximate study duration in years for each research site.

reported that study duration did not significantly affect total or cultural (as defined here) tool repertoires.

In contrast to the species-wide total numbers mentioned above, we used *average* numbers per site to compare (sub)species and populations to ensure a fair comparison with the comparative data within orangutans. Figure 9.3 shows the tool repertoire size, and the number of extractive-foraging, cultural, complex and terrestrial tool variants (Figure 9.3a), as well as the number of subsistence, physical-comfort and communication tool variants (Figure 9.3b), for orangutans compared with chimpanzees.

Chimpanzees had significantly more complex (MWU = 20; $p = 0.036$), extractive-foraging (MWU = 6; $p = 0.001$) and terrestrial (MWU = 0; $p < 0.001$), but not cultural (MWU = 30; $p = 0.390$) tool variants per long-term study site compared with orangutans. The contexts in which tool variants were used also tended to differ between the two species ($\chi^2(2) = 5.58$; $p = 0.061$). Comparing each context separately, we found that chimpanzees had significantly more subsistence tool variants (MWU = 18.5; $p = 0.050$) than orangutans, significantly fewer physical comfort tool variants (MWU = 9; $p = 0.004$) and no substantial differences in the number of communication tool variants (MWU = 38; $p = 0.861$).

Variation in tool-repertoire composition

To establish the main components distinguishing tool repertoires across sites, and to gain some insight into the level of interdependency between outcome variables, we conducted a PCA. The Kaiser–Meyer–Olkin value of the combined set of variables indicated an

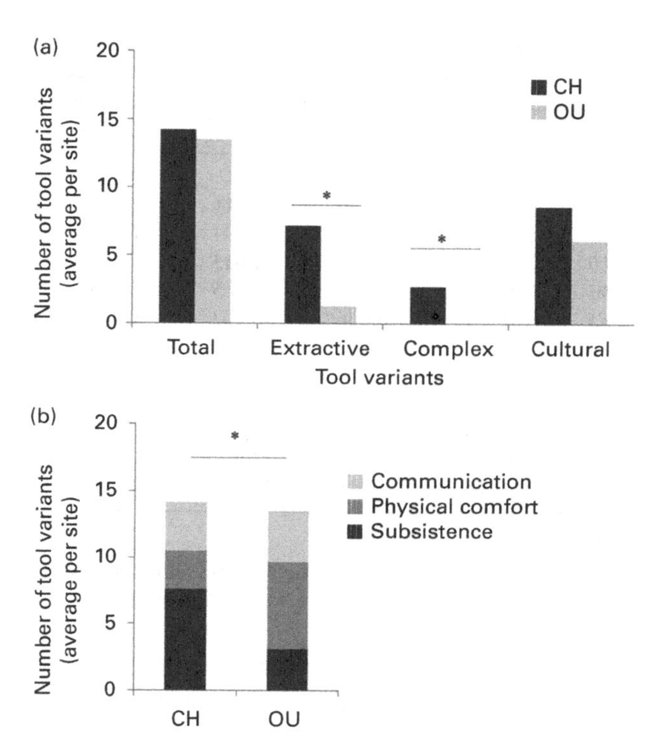

Figure 9.3 The tool repertoires of orangutans and chimpanzees compared. Average numbers for each subset of tool variants are reported to control for the number of study sites and the variation between them. In (a) the average number of tool variants in total, related to extractive foraging, ones that are potentially cultural according to the geographic method, or complex are reported for each of the two great ape species. (b) shows the average distribution of tool variants over the three contexts for the two great ape species. Chimpanzees had significantly more complex (MWU = 20; $p = 0.036$) and extractive-foraging (MWU = 6; $p = 0.001$), but not cultural (MWU = 30; $p = 0.390$) tool variants compared to orangutans. Also, the context in which tool variants were used did tend to differ between the two species ($\chi^2(2) = 5.58$; $p = 0.061$), with chimpanzees having significantly more subsistence tool variants (MWU = 18.5; $p = 0.050$) and significantly fewer physical-comfort tool variants (MWU = 9; $p = 0.004$) relative to orangutans. No substantial differences were found with regard to the number of communication tool variants (MWU = 38; $p = 0.861$).

adequate sampling (when communication variants were excluded – see discussion) for the analysis (KMO = 0.765), although not all KMO values for the individual outcome variables of the tool repertoire were above the acceptance limit of 0.5. Bartlett's test of sphericity ($\chi^2(21) = 177.196$; $p < 0.001$) indicated that correlations between the different outcome variables of the tool repertoire were sufficiently large for a PCA.

Two components had eigenvalues larger than 1 (Kaiser's criterion) and in combination explained 92.9% of the variance (retaining two components was supported by the scree plot). The first component contained the total number of tool variants, the number of extractive foraging, cultural, complex, terrestrial and subsistence tool variants, whereas the second contained the total number of tool variants and physical-comfort tool variants (Table 9.3). In agreement with the predictions based on the refined model, the items that loaded highly on

Table 9.3 Results of a principal components analysis (PCA) of tool-repertoire outcome variables ($n = 18$ study sites).

Tool repertoire (outcome variables)	Component 1 (Foraging related)		Component 2 (Comfort related)		Component 3 (Communication related)
Rotated factor loadings	−N−C	+N+C	−N−C	+N+C	+N+C
Extractive variants	*0.987*	*0.854*	−0.074	**−0.411**	−0.241
Subsistence variants	*0.968*	*0.925*	0.099	−0.191	0.213
Terrestrial variants	*0.920*	*0.730*	−0.269	**−0.615**	0.194
Cultural variants	*0.893*	*0.900*	0.357	0.115	0.339
Complex variants	*0.885*	*0.952*	0.066	−0.103	0.001
Total # variants	*0.776*	*0.771*	*0.602*	*0.516*	−0.122
Physical-comfort variants	−0.075	0.069	*0.987*	*0.952*	0.206
Nest element variants		−0.234		*0.937*	0.088
Communication variants		0.101		0.157	*0.955*
Eigenvalues	4.946	4.499	1.555	2.681	1.234
Percentage of variance	70.657	49.988	22.215	29.793	13.706

Note

N = nest tool variants, C = communication tool variants, − = excluding, + = including. Positive factor loadings above 0.4 are printed in italic. Negative factor loadings below −0.4 are printed in bold.

the same components suggest that component 1 represents a general proficiency for the use of foraging tools, whereas component 2 reflects a propensity for using comfort tools.

Testing the four factors of the model

Intelligence and the number of non-extractive and complex tool variants

The first prediction concerning the effect of intelligence was that the repertoire of non-extractive tool variants of Bornean *P. p. morio* is smaller than that of the other orangutans (Table 9.1). We compared the mean number of non-extractive tool variants for the two *P. p. morio* sites with that for all six other orangutan sites (Figure 9.4a). Although, as predicted, *P. p. morio* tended to have fewer non-extractive tool variants than the other orangutans, this difference was not significant (MWU = 0; $p = 0.071$). We also tested the between-species component of this prediction, namely that non-*morio* orangutans have equal-sized repertoires of non-extractive tool variants as chimpanzees (Figure 9.4b). However, the results show that non-*morio* orangutans have significantly more such tool variants per site than chimpanzees (MWU = 2; $p = 0.001$). Including *P. p. morio* in the analysis did not affect the results (MWU = 7; $p = 0.002$). We will examine this unexpected result in the discussion below.

The second prediction was that tool complexity of *P. p. morio* should be less than that of the other orangutans. Because the orangutan tool repertoire did not include any complex tool variants, it is not surprising that we did not detect any differences in tool complexity between *P. p. morio* and the other orangutans (MWU = 6; $p = 1.000$; Figure 9.4a). The between-species comparison, however, showed that the various chimpanzee populations show a trend of having more complex tool variants than populations of non-*morio* orangutans (Figure 9.4b; MWU = 15; $p = 0.093$). When we included *P. p. morio* in the analysis, this result became significant (MWU = 20; $p = 0.036$).

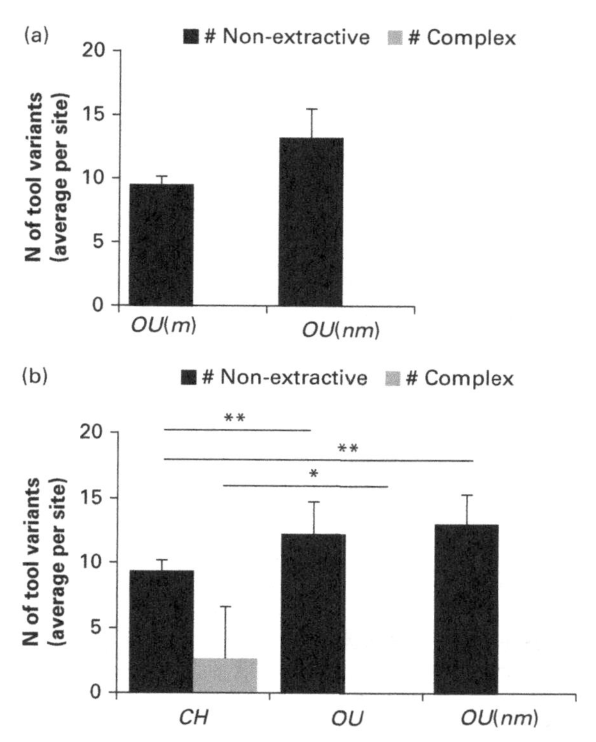

Figure 9.4 The role of intelligence on tool use reflected by the number of non-extractive tool variants and tool complexity. (a) and (b) show the number of non-extractive tool variants as an indicator of innovative abilities, comparing (a) *P. p. morio* and non-*morio* orangutans ($p = 0.071$) and (b) chimpanzees and (non-*morio*) orangutans ($p = 0.001$) or orangutans ($p = 0.002$). The results indicate that *P. p. morio* shows a trend of having less non-extractive tool variants in the repertoire, and that orangutans have significantly more tool innovations in the non-extractive foraging context, compared to chimpanzees. Additionally, complex tool variants are compared between (a) *P. p. morio* and non-*morio* orangutans and (b) (non-*morio*) orangutans and chimpanzees. The mean number of complex tool variants did not differ significantly between the orangutan populations ($p = 1.000$), but a trend was shown when comparing chimpanzees to non-*morio* orangutans ($p = 0.093$), which reached significance when *P. p.morio* were included ($p = 0.036$); in favor of the chimpanzees).

These findings either suggest that innovative ability (as indexed by brain size) does not affect complex tool use, or, more plausibly, reflect the combined effect of tolerant proximity and terrestriality (see below).

Extractive foraging and the number of extractive-foraging tool variants

We predicted more extractive-foraging tool variants for Sumatran versus Bornean orangutans and similar numbers for non-*morio* orangutans and chimpanzees. The number of extractive-foraging tool variants in the local repertoire did not differ significantly (MWU = 0; $p = 0.133$) between non-*morio* Bornean orangutans and the Sumatrans (Figure 9.5a), probably due to the small sample size ($n = 7$). The Sumatra–Borneo difference did become significant when *P. p. morio* were included (MWU = 0;

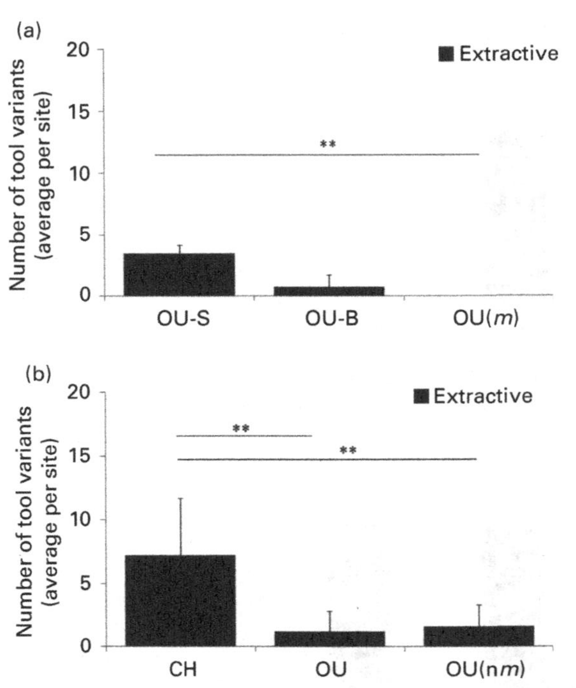

Figure 9.5 The role of extractive foraging on tool use. The number of extractive tool variants are compared between (a) Sumatran orangutans (*P. abelii*) and non-*morio* Bornean orangutans (*P. p. wurmbii*: $p=0.133$, or $p=0.036$ when *P. p. morio* were included), and (b) chimpanzees and non-*morio* orangutans (*P. abelii, P. p. wurmbii*; $p=0.007$) or orangutans (including *P. p. morio*: $p=0.001$), to evaluate the effect of extractive foraging tendencies. Thus, both Sumatran orangutans and chimpanzees have significantly more extractive-foraging tool variants in their repertoire than non-*morio* Bornean orangutans (*P. p. wurmbii*) and non-*morio* orangutans (*P. p. wurmbii* and *P. abelii*), respectively. However, this difference was only significant for the orangutan chimpanzee comparison, probably because of the small number of extractive tool variants within the orangutan tool repertoire (7 out of 38).

$p=0.036$), suggesting that indeed the abundance of opportunities for extractive foraging may have some effect on the innovation of the relevant tools. However, orangutans showed significantly smaller local repertoires of extractive tool variants than chimpanzees (Figure 9.5b; MWU = 6; $p=0.007$ for non- *morio* orangutans; MWU = 6; $p=0.001$ for when *P. p. morio* were included). This pattern can be explained by taking the terrestriality effect into account (see below).

Social tolerance and the number of cultural tool variants

The number of cultural tool variants in the local repertoire may be smaller among non-*morio* Bornean orangutans than the Sumatrans (Figure 9.6a), but this difference was not significant (MWU = 1.5; $p=0.400$ when *P. p. morio* were excluded; $p=0.143$ when they were included). Similarly, the non-*morio* orangutans did not seem to differ in the number of cultural tool variants from the more gregarious chimpanzees (Figure 9.6b; MWU = 26.5; $p=0.728$ when *P. p. morio* were excluded; MWU = 30; $p=0.390$ when they were included).

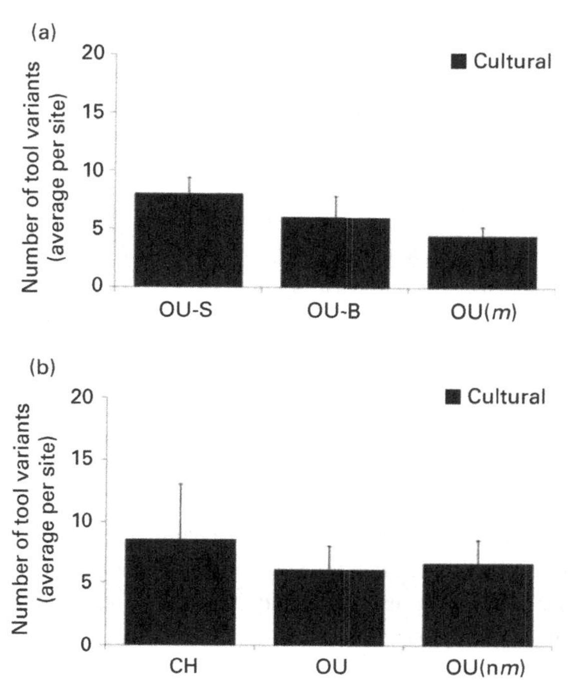

Figure 9.6 The role of social learning opportunities (based on the level of gregariousness), as reflected by the number of cultural tool variants. (a) The non-*morio* Bornean orangutans (*P. p. wurmbii*) have fewer cultural tool variants than the more gregarious Sumatran orangutans (*P. abelii*), although the difference was not significant ($p=0.400$ when *P. p. morio* were excluded, or $p=0.143$ when *P. p. morio* were included). (b) The more gregarious chimpanzees also tend to have more cultural variants compared to the less gregarious non-*morio* orangutans ($p=0.728$) or orangutans ($p=0.390$), but this difference was again not significant. Hence, social learning opportunities based solely on levels of gregariousness do not explain the variation in the number of tool variants in the cultural tool repertoire.

These results, therefore, do not support the contention that increasing opportunities for social learning positively affect the size of the cultural tool repertoire, or at least not to the extent we predicted based on the degree of gregariousness, but do make sense when the effect of terrestriality is included (see below).

The effect of terrestriality

We predicted an effect of terrestriality, especially on the number of extractive (cultural) and complex (cultural) tool variants in favor of the more terrestrial chimpanzees in comparison to the arboreal orangutans (see Table 9.1). Above, we already found that chimpanzees indeed exceed orangutans in the number of extractive and complex tool variants. Likewise, chimpanzees surpassed orangutans in the number of extractive cultural (MWU$=6.5$; $p=0.009$ when *P. p. morio* were excluded; MWU$=7.5$; $p=0.002$ when *P. p. morio* were included) and complex cultural (MWU$=3$; $p=0.002$ when *P. p. morio* were excluded; MWU$=4$; $p<0.001$ when *P. p. morio* were included) tool variants (Figure 9.7). Complex tool variants were, in addition, more often used terrestrially (MWU$=225.5$; $p<0.001$) and extractively

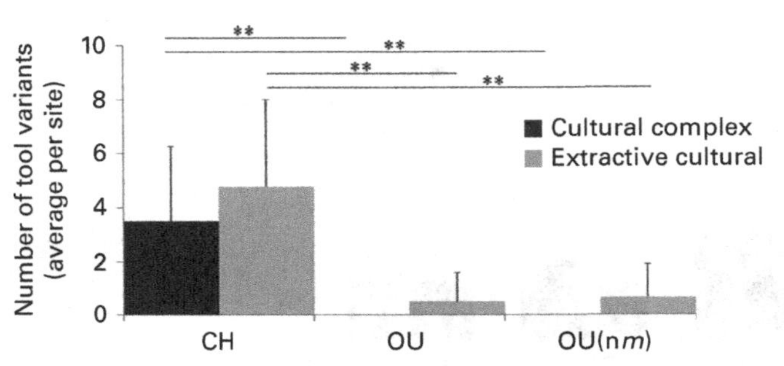

Figure 9.7 Effects of terrestriality on the cultural repertoire of (a) extractive-foraging tools and (b) complex tool variants; compared between non-*morio* orangutans (*P. abelii*, *P. p. wurmbii*) and chimpanzees. Relative to orangutans, chimpanzees have significantly more cultural tool variants for extractive foraging ($p=0.009$ when *P. p. morio* were excluded; or $p=0.001$ when *P. p. morio* were included) and complex cultural tool variants ($p=0.002$ when *P. p. morio* were excluded, or $p=0.001$ when included).

(MWU = 225.5; $p<0.001$) than non-complex tool variants. Including terrestriality as a factor thus explains why the effect of intelligence, extractive foraging tendencies and gregariousness *per se*, on the number of complex, extractive and cultural tool variants, is so limited when comparing orangutans and chimpanzees.

Discussion

Support for the refined model

When limiting the comparisons to orangutans only, the results support the original socioecological model (as well as the refined model, because orangutans are rarely terrestrial). First, the various outcome variables of the tool repertoire were always in favor of *P. abelii* and against *P. p. morio*, with *P. p. wurmbii* being intermediate (although mostly non-significant probably due to the small samples of tool variants in a specific category). The exception was tool complexity, because no orangutan population had complex tool variants. However, flexible and habitual use suggestive of social acquisition have only been reported for seed-extraction and tree-hole tool use, which are both exhibited only by (non-*morio*) Sumatran orangutans (van Schaik *et al.*, 1996, 2003b; van Schaik & Knott, 2001).

With regard to the results from the orangutan–chimpanzee comparisons, we found a significant bias toward extractive-foraging, complex, subsistence and terrestrial tool variants in favor of chimpanzees, whereas the bias was in favor of orangutans for the number of non-extractive and physical-comfort tool variants. These results did not support the original socioecological model and underlined the need for invoking a role of terrestriality. This, because the inclusion of terrestriality in the model, improves the fit between (1) opportunities for extractive foraging and the number of extractive tool variants; (2) brain size (as a predictor for intelligence) and the number of complex tool variants; and (3) opportunities for social learning and the cultural tool repertoire. The PCA additionally illustrated a high correlation between terrestrial tool variants into the first component in which also cultural, complex, extractive-foraging and subsistence tool variants were clustered. In a previous study we furthermore found that, within chimpanzees, terrestrial extractive tool variants are more complex than arboreal tool variants (Meulman *et al.*, 2012). These results therefore suggest that terrestriality positively affects tool innovations, especially within the extractive foraging context; and, second, that, whereas opportunities for social learning are a necessary precondition for cultural tool variants, terrestriality is additionally needed to increase tool complexity.

Terrestriality versus complex arboreal tool use and nest building

We argue that the terrestriality effect is largely mediated by its effect on opportunities for social learning. A similar effect on tool complexity can therefore be expected for the arboreal honey extraction in chimpanzees when using pounding tools (Meulman *et al.*, 2012). Likewise, nest building can create a similar effect (again, see Meulman *et al.*, 2012). Strictly speaking, some variants of orangutan nest building should be regarded as tool use because they involve the detachment of vegetative material(s) from a fixed substrate (see Hansell & Ruxton, 2008 for a more detailed discussion). Interestingly, were we to consider nest-building variants tool use, we would indeed find complex tool use in wild orangutans.

One can also run the same argument in reverse. When a terrestrial context is not associated with enhanced opportunities for social learning, we should not expect to find complex tool use. For instance, male Bornean orangutans are fairly terrestrial but also almost exclusively solitary (apart from brief consortships with females, cf. Utami-Atmoko *et al.*, 2009). Indeed, we do not find any complex tool use in male Bornean orangutans.

The importance of species-specific innovation biases

The increased tool complexity in orangutans when including nest variants are in line with the differential challenges faced by orangutans and chimpanzees and the resulting innovation biases of orangutans toward comfort tools and of chimpanzees toward subsistence tools (see also van Schaik *et al.*, 2006). These innovation biases also explain the higher number of non-extractive tool variants for orangutans relative to chimpanzees that we could not explain with the model (original or revised).

When we include the five nest elements as tool variants and redo the analyses we indeed find some interesting results. First, the difference between tool complexity in orangutans and chimpanzees, which was significant before, becomes non-significant (MWU = 25, $p = 0.192$ including *P. p. morio*; MWU = 21, $p = 0.357$ excluding *P. p. morio*). In addition, including the five nest elements changed the results of the PCA (KMO = 0.633; Bartlett's test of sphericity: $\chi^2(36) = 205.556l$, $p < 0.001$), so that extractive-foraging and terrestrial tool variants loaded strongly negatively on the physical-comfort component, which includes the nest elements (see Table 9.3). Moreover, inclusion of nest elements furthermore changed the percentages of variance explained by the foraging- (50% versus 70% when excluding nests) and comfort-related (30% versus 20% when excluding nests)

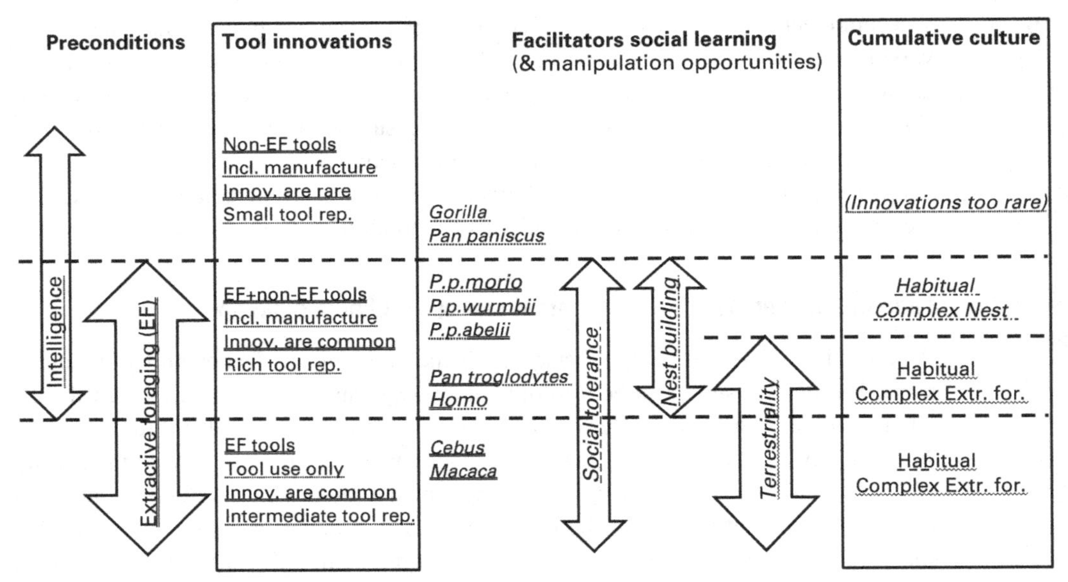

Figure 9.8 Diagram visualizing the five main evolutionary constraints acting on primate tool use. Predictor variables (or evolutionary constraints) are indicated by arrows. The gray rectangles describe the features of the tool repertoires based on these predictor variables. The width of the arrow is associated with its relative importance for the tool repertoire. The type of underlining indicates the link between the predictor variables and (1) the different features of the tool repertoire, as well as (2) how much they are represented in the different species. The dashed lines designate differences in predictor variables that affect the tool repertoires. Extractive foraging (indicated by EF or Extr. for.) seems to be the main driver for tool innovations (which become more common and diverse), although intelligence may compensate to some extent by its effect on innovative tendencies (also in the non-extractive-foraging context) and the complexity of these innovations (i.e., manufacture vs. use). Opportunities for social learning subsequently determine whether innovations may persist in the repertoire. Additionally, contexts such as terrestriality and nest building enable more complex manipulations and socially facilitated affordance learning, crucial for the manifestation of cumulative material cultures.

components. Thus, although the foraging-related component still clearly outweighs the comfort-related component, the separation has become less strict. In conclusion, the lower tool complexity in wild orangutans may therefore largely be due to the innovation bias toward arboreal settings and the physical-comfort context.

The evolution of tool use in primates revisited

Based on the variation in orangutan and chimpanzee tool repertoires we can now extrapolate and see what these findings may mean for the evolution of tool use in primates (see Figure 9.8 for a schematic overview). The factors postulated by the original model (extractive foraging, innovative ability [i.e., intelligence] and opportunities for social learning [i.e., social tolerance]) remain relevant, but the effect of intelligence, extractive foraging and social tolerance is strongly affected by terrestriality, which therefore must be seen as an essential ingredient of the model.

Extractive foraging remains the basic precondition for tool innovations (Parker & Gibson, 1977; van Schaik *et al.*, 1999; Panger, 2007). Apes with extractive foraging have larger tool repertoires than the other apes (bonobos, gorillas) (McGrew *et al.*, 2007; Deblauwe & Janssens, 2008; Deblauwe, 2009; Lonsdorf *et al.*, 2009; Bentley-Condit & Smith, 2010), and among monkeys the only taxa with habitual tool use comprise extractive foragers that use tools mainly extractively (long-tailed macaques: Gumert *et al.*, 2009; capuchins: Visalberghi, 2009). Intelligence may compensate to some extent for the lack of an innovation bias toward the extractive-foraging context, but it has a limited effect on the occurrence of tool innovations in general, and even less on the occurrence of habitual tool use or tools used for extractive foraging. This is reflected in the presence of non-extractive tool variants in all great apes (gorilla and bonobo tool use is almost exclusively non-habitual, non-extractive or even non-foraging related), whereas monkeys have virtually nothing in this regard (Ottoni & Izar, 2008; Gumert *et al.*, 2009; Bentley-Condit & Smith, 2010; Shumaker *et al.*, 2011).

Opportunities for social transmission determine subsequently whether tool innovations promoted by extractive foraging and intelligence can be maintained in the behavioral repertoire. Although social tolerance explains the presence of habitual tool variants (Whiten *et al.*, 2001; van Schaik *et al.*, 2003a; Leca *et al.*, 2007; Mannu & Ottoni, 2009), it is less important than terrestriality. Terrestriality is important because it facilitates an increased potential for complex innovations and skill acquisition through social learning. Terrestriality (and to a lesser extent also nest building) can therefore additionally explain the "orangutan tool paradox," and the occurrence of habitual and complex tool use in primates in general (for further discussion, see Meulman *et al.*, 2012).

Hominin evolution

The new version of the model (Figure 9.8) can also account for the flourishing of tool use into elaborate lithic technologies during hominin evolution. Since the emergence of the Oldowan, around 2.5 mya, hominins were at least partially terrestrial, and used tools in terrestrial contexts (Plummer, 2004; Foley & Gamble, 2009). In addition, higher sociability (tolerant proximity), as implied by hunting of large game, allowed for more efficient information transfer. The rise of teaching following the adoption of alloparental care (cf. Burkart *et al.*, 2009) must have made transfer of technology to subsequent generations more efficient still. Thus, terrestriality, in combination with greater opportunities for social learning, afforded by greater sociability and teaching, goes far to explain the technological differences between great apes and humans (see also Meulman *et al.*, 2012).

Future directions

Although the new model can encompass the findings of this study, as well as the occurrence of true, habitual and complex tool use in general, the quality of the data could be improved to enable more quantitative analyses. First, more quantitative data on the frequency of extractive foraging (perhaps even classifying whether tool innovations would be required or not) are important to more quantitatively assess the role of extractive foraging on the evolution of tool use. Second, it may be commendable to distinguish in the future between mere gregariousness and actual opportunities for social learning through observational learning, direct tool transfers and/or indirect tool transfers (or stimulus enhancement). Third, more specific data on variation in the level of terrestriality among populations and its consequences for social proximity and tool affordance learning would be crucial to confirm the importance of terrestriality for the evolution of cumulative technology and cultural intelligence. Especially data on social tolerance levels and socially facilitated skill acquisition in terrestrial contexts versus arboreal contexts would provide us with crucial empirical data in this regard. Likewise, more quantitative data are needed regarding the effect of terrestriality on the occurrence of tool innovations and complex manipulations. Systematic comparisons of complex technology within the nest-building context (when socially learned) could similarly be very interesting and moreover provide more insight on the importance of innovation biases (e.g., chimpanzee versus orangutan nest building). Finally, more species and populations need to be included to confirm our conclusions for primates in general.

Acknowledgments

We thank the State Ministry of Research and Technology (Ristek) and the Indonesian Institute of Sciences (LIPI) for permission to work in Indonesia and the Ministry of Forestry for permission to work in the Gunung Leuser National Park. Additionally, we acknowledge PanEco (Switzerland) and the A. H. Schultz Foundation for their financial support. For mental and logistic support, we would like to thank the Sumatran Orangutan Conservation Programme (SOCP), Fakultas Biologi Universitas Nasional (UNAS-Jakarta) and the Anthropological Institute of the University of Zürich. Furthermore, we would like to express a special thanks to the following people, for all the help, fruitful discussions and advice that helped us produce this chapter: Syamsuar, Ishak, Zulkifli, Izumi, Mahmuddin, Edit, Zulfikar, Toni, Armas, Ari, Rustam, Syafi'i, Syahrul, Jak, Azhar, Santi and Asril (Indonesian counterparts and field assistants), Ian Singleton and Regina Frey (PanEco), Crickette Sanz (Washington University) and Karin Isler, Maria van Noordwijk, Erik Willems, Adrian Jaeggi, Simon Townsend, Sofia Forss and Andrea Gibson (University of Zürich). We would also like to thankfully acknowledge the organizers and participants of the workshop on "Understanding tool use" at the Max Planck Institute for Evolutionary Anthropology in Leipzig, which inspired us to write this chapter.

References

Arora, N., Nater, A., van Schaik, C. P., *et al.* (2010). Effects of Pleistocene glaciations and rivers on the population structure of Bornean orangutans (*Pongo pygmaeus*). *Proceedings of the National Academy of Sciences USA*, **107**(50), 21376–21381.

Beck, B. B. (1980). *Animal Tool Behavior: The Use and Manufacture of Tools by Animals*. New York: Garland STPM Publishers.

Bentley-Condit, V. K. & Smith, E. O. (2010). Animal tool use: current definitions and an updated comprehensive catalog. *Behaviour*, **147**(2), 185–221.

Boesch, C. & Boesch, H. (1990). Tool use and tool making in wild chimpanzees. *Folia Primatologica*, **54**, 86–99.

Boesch-Achermann, H. & Boesch, C. (1994). Hominization in the rainforest: the chimpanzee's piece of the puzzle. *Evolutionary Anthropology*, **3**(1), 9–16.

Burkart, J. M., Hrdy, S. B. & van Schaik, C. P. (2009). Cooperative breeding and human cognitive evolution. *Evolutionary Anthropology*, **18**(5), 175–186.

Byrne, R. W. (1995). *The Thinking Ape: Evolutionary Origins of Intelligence*. Oxford and New York: Oxford University Press.

Coussi-Korbel, S. & Fragaszy, D. (1995). On the relation between social dynamics and social learning. *Animal Behaviour*, **50**, 1441–1553.

Deaner, R. O., Isler, K., Burkart, J., *et al.* (2007). Overall brain size, and not encephalization quotient, best predicts cognitive ability across non-human primates. *Brain Behavior and Evolution*, **70**(2), 115–124.

Deblauwe, I. (2009). Temporal variation in insect-eating by chimpanzees and gorillas in Southeast Cameroon: extension of niche differentiation. *International Journal of Primatology*, **30**(2), 229–252.

Deblauwe, I. & Janssens, G. P. J. (2008). New insights in insect prey choice by chimpanzees and gorillas in southeast Cameroon: the role of nutritional value. *American Journal of Physical Anthropology*, **135**(1), 42–55.

Foley, R. & Gamble, C. (2009). The ecology of social transitions in human evolution. *Philosophical Transactions of the Royal Society of London B: Biological Sciences*, **364**(1533), 3267–3279.

Fox, E. A. & Bin'Muhammad, I. (2002). Brief communication: new tool use by wild Sumatran orangutans (*Pongo pygmaeus abelii*). *American Journal of Physical Anthropology*, **119**(2), 186–188.

Fragaszy, D. M. & Perry, S. (2003). *The Biology of Traditions: Models and Evidence*. Cambridge: Cambridge University Press.

Groves, C. P. (2001). *Primate Taxonomy*. Washington, DC: Smithsonian Institution Press.

Gumert, M. D., Kluck, M. & Malaivijitnond, S. (2009). The physical characteristics and usage patterns of stone axe and pounding hammers used by long-tailed macaques in the Andaman Sea region of Thailand. *American Journal of Primatology*, **71**(7), 594–608.

Haidle, M. N. (2010). Working-memory capacity and the evolution of modern cognitive potential implications from animal and early human tool use. *Current Anthropology*, **51**, S149–S166.

Hansell, M. & Ruxton, G. D. (2008). Setting tool use within the context of animal construction behaviour. *Trends in Ecology and Evolution*, **23**(2), 73–78.

Huang, C. T. & Charman, T. (2005). Gradations of emulation learning in infants' imitation of actions on objects. *Journal of Experimental Child Psychology*, **92**(3), 276–302.

Humle, T. & Matsuzawa, T. (2009). Laterality in hand use across four tool-use behaviors among the wild chimpanzees of Bossou, Guinea, West Africa. *American Journal of Primatology*, **71**(1), 40–48.

Isler, K., Kirk, E. C., Miller, J. M. A., *et al.* (2008). Endocranial volumes of primate species: scaling analyses using a comprehensive and reliable data set. *Journal of Human Evolution*, **55**(6), 967–978.

Kelley, J. (2002). The hominoid radiation in Asia. In W. C. Hartwig (ed.) *The Primate Fossil Record* (pp. 369–384). Cambridge: Cambridge University Press.

Krützen, M., Willems, E. P. & van Schaik, C. P. (2011). Culture and geographic variation in orangutan behavior. *Current Biology*, **21**(21), 1808–1812.

Leca, J. B., Gunst, N. & Huffman, M. A. (2007). Japanese macaque cultures: inter- and intra-troop behavioural variability of stone handling patterns across 10 troops. *Behaviour*, **144**, 251–281.

Lonsdorf, E. V., Ross, S. R., Linick, S. A., *et al.* (2009). An experimental, comparative investigation of tool use in chimpanzees and gorillas. *Animal Behaviour*, **77**(5), 1119–1126.

Mannu, M. & Ottoni, E. B. (2009). The enhanced tool-kit of two groups of wild bearded capuchin monkeys in the Caatinga: tool making, associative use, and secondary tools. *American Journal of Primatology*, **71**(3), 242–251.

Marshall, A. J., Ancrenaz, M., Brearley, F. Q., *et al.* (2009). The effects of forest phenology and floristics on populations of Bornean and Sumatran orangutans: are Sumatran forests better orangutan habitat than Bornean forests? In S. A. Wich, S. S. Utami Atmoko, T. Mitra Setia & C. P. van Schaik (eds.) *Orangutans: Geographic Variation in Behavioral Ecology and Conservation* (pp. 97–117). New York: Oxford University Press.

McGrew, W. C. (1992). Tool-use by free-ranging chimpanzees: the extent of diversity. *Journal of Zoology*, **228**(4), 689–694.

McGrew, W. C. (2004a). *The Cultured Chimpanzee: Reflections on Cultural Primatology*. Cambridge: Cambridge University Press.

McGrew, W. C. (2004b). Primatology: advanced ape technology. *Current Biology*, **14**(24), R1046–R1047.

McGrew, W. C. (2010). In search of the last common ancestor: new findings on wild chimpanzees. *Philosophical Transactions of the Royal Society of London B: Biological Sciences*, **365**(1556), 3267–3276.

McGrew, W. C. & Marchant, L. F. (1997). Using the tools at hand: manual laterality and elementary technology in *Cebus spp*. and *Pan spp*. *International Journal of Primatology*, **18**(5), 787–810.

McGrew, W. C., Marchant, L. F., Beuerlein, M. M., *et al.* (2007). Prospects for bonobo insectivory: Lui Kotal, Democratic Republic of Congo. *International Journal of Primatology*, **28**(6), 1237–1252.

Meulman, E. J. M., Sanz, C. M., Visalberghi, E., *et al.* (2012). The role of terrestriality in promoting primate technology. *Evolutionary Anthropology*.

Morrogh-Bernard, H. C., Husson, S. J., Knott, C. D., *et al.* (2009). Orangutan activity budgets and diet: a comparison between species, populations and habitats. In S. A. Wich, S. S. Utami Atmoko, T. Mitra Setia & C. P. van Schaik (eds.) *Orangutans: Geographic Variation in Behavioral Ecology and Conservation* (pp. 97–117). New York: Oxford University Press.

Ottoni, E. B. & Izar, P. (2008). Capuchin monkey tool use: overview and implications. *Evolutionary Anthropology*, **17**(4), 171–178.

Panger, M. (2007). Tool use and cognition in primates. In C. Campbell, S. Bearder, A. Fuentes, *et al.* (eds.) *Primates in Perspective* (pp. 665–677). Oxford: Oxford University Press.

Parker, S. T. & Gibson, K. R. (1977). Object manipulation, tool use and sensorimotor intelligence as feeding adaptations in *Cebus* monkeys and great apes. *Journal of Human Evolution*, **6**(7), 623–641.

Plummer, T. (2004). Flaked stones and old bones: biological and cultural evolution at the dawn of technology. *American Journal of Physical Anthropology*, **39**, 118–164.

Pradhan, G. R., Tennie, C. & van Schaik, C. P. (2012). Social organization and the evolution of cumulative technology in apes and hominins. *Jounal of Human Evolution*.

Raaum, R. L., Sterner, K. N., Noviello, C. M., *et al.* (2005). Catarrhine primate divergence dates estimated from complete mitochondrial genomes: concordance with fossil and nuclear DNA evidence. *Journal of Human Evolution*, **48**(3), 237–257.

Reader, S. M. & Laland, K. N. (2002). Social intelligence, innovation, and enhanced brain size in primates. *Proceedings of the National Academy of Sciences USA*, **99**(7), 4436–4441.

Reader, S. M., Hager, Y. & Laland, K. N. (2011). The evolution of primate general and cultural intelligence. *Philosophical Transactions of the Royal Society of London B: Biological Sciences*, **366**(1567), 1017–1027.

Russon, A. E., van Schaik, C. P., Kuncoro, P., *et al.* (2009). Innovation and intelligence in orangutans. In S. A. Wich, S. S. Utami Atmoko, T. Mitra Setia & C. P. van Schaik (eds.) *Orangutans: Geographic Variation in Behavioral Ecology and Conservation* (pp. 279–298). Oxford: Oxford University Press.

Sanz, C. M. & Morgan, D. B. (2007). Chimpanzee tool technology in the Goualougo Triangle, Republic of Congo. *Journal of Human Evolution*, **52**(4), 420–433.

Sanz, C. M. & Morgan, D. B. (2009). Flexible and persistent tool-using strategies in honey-gathering by wild chimpanzees. *International Journal of Primatology*, **30**(3), 411–427.

Sanz, C. M. & Morgan, D. B. (2010). The complexity of chimpanzee tool-use behaviors. In E. V. Lonsdorf, S. R. Ross & T. Matsuzawa (eds.) *The Mind of the Chimpanzee: Ecological and Experimental Perspectives* (pp. 127–140). Chicago, IL and London: University of Chicago Press.

Sanz, C. M., Schoning, C. & Morgan, D. B. (2010). Chimpanzees prey on army ants with specialized tool set. *American Journal of Primatology*, **72**(1), 17–24.

Shumaker, R. W., Walkup, K. R. & Beck, B. B. (2011). *Animal Tool Behavior: The Use and Manufacture of Tools by Animals*. Baltimore, MD: Johns Hopkins University Press.

Sitompul, A. F. I. (1995). Penggunaan alat pada orangutan Sumatra (*Pongo pygmaeus abelii*, Lesson 1827) dalam memanfaatkan sumber pakan serangga sosial di Suaq Balimbing, Kluet, Taman Nasional Gunung Leuser. Bachelor (S1), Universitas Indonesia.

Spagnoletti, N., Izar, P. & Visalberghi, E. (2009). Tool use and terrestriality in wild bearded capuchin monkey (*Cebus libidinosus*). *Folia Primatologica*, **80**(2), 142.

Stumpf, R. M. (2007). Chimpanzees and bonobos: diversity within and between species. In C. B. Campbell, A. Fuentes, K. C. MacKinnon, M. Panger & S. Bearder (eds.) *Primates in Perspective* (pp. 321–344). Oxford: Oxford University Press.

Taylor, A. B. & van Schaik, C. P. (2007). Variation in brain size and ecology in *Pongo*. *Journal of Human Evolution*, **52**(1), 59–71.

Utami-Atmoko, S. S., Singleton, I., van Noordwijk, M. A., *et al.* (2009). Male–male relationships in orangutans. In S. A. Wich, S. S. Utami Atmoko, T. Mitra Setia & C. P. van Schaik (eds.) *Orangutans: Geographic Variation in Behavioral Ecology and Conservation* (pp. 225–233). Oxford: Oxford University Press.

van Noordwijk, M. A., Sauren, S. E. B., Nuzuar, *et al.* (2009). Development of independence. Sumatran and Bornean orangutans compared. In S. A. Wich, S. S. Utami Atmoko, T. Mitra Setia & C. P. van Schaik (eds.) *Orangutans: Geographic Variation in Behavioral Ecology and Conservation* (pp. 189–203). Oxford: Oxford University Press.

van Schaik, C. P. (1999). The socioecology of fission-fusion sociality in Sumatran orangutans. *Primates*, **40**, 73–90.

van Schaik, C. P. (2004). *Among Orangutans: Red Apes and the Rise of Human Culture*. Cambridge, MA and London: Belknap Press of Harvard University Press.

van Schaik, C. P. (2006). Why are some animals so smart? *Scientific American*, **294**(4), 64–71.

van Schaik, C. P. & Knott, C. D. (2001). Geographic variation in tool use on *Neesia* fruits in orangutans. *American Journal of Physical Anthropology*, **114**(4), 331–342.

van Schaik, C. P., Fox, E. A. & Sitompul, A. E. (1996). Manufacture and use of tools in wild Sumatran orangutans: implications for human evolution. *Naturwissenschaften*, **83**, 186–188.

van Schaik, C. P., Deaner, R. O. & Merrill, M. Y. (1999). The conditions for tool use in primates: implications for the evolution of material culture. *Journal of Human Evolution*, **36**(6), 719–741.

van Schaik, C. P., Ancrenaz, M., Borgen, G., *et al.* (2003a). Orangutan cultures and the evolution of material culture. *Science*, **299**(5603), 102–105.

van Schaik, C. P., Fox, E. A. & Fechtman, L. T. (2003b). Individual variation in the rate of use of tree-hole tools among wild orang-utans: implications for hominin evolution. *Journal of Human Evolution*, **44**(1), 11–23.

van Schaik, C. P., Fragaszy, D. M. & Perry, S. (2003c). *Local Traditions in Orangutans and Chimpanzees: Social Learning and Social Tolerance – the Biology of Traditions*. Cambridge: Cambridge University Press.

van Schaik, C. P., van Noordwijk, M. A. & Wich, S. A. (2006). Innovation in wild Bornean orangutans (*Pongo pygmaeus wurmbii*). *Behaviour*, **143**, 839–876.

van Schaik, C. P., Ancrenaz, M., Djojoasmoro, R., *et al.* (2009). Orangutan cultures revisited. In S. A. Wich, S. S. Utami Atmoko, T. Mitra Setia & C. P. van Schaik (eds.) *Orangutans: Geographic Variation in Behavioral Ecology and Conservation* (pp. 299–309). New York: Oxford University Press.

Visalberghi, E. (2009). Wild capuchin monkeys use tools: why and how it challenges our ideas about tool use in human evolution. *Folia Primatologica*, **80**(2), 108.

Visalberghi, E., Fragaszy, D. M., Izar, P., *et al.* (2005). Terrestriality and tool use. *Science (Letters)*, **308**(5724), 951–952.

Visalberghi, E., Addessi, E., Truppa, V., *et al.* (2009). Selection of effective stone tools by wild bearded capuchin monkeys. *Current Biology*, **19**(3), 213–217.

Warren, K. S., Verschoor, E. J., Langenhuijzen, S., *et al.* (2001). Speciation and intrasubspecific variation of Bornean orangutans, *Pongo pygmaeus pygmaeus*. *Molecular Biology and Evolution*, **18**(4), 472–480.

Whiten, A., Goodall, J., McGrew, W. C., *et al.* (1999). Cultures in chimpanzees. *Nature*, **399**, 682–685.

Whiten, A., Goodall, J., McGrew, W. C., *et al.* (2001). Charting cultural variation in chimpanzees. *Behaviour*, **138**(11/12), 1481–1516.

Whiten, A., Schick, K. & Toth, N. (2009). The evolution and cultural transmission of percussive technology: integrating evidence from palaeoanthropology and primatology. *Journal of Human Evolution*, **57**(4), 420–435.

Wich, S. A., Utami-Atmoko, S. S., Mitra-Setia, T., *et al.* (2009). *Orangutans: Geographic Variation in Behavioral Ecology and Conservation*. New York: Oxford University Press.

Xu, X. F. & Arnason, U. (1996). The mitochondrial DNA molecule of Sumatran orangutan and a molecular proposal for two (Bornean and Sumatran) species of orangutan. *Journal of Molecular Evolution*, **43**(5), 431–437.

10 The Etho-*Cebus* Project: Stone-tool use by wild capuchin monkeys

Elisabetta Visalberghi
Istituto di Scienze e Tecnologie della Cognizione, Consiglio Nazionale delle Ricerche

Dorothy Fragaszy
Psychology Department, University of Georgia

Tool use, according to St Amant and Horton (2008: 1203),[1] is the "exertion of control over a freely manipulable external object (the tool) with the goal of altering the physical properties of another object, substance, surface or medium (the target, which may be the tool user or another organism), via a dynamic mechanical interaction." Among wild great apes, only chimpanzees use tools habitually, in many varied formats across their geographical distribution, and for diverse purposes (see McGrew, 1992; Yamakoshi, 2004 for a review; Boesch *et al.*, 2009; and this volume). Tool use is observed much less often in wild Sumatran orangutans (*Pongo abelii*), and even more rarely in the other wild great apes (Western gorillas, *Gorilla gorilla*, and bonobos, *Pan paniscus*), though in captivity all the great ape species use tools spontaneously in flexible and diverse ways. Among monkeys, very few species use tools in natural settings (*Macaca fascicularis*: Malaivijitnond *et al.*, 2007; Gumert *et al.*, 2009; *Cebus*[2] *libidinosus* and *C. xanthosternos*: Canale *et al.*, 2009; for a comprehensive review concerning the genus *Cebus*, see Ottoni & Izar, 2008), although many species occasionally use tools in captivity (for reviews, see Anderson, 1996; Panger, 2007; Bentley-Condit & Smith, 2010; Shumaker *et al.*, 2011). Among monkeys, the capuchins (species belonging to the newly identified genus *Sapajus*) excel in all respects (Bentley-Condit & Smith, 2010) and their tool use fully fits St Amant and Horton's (2008) definition.

Tool-using skills of captive capuchins were reported in Europe in the sixteenth century (see Visalberghi & Fragaszy, 2012), long before the first illustration of a Liberian chimpanzee digging for termites appeared in a stamp issued in 1906 (Whiten & McGrew, 2001)

[1] We cite this definition of tool use, among several available, because it highlights the requirement that the actor must accomplish a mechanical interaction with the target. Use of a stone as a percussor epitomizes this aspect of tool use.

[2] Recent molecular analysis has revealed that capuchin monkeys, formerly identified as the single genus *Cebus*, are two genera, with the robust forms (including *libidinosus*, *xanthosternos* and several other species) now recognized as the genus *Sapajus*, and the gracile forms retained as the genus *Cebus* (Lynch Alfaro *et al.*, 2011, 2012). To date, tool use has been observed in some species of wild *Sapajus*, but no species of wild *Cebus*. In this chapter we retain *Cebus* when citing published findings using the older nomenclature and *Sapajus (Cebus)* to refer to the taxa in general and new findings about species in this newly recognized genus.

Tool Use in Animals: Cognition and Ecology, eds. Crickette M. Sanz, Josep Call and Christophe Boesch.
Published by Cambridge University Press.

or reached the scientific community through the work of Jane Goodall (1964). However, first-hand published reports on tool use by wild capuchins are relatively recent. Fernandes (1991) published the very first account of direct observation of tool use. He observed a wild capuchin (*Cebus apella*) using a broken oyster shell to strike oysters still attached to the substrate and successfully breaking them open. Boinski (1988) observed a wild male white-faced capuchin (*C. capucinus*) hitting a snake with a branch obtained from nearby vegetation. These observations each concerned one individual and one event. Habitual tool use (i.e., by several individuals over a period of time) in wild capuchins has been discovered and investigated only in the present millennium.

Surprising to some, the phenomenon of tool use by capuchin monkeys appears to be geographically widespread in wild populations of bearded capuchins (*Sapajus (Cebus) libidinosus*) and to encompass a range of materials, methods and goals. The reports come from seasonally dry Cerrado and Caatinga habitats in the north and east of Brazil. At Fazenda Boa Vista, State of Piauí (hereafter FBV; Fragaszy *et al.*, 2004a) bearded capuchin monkeys crack open palm nuts and other encased foods with stone hammers and anvils. In the Serra da Capivara National Park, also in the State of Piauí, bearded capuchins use stones to access embedded food by percussion and by scraping, and sticks to probe for honey and to flush vertebrate prey (Moura & Lee, 2004; Mannu & Ottoni, 2009). At the Agua Mineral National Park (Federal District, Brazil) and in the State of Rio Grande do Norte they crack encased foods with stone tools (Waga *et al.*, 2006; Ferreira *et al.*, 2010) (see Figure 10.1). Elsewhere researchers have found indirect evidence of tool use by (presumably) bearded capuchins (i.e., stone tools and palm shells on hard substrates; e.g., Canale *et al.*, 2009). Finally, semi-free-ranging capuchins (that may be mixtures of species and hybrids) have been observed to crack nuts in several sites (for a review see Ottoni & Izar, 2008).

Nut cracking is an integrated dynamic system with biomechanical and morphological components (related to the monkeys' postcranial morphology) and with environmental components (including the mass and material of the hammer stones and of the anvil site, and the material and physical properties of the nut). Using a stone (or log) to pound open or otherwise breach an encased food item placed on a solid surface (an "anvil") is considered the most complex form of tool use by non-human species routinely seen in nature, because it involves producing two spatial relations in sequence (between nut and anvil, and between pounding tool and nut) (Matsuzawa, 2001; Fragaszy *et al.*, 2004b). Furthermore, transporting the food item and sometimes the percussor involves costs (time and energy, among others) and may present cognitive challenges, such as anticipating future needs, recalling elements that are out of sight and planning the course of action. Transporting food resources and repetitive visits to specific places on the landscape to process foods are associated with early *Homo* and are thought to be important innovations of the Oldowan (e.g., Binford, 1981; Isaac, 1984; Potts, 1991).

Our goal here is to provide an overview of the discoveries made by the Etho-*Cebus*[3] team since the use of stone hammers and anvils by wild bearded capuchin monkeys at

[3] In 2005 Dorothy Fragaszy (Department of Psychology, University of Georgia, Athens, GA, US), Elisabetta Visalberghi (Istituto di Scienze e Tecnologie della Cognizione, Consiglio Nazionale delle Ricerche, Rome, Italy), Eduardo Ottoni and Patricia Izar (Department of Experimental Psychology, University of São Paulo,

Figure 10.1 Sites for which there are published observations of the use of percussive tools by wild capuchins. The stars indicate Fazenda Boa Vista (FBV, PI, Brazil) and Serra da Capivara National Park (SdC, PI, Brazil). The circle indicates the Agua Mineral Park, in the Brasilia National Park (DF, Brazil).

Fazenda Boa Vista (Gilbuès, State of Piauí, Brazil) was first reported in 2004. We also briefly compare some of our findings on stone-tool use in bearded capuchins, such as frequency of tool use, sex differences and selection of tools, with those reported for chimpanzees.

Study area and monkeys

Our study area is located at Fazenda Boa Vista (9°39′ S, 45°25′ W; see Figure 10.1), privately owned land 21 km northwest of the town of Gilbués (PI). The physical geography of the field site is a sandy plain at approximately 420 m above sea level, punctuated by

São Paulo, Brazil) started the Etho-*Cebus* Project to investigate the nut-cracking behavior of the FBV capuchins in ecological, developmental, social, physical and historical context. Barth Wright and Kristin Wright, both at Kansas City University of Medicine and Biosciences, have since joined the team. For more information, see http://www.Ethocebus.org

sandstone ridges, pinnacles and plateaus surrounded by cliffs composed of sedimentary rock rising steeply to 20–100 m above the plain. The cliff and plateau consist of interbedded sandstone, siltstone and shale. Rock faces often break off these formations and fall to the base of the cliff close to the plain, shattering into boulders with planar surfaces, and forming a talus (for further information about the geology of FBV, see Visalberghi *et al.*, 2007 and video 1 in the electronic supplementary material at www.cambridge.org/9781107011199). The sandstone cliffs and plateaus are heavily eroded and there are ephemeral water courses that have running water briefly after rainfall. The area is lightly populated by humans, and contains cultivated areas, wetlands, grazing land and some less disturbed woodland areas. The flat areas are open woodland, whereas the slopes of the ridges are more heavily wooded. Palms are abundant in the open woodland. The climate is seasonally dry. Over a 20-year period (1971–1990), the average annual rainfall was 1156 mm; 230 mm of that amount fell during the dry season (May–September) (Brazilian Agricultural Research Corporation, Embrapa, www.embrapa.br).

FBV is located in the transition area between the *Cerrado* and the *Caatinga* domains. It presents four types of vegetation physiognomies according to the terrain and the availability of water. The sandy plain is characterized by a high abundance of palms with subterranean stems and medium-height trees like *Eschweilera nana* and *Hymaenaea courbaril*. The vegetation in wetter areas is characterized by a higher diversity of trees forming gallery forests and a high density of the tall palm tree *Mauritia flexuosa*. Shrubs and small trees dominate the cliff and the talus, whereas in the plateau herbaceous vegetation dominates, especially bromeliads and cactus.

We have studied two groups of wild bearded capuchins at FBV: the C group and the Z group. In 2003, when we first visited FBV, the C group was already habituated to human presence and provisioned in support of ecotourism being promoted in that area. During the year in which data on tool use were systematically collected by Spagnoletti for her PhD (June 2006 to May 2007; Spagnoletti, 2009) the C group received a supplement of water and food (on average 3082 ± 908 kcal per day for the whole group consisting of 12–18 individuals). The C group visited the provisioned area to eat in 53 of the 91 days (58%) over which behavioral observations were carried out on that group. Visits to the provisioned area did not occur regularly and sometimes capuchins did not visit for many days in a row (for example, in February 2007 and February 2009 they did not visit for 11 and 12 days consecutively). Our team habituated Z group to human presence in 2005 and this group was not provisioned. The home range (MCP, minimum convex polygon) of Z group was 4.9 km^2 and that for C group 6.9 km^2, with an overlap of 3.0 km^2 during the period of Spagnoletti's study (Spagnoletti, unpublished data).

Between 2004 and 2009 the Etho-*Cebus* team conducted several behavioral studies, some of which are ongoing at the time of writing. Here, we discuss observations on the 28 bearded capuchins belonging to two groups (C and Z) carried out by Spagnoletti *et al.* (2011, 2012) and a series of field experiments conducted with the C group carried out by Visalberghi, Fragaszy and colleagues.

Palm nuts

In the open woodland of FBV palms are abundant and produce fruit at ground level. The four species of palm nuts most commonly eaten by capuchins at Boa Vista are: tucum (*Astrocaryum campestre*), catulè (*Attalea barreirensis*), piassava (*Orbignya* sp.) and catulí (*Attalea* sp.) (see Figures 1–3 in the electronic supplementary material at www.cambridge.org/9781107011199). Behavioral and phenological data collected in Boa Vista show that catulè nuts are more abundant in the dry season, whereas the other species, and especially piassava, are more uniformly available across the year (Spagnoletti *et al.*, 2012).

Typically, the capuchins collect the palm nuts by plucking one nut from the cluster, pulling and turning it until it comes loose. The mesocarp of catulè, catulí and piassava (but not of tucum) is edible, and capuchins usually eat this layer until the woody endocarp of the nut is exposed, as do cattle and other animals in the area. At this point they immediately look for an anvil site to crack the nut, or abandon the nut on the ground. Capuchins will also pick up nuts that have had the mesocarp removed previously and transport them to an anvil site.

Visalberghi *et al.* (2008) characterized the four species of nuts cracked by capuchins and assessed their peak-force-at-failure. The resistance of the structure of the nuts differed across species and was correlated positively with weight (and volume). In particular, the mean peak-force-at-failure values were 5.15 kN for catulè, 5.57 kN for tucum, 8.19 kN for catulí and 11.50 kN for piassava. Since catulè and tucum values were significantly lower than those of catulí and piassava, nuts were categorized as low-resistance (catulè and tucum) and high-resistance (piassava and catulí). To put these findings into context, all four species of palm nuts exploited by capuchins are at least 13 times more resistant than walnuts (*Juglans regia*) and between two (catulè, tucum, catulí) and five times (piassava) more resistant than macadamia nuts. Indeed, the piassava nuts are approximately as resistant as panda nuts, the most resistant nuts cracked open by chimpanzees (see Table 10.1).

Stones used as hammers and their availability in FBV

Nut cracking by capuchins leaves physical evidence, such as distinctive shallow depressions (pits) on the surface of both wooden anvils and stone anvils, cracked shells and stone hammer(s) on the anvil (see Figure 4 in the electronic supplementary material at www.cambridge.org/9781107011199). Visalberghi *et al.* (2007) used these diagnostic physical remains to infer the occurrence of nut cracking in the area of FBV. By surveying a sample of these anvil sites they found that: (1) anvils (boulders and logs containing shallow, hemi-spherical pits) were located predominantly in the transition zone between the flat open woodland and the ridges, in locations that offered some overhead coverage, and with a tree nearby, but not necessarily near palm trees; (2) hammer stones represented a diverse assemblage of ancient rocks that were much harder than the sedimentary rock (sandstone)

Table 10.1 Summary of the main characteristics of stone-tool use in bearded capuchins living at FBV and in chimpanzees in the Taï National Park (TNP, Ivory Coast). The references are in the text unless indicated.

	Cebus libidinosus	*Pan troglodytes verus*
Customary and habitual	Yes	Yes
Seasonal	No	Yes
Hammers		
Stone	Yes	Yes
Wood	Unsuitable	Yes
Frequency of suitable hammer stones	17.5 stones per hectare (0.3–3 kg)	0.09 hammers (>1 kg) per hectare[1]
Peak-force-at-failure of the hardest species of nuts[2]	*Orbignya* sp., average 11.5 kN	*Panda oleosa*, range: 9.6, 12.2 kN
Hammer selectivity on the basis of material and weight	Yes[*]	Yes
Anvils		
Material	Stone and wood	Stone and wood
Reuse of the same anvil site	Yes	Yes
Anvil transport	No impossible	No, but common in Bossou (pers. obs.)
Anvil selection	Yes	Yes[2]
Anvil pits selection	Yes[*3]	In progress[4]
Frequency of tool use	Males > females	Males < females
Body position when cracking	Bipedal	Seated
Tool use in trees	Yes[5]	Yes
Age of nut cracking acquisition	2–3 years	3.5–5 years[6]

Notes
[1] Boesch & Boesch, 1983.
[2] Carvalho *et al.*, 2009.
[3] Liu *et al.*, 2011.
[4] Carvalho *et al.* in progress.
[5] Etho-*Cebus* team, unpublished.
[6] Inoue-Nakamura and Matsuzawa, 1997.
[*] Experimental confirmation.

prevailing in FBV out of which they eroded; and (3) hammers were mostly cobbles eroded from the few conglomerate layers present in the local stratigraphy.

The stones found on the anvils (or within 3 m of them) were predominantly quartzite (the hardest rock in the area), siltstone, ironstone and sandstone that underwent metamorphism under higher temperature and pressure, becoming harder and less porous than the prevailing sandstone. Overall, the weight of these stones averaged 1096 g ($n = 62$, SD = 462.78; max weight 2530 g, min weight 140 g). Therefore, it appeared that sufficient hardness and weight were requirements for stones to be used as hammers. Since hard stones that make durable, effective hammers were significantly more frequently found on the anvil, or within 30 cm of it, than in the corona (the zone surrounding the base of the anvil, 30–300 cm from the base), Visalberghi *et al.* (2007) suggested that capuchins transport hammer stones to the anvils.

To investigate the latter point, Visalberghi *et al.* (2009a) estimated the occurrence of surfaces suitable as anvils, stones suitable as hammers and palms in the home range of our two study groups of capuchins by counting their frequencies in 40 plots (each measuring 10 m^2 located along a 3 km line transect crossing four different physiognomies (the marsh, the plain, the talus, i.e., the area of transition between the plain and the cliff, and the cliff-plateau). The transect census showed that palms and anvil-like surfaces were relatively common, whereas stones large enough and hard enough to use as hammers were rare. Overall, in the 40 plots (totaling 400 m^2 of surface area) there were only seven hammer-like stones which were found in two plots located at the cliff-plateau and in one plot located at the talus. The low number of hammer-like stones (17.5 stones per hectare) contrasts with the abundance of anvil-like surfaces, which in many areas are present as boulder fields at the foot of the cliffs. Both hammer-like stones and anvil-like surfaces were found more frequently in the talus and in the cliff-plateau areas than elsewhere, whereas palms were common everywhere except in the cliff-plateau. In short, the elements indispensable for tool use to crack nuts (i.e., hammer-like stones, anvils and palms, and therefore nuts) co-occur only in the cliff-plateau and in the talus; only the talus is close to the plain where palms are abundant. This picture confirms Visalberghi *et al.*'s (2007) report that active anvil sites are located at the transition zone between the cliff and the flat open woodland. Finally, the overall abundance of ephemeral water courses and direct observation of stones and tree trunks moved by water during heavy rainfalls support the hypothesis that when the conglomerate beds are weathered and eroded, the quartzite pebbles become loosened from the surrounding rock matrix, and are carried from the cliff-plateau to the talus below by water. Thus they become available to the capuchins as loose stones on an unpredictable basis, temporally and spatially.

Behavioral observations of tool use

Tool use occurred all year round at equivalent rates; monthly frequencies did not differ between groups or seasons (Spagnoletti *et al.*, 2011, 2012). Adult capuchins used tools to exploit palm nuts in 87% of the tool-use episodes and to exploit other encased fruits in the remaining episodes. One-third of the episodes of nut cracking targeted high-resistance nuts. Informal observations by our research team indicate that infants strike nuts on surfaces and strike nuts with other objects during the first year of life, but they do not open a whole nut or any part of a nut for many more months, despite persistent practice. Some individuals have been observed using effective strikes in their second year, but others did not do so until sometime in their third year or even later; Visalberghi and Fragaszy, unpublished data). Of the 23 physically normal capuchins observed by Spagnoletti at FBV that were at least three years old, only one female was never seen to crack nuts using a stone tool. Her health did not seem compromised by this deviation. Indeed, she was the heaviest adult female in her group and she bore several viable infants between 2004 and 2009.

In contrast with the pattern observed in chimpanzees, where females use tools more frequently than males (see McGrew, 1992 for a review; Lonsdorf *et al.*, 2004), adult male capuchins used tools about three times more often than females did to crack palm nuts (but not other food items) (Spagnoletti *et al.*, 2011). When the weights of the hammer stones are considered in relation to the body mass of the tool users (weighed in the field with a scale by Fragaszy *et al.*, 2010), it emerges that to crack open nuts the two sexes use stones of roughly equal mass. Adult female capuchins used stones weighing on average 978 g (i.e., 44–51% of their body mass, depending on the individual), whereas adult males used stones weighing on average 1072 g (i.e., 24–40% of their body mass, depending on the individual) (see Spagnoletti *et al.*, 2011; Fragaszy *et al.*, 2010 for further details). Overall, females used on average 14 ± 3.2 strikes to open a nut and males 10 ± 1.4 strikes (Figure 10.2).

Why does the frequency of nut cracking vary between male and female capuchins? Females have higher energetic costs of reproduction than males, but if males are sufficiently larger than females they may have higher energetic requirements than females (Key & Ross, 1999). In terms of daily energy expenditure, male and female capuchins appear close to even and should be expected to exploit nuts to similar extents. However, since at our site female capuchins were 36% smaller than males, and use more strikes to open a nut than males, nut cracking is more energy-demanding, more time-consuming and/or a less reliable method of producing food for them than for males. Additionally, females may be more sensitive than males to competitive costs, or costs related to risk of predation, such as spending time on the ground (Fragaszy, 1990; Rose, 1994). However, this does not seem to be the case in FBV, where the daily time spent on the ground by adult males and females is 29% and 31%, respectively (Spagnoletti, 2009). Thus, cracking nuts might entail a different set of costs and potential benefits for males than for females. Any and all of these factors could make tool use more advantageous for male than for female capuchins (Spagnoletti *et al.*, 2011). In contrast, for chimpanzees nut cracking takes little time or effort and is a reliable method of foraging (Boesch & Boesch-Achermann, 2000); physical sexual dimorphism is also less pronounced in chimpanzees than in bearded capuchins (compare Key & Ross, 1999 with Fragaszy *et al.*, 2010).

Spagnoletti *et al.* (2011) report that capuchins were successful in 84% of the episodes with high-resistance palm nuts, 91% of the episodes with low-resistance palm nuts and 99% of the episodes involving other encased foods. The rate of success is similar for females and males, and both sexes cracked low-resistance palm nuts more often than high-resistance nuts. This pattern of food choice may have resulted in equivalent reliability of energy gain for all animals, or may have enhanced reliability of energy gain for females compared to males. Interestingly for the discussion concerning tool selectivity, female capuchins needed significantly more strikes than males to crack low-resistance palm nuts, but not high-resistance palm nuts. Though this might appear contradictory, females used significantly heavier hammers to crack open high-resistance palm nuts than low-resistance palm nuts, whereas males did not.

Efficiency in cracking nuts with tools varies widely among wild capuchins, even when the same hammer stone and the same anvil are used to crack open nuts of the same palm species (Fragaszy *et al.*, 2010). In Fragaszy *et al.*'s sample, the most efficient monkey opened on average 15 nuts per 100 strikes (6.6 strikes per nut). The least efficient monkey

Figure 10.2 An adult male (a) and an adult female (b) bearded capuchin using a 3.5 kg stone to crack a piassava nut. The height of the female's strike is much lower than that of the male. The male weighs 4.2 kg; the female, 2.2 kg. This stone is heavier than those normally used by the monkeys (photos by E. Visalberghi).

opened on average 1.32 nuts per 100 strikes (more than 75 strikes per nut). They also report that the efficiency of one physically fit adult human male (20 years, 185 cm, 78 kg) striking the same species of nuts with the same stone was 16.1 nuts per 100 strikes (6.2 strikes per nut).

Logistic regression revealed that the monkey's body weight and diameter of the nut best predicted whether a monkey would crack a nut on a given strike. Increasing body weight improved the likelihood of cracking the nut; increasing diameter of the nut decreased the likelihood. In fact, smaller monkeys (females and youngsters) often fail to crack whole nuts even after numerous strikes with a stone that weighs proportionally more than 50% of their body weight.

Whereas chimpanzees are seated when cracking nuts, wild capuchins most often adopt a bipedal stance, raising and rapidly lowering the hammer by flexing the lower extremities and the hip. By studying the kinematics and energetics of nut cracking of two adult males and two adult females, Liu *et al.* (2009) demonstrated that the two males achieved greater maximum downward vertical velocities with the stone than the females (mean = $3.81\ m\,s^{-1}$ vs. $3.16\ m\,s^{-1}$; males and females, respectively). Therefore, the males generated higher maximum kinetic energy than the females. As the males lifted the stones to a higher maximum vertical height (in accord with their longer body length), the potential energy that they generated was also higher than the potential energy generated by females. All the monkeys produced work in the downward phase; that is, they added energy to the stone in the downward direction. Males produced nearly twice the work that females produced (mean = 5.61 J vs. 2.89 J; males and females, respectively). The overall consequence of the above differences is that large monkeys (adult males) crack nuts with fewer strikes than small monkeys. The differential cost of cracking for small and large animals is probably one of the factors accounting for the lower frequency of tool use in females than in males (Spagnoletti *et al.*, 2011).

Table 10.1 summarizes some of the above results and provides a comparison between our findings and those reported for chimpanzees. The findings reported so far provide enough information to draw two conclusions. First, most adult capuchins at FBV use tools throughout the year, fulfilling McGrew's (1992) definition of a habitual behavior (pattern of behavior shown repeatedly by several members of the group) and Whiten *et al.*'s (1999) definition of customary behavior (behavior that occurs in all or most able-bodied members of at least one age and sex class). Second, cracking nuts requires the use of stones of suitable weight and material, which are not very common in FBV. This conclusion prompted us to expect that capuchins select appropriate stones and transport them (as well as nuts or other encased hard fruits) to anvils. We turn now to what we have learned about anvil sites and the transport of nuts and stones to these sites.

Anvil sites as indirect evidence for tool use

Capuchins use anvil sites habitually. Spagnoletti *et al.* (2011) report that over a 12-month period (1709 hours of observation) capuchins cracked palm nuts on 116 different anvils throughout an area of 9 km^2, and that these anvils accommodated 607 episodes of tool use

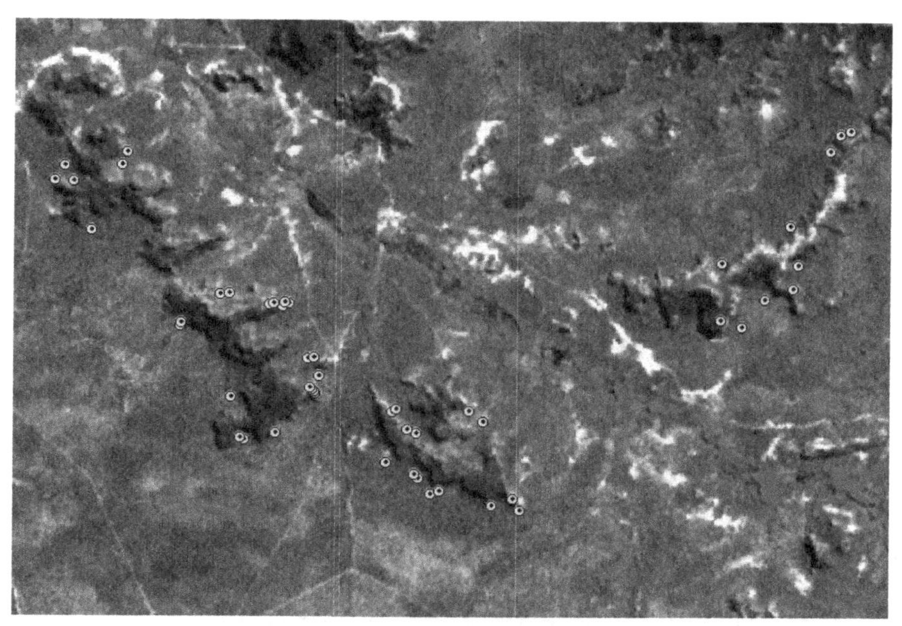

Figure 10.3 Map of the anvil sites surveyed monthly. Note that they are located in the transition area between the plain and the ridges. The two groups of capuchins we observed used the anvil sites located on the center-left.

(5.2 tool episodes per anvil) (Figure 10.3). More comprehensive evidence of repeated use of familiar anvil sites comes from a longitudinal survey conducted monthly by the Etho-*Cebus* team of 40 anvil sites (from February 2005 to January 2006) and 58 anvil sites (which included the previous 40) for two additional years (from February 2006 to January 2008; Visalberghi *et al.*, submitted). As illustrated in Figure 10.4, in each monthly visit we scored (a) displacement of the hammer stone(s), compared to its placement the previous month, and (b) the presence of nut shells on the anvil. The joint occurrence of hammer displacement and presence of shells was considered as strong evidence of tool use, displacement of the hammer stone without shells was taken as weak evidence and the presence of shells without displacement of the hammer position was not considered evidence of tool use. We recorded strong evidence of tool use on an average of 34% of the anvil sites each month, and weak evidence of use on an additional 7% of anvil sites.

During the three years of the study 17 hammer stones went missing from their original anvil sites; two were brought back to the same anvil site by capuchins five months and one month after disappearance. On nine occasions we found a new stone at an anvil site, all of a suitable weight, and seven of these were suitably fracture-resistant to serve as hammers. The other two were weathered sandstone, which breaks when used to strike nuts (Visalberghi *et al.*, 2007). The frequencies with which shells of piassava and catulé nuts were found on the anvils in the survey match the relative frequencies of directly observed tool-use episodes involving these two species of nuts (Spagnoletti, 2009). The similarity between direct and indirect assessments indicates surveying anvil sites is a

Figure 10.4 Procedure for the survey of the anvil sites. Each month we visited each anvil site to gather indirect evidence of tool use by scoring whether the hammer had been displaced from its initial position (a), and whether shells from nuts or remains from other encased fruits were present on the anvil. The comparison between these pictures shows that the anvil site has been used to crack nuts between when photo a was taken and 30 days later, when photo b was taken. In particular, in the later photo b, the hammer has been moved from its initial position (line in the front) to a new one (indicated by the arrow) and nut shells (circled with dotted lines) are present. Close to the nut shell on the far left there is a new small black pebble, presumably brought by capuchins (photos by E. Visalberghi).

reliable way to assess tool use in capuchins. In short, the evidence shows that (1) capuchins use anvils habitually; (2) capuchins transport new hammer stones to anvils and familiar stones among anvils; and (3) tool use to crack nuts by capuchins can be inferred reliably by surveying anvil sites, lending support to the use of this type of indirect evidence in areas where the monkeys are not habituated to human observers and/or the area is unfamiliar to the researchers, as has been done by Ferreira *et al.* (2010), Canale *et al.* (2009) and Langguth and Alonso (1997).

Archaeological excavations at abandoned nut-cracking sites of chimpanzees (Mercader *et al.*, 2002, 2007) and combining classical archaeological approaches with direct observation of chimpanzees' nut cracking are promising avenues for archaeologists to infer more accurately the function of pounding tools coming from Plio-Pleistocene assemblages (Mercader *et al.*, 2007; Carvalho *et al.*, 2008, 2009). Studies of this type with capuchin monkeys would allow interesting comparisons among closely and less closely related species that use percussive tools. We expect to begin studies along these lines with capuchins at Boa Vista, such as examining microwear produced by capuchins on percussors of high-quality material (e.g., lava basalt).

Transport and selectivity of hammer stones: behavioral observations and experimental evidence

Although both adult and juvenile capuchins spontaneously transport nuts and other encased foods and hammers to anvil sites (Visalberghi *et al.*, 2009a), only 3.4% of the tool-use episodes observed by Spagnoletti (2009) included stone hammer transport. This low rate of occurrence reflects the rarity of suitable stones in the general habitat, the presence of hammers on anvil in use and, at least for small individuals, the high cost of transport (Visalberghi *et al.*, 2007; Massaro *et al.*, 2012).

Of the 22 tool-using capuchins observed by Spagnoletti (2009), 12 spontaneously transported a hammer to an anvil at least once; frequencies of transport did not differ between sex and age classes. Adult capuchins transported heavier hammer stones than juveniles; an adult female transported the heaviest hammer stone, weighing 1600 g, for 6 m. If capuchins take into account the resistance of nuts when looking for a hammer, then they should transport stones suitable to overcome the nuts' resistance, such as quartzites and siltstones, significantly more often than unsuitable ones, such as weathered sandstones. To exploit nuts adult males carried suitably hard stones in 15 cases and unsuitably soft stones in four cases; adult females carried suitably hard stones in all five cases; and juveniles transported suitably hard stones twice and unsuitably soft stones five times (Visalberghi *et al.*, 2009a). Conversely, when exploiting other encased foods (less resistant than the nuts), adults transported soft stones in four cases out of five, and juveniles transported a soft stone in one out of two cases. Although the sample size is small, our observations suggest that adult capuchins take into account the resistance of the food item to be cracked when transporting and using a stone as a hammer (Spagnoletti, 2009; Visalberghi *et al.*, 2009a).

To confirm that the monkeys selected stones for their weight and friability (resistance to crumbling), we carried out the first experimental study on this topic with wild animals (Visalberghi *et al.*, 2009b)[4] by repeatedly providing individuals with sets of stones varying in specific properties. We presented to eight capuchins choices between two (or among three) stones differing in functional features. In the first two conditions, when no other

[4] Video clips can be viewed at http://www.cell.com/current-biology/supplemental/S0960-9822%2808%2901624-2

stones were available in the area, subjects chose between novel natural stones, similar to those they usually encounter in their habitat, differing in friability (sandstone vs. siltstone), or in size and weight (small vs. large quartzite stones). One of the stones in each set was deemed "functional"; i.e., it weighed 500 g or more and would not fracture when used to strike a nut. In both the above conditions, all subjects first touched, transported and used the functional stone significantly more often than expected by chance.

In the next three conditions subjects chose between novel manufactured stones (composed of ground stone and resin of the same color and material), the weight (an "invisible" feature) of which did not correlate with size (contrary to the usual case). Capuchins had to choose between stones of the same size and different weight, between a light large stone and a heavy small stone, and among a light large stone, a light small stone and a heavy large stone. Again, each subject demonstrated a significant bias to transport and use the heavier stone in all conditions (except one monkey in one condition). They always used the stone they chose first and never modified their initial choice after the first strike(s).

Interestingly, when visual cues were available and reliable, capuchins always touched the functional stone first, suggesting they discriminated the volume of the stones by sight, and they did not tap either stone. In contrast, when visual cues were not predictive or were conflicting, individuals gained information about the mass of the experimental stones by moving, lifting and/or tapping them. Tapping may have allowed them to infer the density of the stones (see also Schrauf *et al.*, 2008). Fragaszy *et al.* (2010) showed that the monkeys preferentially select the heavier stone when they have a choice of two manufactured stones of equal volume, even when the difference in weight between the two stones is a small fraction of the total weight. Overall, naturalistic observations and field experiments provide compelling evidence that when wild capuchins encounter stones that differ in friability and weight, they choose, transport and use the more effective stones to crack open nuts. Moreover, when weight cannot be judged by visual attributes, capuchins act to gain information to guide their selection. In all these ways, capuchins evidence planning and skill in tool use.

Final remarks

Capuchins from very early in life pound objects on surfaces (Adams-Curtis & Fragaszy, 1994) and all over their wide geographical distribution they often pound encased foods, such as hard fruits, seeds or shelled invertebrates, on hard surfaces in order to get access to the inner parts (for a review, see Fragaszy *et al.*, 2004b). Although captive capuchins spontaneously learn to use hammer stones to crack open nuts (for a review, see Visalberghi & Fragaszy, 2012), percussive tool use in the wild seems a geographically circumscribed phenomenon reported only in some populations of *Cebus libidinosus* and *C. xanthosternos* living in northeastern Brazil (Ottoni & Izar, 2008; Canale *et al.*, 2009; Ferreira *et al.*, 2010). Why might this be so?

Three hypotheses have been proposed to account for the appearance of percussive tool use in some populations of capuchins and not in others. According to Moura and Lee

(2004), the occurrence of tool use is related to energy bottlenecks. Tool use is thought to allow the monkeys to exploit tough or hidden foods in regions where they experience periods of extreme food scarcity. According to the second hypothesis, terrestriality is the key factor promoting the emergence of tool use (see Chapter 9; Meulman *et al.*, 2012). The reasoning here is that foraging and traveling on the ground increase an individual's chances of discovering and practicing percussive tool use, because nuts, anvils and potential tools are all found on the ground (Visalberghi *et al.*, 2005). These materials can also be combined more readily on the ground. For example, on the ground, but not in trees, nuts can be placed on flat surfaces and they can be recovered if displaced or dropped. Resende *et al.* (2008) suggested that placing the nut on an anvil and releasing it is one of the main challenges facing young capuchins learning to crack nuts; practicing positioning nuts and other objects on flat surfaces on the ground (but not in trees) may help to overcome this reluctance, because the monkey can learn that a released object remains in arm's reach. The third hypothesis is that the abundance of tough foods in the diet can increase the dietary benefits of tool use to access tough foods (independent of food abundance or terrestriality) (Wright *et al.*, 2009).

It appears from estimates of fruit biomass (Verderane *et al.*, unpublished data; see also Spagnoletti *et al.*, 2012) that the monkeys at FBV do not face seasonal food shortages as extreme as those seen in other habitats (e.g., Atlantic coastal forest). Their reproductive performance and general health suggest a consistently good food supply as well. For example, in late 2008, two pairs of twins were born, one pair in each group. Of the four twins, one disappeared at 11 months and the other three are still alive at this writing (at 16–19 months). Twins are not common in capuchins and survival past 30 days is less likely for a twin than for a singleton infant in captivity (Leighty *et al.*, 2004). Time budgets provide another indirect index of food availability. The capuchins at FBV spent 7% of their daylight time in social activities (Spagnoletti, 2009), which is at the upper range for capuchins living in other semi-deciduous dry forest habitats (Fragaszy, 1990; Rose, 1994; Miller, 1997) and substantially higher than for capuchin monkeys living in evergreen forests (Terborgh, 1983), suggesting that the capuchins at FBV manage to find sufficient food quickly enough to have time for "leisure" activities.

However, their diet does comprise a greater proportion of tough items than the diet of monkeys at other sites (Wright *et al.*, 2009). Although most of these items are breached with teeth and hands, the toughest are breached through hammering with a stone. Thus capuchins at FBV are using tools in a relatively food-rich environment.

The monkeys at FBV (similar to other populations of capuchins that use tools – e.g., Mannu & Ottoni, 2009) spend a greater proportion of time on the ground (roughly 26–34.5% of the day; Spagnoletti, 2009) to forage, rest and play than other populations of capuchins for which tool use has not been reported (e.g.: *Cebus olivaceus* 13.4 %, Robinson, 1986; *Sapajus (Cebus) nigritus* 1.9 %, Brandon Wheeler, pers. comm.; see Spagnoletti, 2009; Spagnoletti *et al.*, 2009). Thus the preponderance of the evidence suggests that seasonal food shortages are not an adequate explanation of the appearance of tool use among capuchins in *Cerrado* environments. Instead, a combination of ecological circumstances including spatial location of resources, scarcity of terrestrial predators (influencing the risk of descending to the ground) and mechanical properties of available foods would all seem to play influential roles in the appearance of tool use.

Overall, our findings show that at FBV wild bearded capuchin monkeys are habitual tool users (*sensu* McGrew, 1992) and that the sophistication of their behavior with respect to selection and transport of tools matches that reported for wild chimpanzees using percussive tools (see Table 10.1). This is particularly interesting given the phylogenetic distance between the two species. Although extensive and flexible tool use was once considered a defining human characteristic, discovery of habitual tool use among wild chimpanzees (*Pan troglodytes*) in the 1960s led anthropologists to suggest that the last common ancestor of chimpanzees and humans was a tool user. For many archaeologists and anthropologists, chimpanzees have become the referent for modeling early hominins (Sayers & Lovejoy, 2008). However, wild bearded capuchin monkeys, a species that separated from the human lineage about 35 million years ago, also habitually use tools, but great apes other than chimpanzees rarely do. Thus we need to re-think the accepted explanations of continuity and convergence in primate tool use (Wynn & McGrew, 1989; Byrne 2004; Fox *et al.*, 2004; Davidson & McGrew, 2005; but see Panger *et al.*, 2002; Haslam *et al.*, 2009). Percussive tool use in non-humans has particular importance for the understanding of tool use in early hominins that also used percussive tools to access encased food. As an outgroup species, capuchins illuminate convergences in behavior that suggest ecological foundations for the character of interest (in this case, percussive tool use). A rigorous comparison of percussive tool use in capuchins with tool use in chimpanzees and humans, extant and extinct, will contribute to our understanding of the origins and evolution of this key feature of human behavior.

Acknowledgments

Permission to work in Brazil granted by IBAMA and CNPq to DF and EV. Thanks to the *Familia M* for permission to work at FBV and Jozemar, Arizomar and M. Junior Oliveira for their assistance in the field, Noemi Spagnoletti and the Etho-*Cebus* team for comments on this manuscript. Funded by CNR, EU-Analogy (STREP Contr. No 029088), European project IM-CLeVeR FP7-ICT-IP-231722, the National Geographic Society, and the Leakey Foundation. We also thank Maria Elena Miletto Petrazzini for her help.

References

Adams-Curtis, L. E. & Fragaszy, D. M. (1994). Development of manipulation in capuchin monkeys during the first six months. *Developmental Psychobiology*, **27**, 123–136.

Anderson, J. R. (1996). Chimpanzees and capuchin monkeys: comparative cognition. In A. Russon, K. Bard & S. Parker (eds.) *Reaching into Thought: The Minds of the Great Apes* (pp. 23–55). Cambridge: Cambridge University Press.

Bentley-Condit, V. K. & Smith, E. O. (2010). Animal tool use: current definition and an updated comprehensive catalog. *Behaviour*, **174**, 185–221.

Binford, L. (1981). *Bones: Ancient Men and Modern Myths*. New York: Academic Press.

Boesch, C. & Boesch, H. (1983). Optimisation of nut-cracking with natural hammers by wild chimpanzees. *Behaviour*, **83**, 265–286.

Boesch, C. & Boesch-Achermann, H. (2000). *The Chimpanzees of the Taï Forest*. Oxford: Oxford University Press.

Boesch, C., Head, J. & Robbins, M. M. (2009). Complex tool sets for honey extraction among chimpanzees in Loango National Park, Gabon. *Journal of Human Evolution*, **56**, 560–569.

Boinski, S. (1988). Use of a club by a wild white-faced capuchin (*Cebus capucinus*) to attack a venomous snake (*Bothrops asper*). *American Journal of Primatology*, **14**, 177–180.

Byrne, R. W. (2004). The manual skills and cognition that lie behind hominid tool use. In A. E. Russon & D. R. Begun (eds.) *The Evolution of Thought: Evolutionary Origins of Great Ape Intelligence* (pp. 31–44). Cambridge: Cambridge University Press.

Canale, G. R., Guidorizzi, C. E., Kierulff, M. C. M. & Gatto, C. A. F. R. (2009). First record of tool use by wild populations of the yellow-breasted capuchin monkey (*Cebus xanthosternos*) and new records for the bearded capuchin (*Cebus libidinosus*). *American Journal of Primatology*, **71**(5), 366–372.

Carvalho, S., Cunha, E., Sousa, C. & Matsuzawa, T. (2008). Chaînes opératoires and resource exploitation strategies in chimpanzee nut-cracking (*Pan troglodytes*). *Journal of Human Evolution*, **55**, 148–163.

Carvalho, S., Biro, D., McGrew, W. & Matsuzawa, T. (2009). Tool-composite reuse in wild chimpanzees (*Pan troglodytes*): archaeologically invisible steps in the technological evolution of early hominins? *Animal Cognition*, **12**, 103–114.

Davidson, I. & McGrew, W. C. (2005). Stone tools and the uniqueness of human culture. *Royal Anthropological Institute*, **11**, 793–817.

Fernandes, E. B. M. (1991). Tool use and predation of oysters (*Crassostrea rhizophorae*) by the tufted capuchin, *Cebus apella apella*, in brackish water mangrove swamp. *Primates*, **32**, 529–531.

Ferreira, R. G, Emidio, R. A. & Jerusalinsky, L. (2010). Three stones for three seeds: natural occurrence of selective tool use by capuchins (*Cebus libidinosus*) based on an analysis of the weight of stones found at nutting sites. *American Journal of Primatology*, **72**, 270–275.

Fox, E. A., van Schaik, C. P., Sitompul, A. & Wright, D. N. (2004). Intra and interpopulational differences in orangutan (*Pongo pygmaeus*) activity and diet: implications for the invention of tool use. *American Journal of Physical Anthropology*, **125**, 162–174.

Fragaszy, D. M. (1990). Sex and age differences in the organization of behavior in wedge-capped capuchins, *Cebus olivaceus*. *Behavioral Ecology*, **1**, 81–94.

Fragaszy, D. M., Izar, P., Visalberghi, E., Ottoni, E. B. & de Oliveira, M. G. (2004a). Wild capuchin monkeys (*Cebus libidinosus*) use anvils and stone pounding tools. *American Journal of Primatology*, **64**, 359–366.

Fragaszy, D. M., Visalberghi, E. & Fedigan, L. M. (2004b). *The Complete Capuchin: The Biology of the Genus* Cebus. Cambridge: Cambridge University Press.

Fragaszy, D. M., Pickering, T., Liu, Q., *et al.* (2010). Bearded capuchin monkeys' and a human's efficiency at cracking palm nuts with stone tools: field experiments. *Animal Behaviour*, **79**, 321–332.

Goodall, J. (1964). Tool-using and aimed throwing in a community of free-living chimpanzees. *Nature*, **201**, 1264–1266.

Gumert, M. D., Kluck, M. & Malaivijitnond, S. (2009). The physical characteristics and usage patterns of stone axe and pounding hammers used by long-tailed macaques in the Andaman Sea region of Thailand. *American Journal of Primatology*, **71**, 594–608.

Haslam, M., Hernandez-Aguilar, A., Ling, V., *et al.* (2009). Primate archaeology. *Nature*, **460**, 339–344.

Inoue-Nakamura, N. & Matsuzawa, T. (1997). Development of stone tool use by wild chimpanzees (*Pan troglodytes*). *Journal of Comparative Psychology*, **111**, 159–173.

Isaac, G. L. (1984). The archaeology of human origins: studies of the lower Pleistocene in East Africa, 1971–1981. *Advances in World Archaeology*, **3**, 1–86.

Key, C. & Ross, C. (1999). Sex differences in energy expenditure in non-human primates. *Proceedings of the Royal Society of London B*, **266**, 2479–2480.

Langguth, A. & Alonso, C. (1997). Capuchin monkeys in the Caatinga: tool use and food habits during drought. *Neotropical Primates*, **5**, 77–78.

Leighty, K. A., Byrne, G., Fragaszy, D. M. & Visalberghi, E. (2004). Twinning in tufted capuchins (*Cebus apella*): rate, survivorship, and weight gain. *Folia Primatologica*, **75**(1), 14–18.

Liu, Q., Simpson, K., Izar, P., *et al.* (2009). Kinematics and energetics of nut-cracking in wild capuchin monkeys (*Cebus libidinosus*) in Piauí, Brazil. *American Journal of Physical Primatology*, **138**, 210–220.

Liu, Q., Fragaszy, D., Izar P., Ottoni, E. & Visalberghi, E. (2011). Wild capuchin monkeys (*Cebus libidinosus*) place nuts in anvils selectively. *Animal Behaviour*, **81**, 297–305.

Lonsdorf, E. V., Pusey, A. E. & Eberly, L. (2004). Sex differences in learning in chimpanzees. *Nature*, **428**, 715–716.

Lynch Alfaro, J., Boubli, J., Olson, L., *et al.* (2011). Explosive Pleistocene range expansion leads to widespread Amazonian sympatry between robust and gracile capuchin monkeys. *Journal of Biogeography*, **39**(2), 272–288.

Lynch Alfaro, J., Silva, J. de Sousa & Rylands, A. (2012). How different are robust and gracile capuchin monkeys? An argument for the use of *Sapajus* and *Cebus*. *American Journal of Primatology*, **74**, 273–286.

Malaivijitnond, S., Lekprayoon, C., Tandavanittj, N., *et al.* (2007). Stone-tool usage by Thai long-tailed macaques (*Macaca fascicularis*). *American Journal of Primatology*, **69**, 227–233.

Mannu, M. & Ottoni, E. B. (2009). The enhanced tool-kit of two groups of wild bearded capuchin monkeys in the Caatinga: tool making, associative use, and secondary tools. *American Journal of Primatology*, **71**, 242–251.

Massaro, L., Liu, Q., Visalberghi, E. & Fragaszy, D. (2012). Wild bearded capuchina select hammer tools on the basis of both stone mass and distance from the anvil. *Animal Cognition*, **15**, 1065–1074.

Matsuzawa, T. (2001). Primate foundations of human intelligence: a view of tool use in nonhuman primates and fossil hominids. In T. Matsuzawa (ed.) *Primate Origins of Human Cognition and Behavior* (pp. 3–25). Tokyo: Springer-Verlag.

McGrew, W. C. (1992). *Chimpanzee Material Culture: Implications for Human Evolution*. Cambridge: Cambridge University Press.

Mercader, J., Panger, M. & Boesch, C. (2002). Excavation of a chimpanzee stone tool site in the African rainforest. *Science*, **296**, 1452–1455.

Mercader, J., Barton, H., Gillespie, J., *et al.* (2007). 4,300-year-old chimpanzee sites and the origins of percussive stone technology. *Proceedings of the National Academy of Sciences USA*, **104**, 3043–3048.

Meulman, E. J. M., Sanz, C. M., Visalberghi, E. & van Schaik, C. P. (2012). The role of terrestriality in promoting primate technology. *Evolutionary Anthropology*, **21**, 58–68.

Miller, L. (1997). Methods of assessing dietary intake: a case study from wedge-capped capuchins in Venezuela. *Neotropical Primates*, **5**(4), 104–108.

Moura, A. C. A. & Lee, P. C. (2004). Capuchin stone tool use in Caatinga dry forest. *Science*, **306**, 1909.

Ottoni, E. & Izar, P. (2008). Capuchin monkey tool use: overview and implications. *Evolutionary Anthropology*, **17**, 171–178.

Panger, M. (2007). Tool use and cognition in primates. In C. J. Campbell, A. Fuentes, K. C. MacKinnon, M. Panger & S. K. Bearder (eds.) *Primates in Perspective* (pp. 665–677). New York: Oxford University Press.

Panger, M., Brooks, A., Richmond, B. G. & Wood, B. (2002). Older than the Oldowan? Rethinking the emergence of hominin tool use. *Evolutionary Anthropology*, **11**, 235–245.

Potts, R. (1991). Why the Oldowan? Plio-Pleistocene tool making and the transport of resources. *Journal of Anthropological Research*, **47**, 153–176.

Resende, B. D., Ottoni, E. B. & Fragaszy, D. M. (2008). Ontogeny of manipulative behavior and nut-cracking in young tufted capuchin monkeys (*Cebus apella*): a perception-action perspective. *Developmental Science*, **11**, 828–840.

Robinson, J. G. (1986). Seasonal variation in use of time and space by the wedge-capped capuchin monkey *Cebus olivaceus*: implications for foraging theory. *Smithsonian Contributions to Zoology*, **431**, 1–60.

Rose, L. M. (1994). Sex differences in diet and foraging behavior in white-faced capuchins (*Cebus capucinus*). *International Journal of Primatology*, **15**, 95–114.

Sayers, K. & Lovejoy, C. O. (2008). The chimpanzee has no clothes: a critical examination of *Pan troglodytes* in models of human evolution. *Current Anthropology*, **49**, 87–114.

Schrauf, C., Huber, L. & Visalberghi, E. (2008). Do capuchin monkeys use weight to select hammer tools? *Animal Cognition*, **11**, 413–422.

Shumaker, R., Walkup, K. & Beck, B. (2011). *Animal Tool Behavior: The Use and Manufacture of Tools by Animals*. Baltimore, MD: Johns Hopkins University Press.

Spagnoletti, N. (2009). Tool use in a wild population of *Cebus libidinosus* in Piauí, Brazil. PhD thesis, University La Sapienza of Rome.

Spagnoletti, N., Izar, P. & Visalberghi, E. (2009). Tool use and terrestriality in wild bearded capuchin monkey (*Cebus libidinosus*). *Folia Primatologica*, **80**, 142.

Spagnoletti, N., Visalberghi, E., Ottoni, E., Izar, P. & Fragaszy, D. (2011). Stone tool use by adult wild bearded capuchin monkeys (*Cebus libidinosus*). *Journal of Human Evolution*, **61**, 97–107.

Spagnoletti, N., Visalberghi, E., Verderane, M. P., *et al.* (2012). Stone tool use in wild bearded capuchin monkeys (*Cebus libidinosus*): is it a strategy to overcome food scarcity? *Animal Behaviour*, **83**, 1285–1294.

St Amant, R. & Horton, T. E. (2008). Revisiting the definition of animal tool use. *Animal Behavior*, **75**, 1199–1208.

Terborgh, J. (1983). *Five New World Primates: A Study in Comparative Ecology*. Princeton, NJ: Princeton University Press.

Visalberghi, E. & Fragaszy, D. (2012). What is challenging about tool use? The capuchins' perspective. In T. R. Zentall & E. A. Wasserman (eds.) *Comparative Cognition: Experimental Explorations of Animal Intelligence* (pp. 777–799). Oxford: Oxford University Press.

Visalberghi, E., Fragaszy, D., Izar, P. & Ottoni, E. B. (2005). Terrestriality and tool use. *Science*, **308**, 951.

Visalberghi, E., Fragaszy, D., Ottoni, E., *et al.* (2007). Characteristics of hammer stones and anvils used by wild bearded capuchin monkeys (*Cebus libidinosus*) to crack open palm nuts. *American Journal of Physical Anthropology*, **132**, 426–444.

Visalberghi, E., Haslam, M., Spagnoletti, N. & Fragaszy, D. (submitted). Use of stone hammer tools and anvils by bearded capuchin monkeys over time and space: dynamic construction of an archeological record of tool use. *Journal of Archaeological Science*.

Visalberghi, E., Sabbatini, G., Spagnoletti, N., *et al.* (2008). Physical properties of palm fruits processed with tools by wild bearded capuchins (*Cebus libidinosus*). *American Journal of Primatology*, **70**, 1–8.

Visalberghi, E., Spagnoletti, N., Ramos Da Silva, E., *et al.* (2009a). Distribution of potential hammers and transport of hammer tools and nuts by wild capuchin monkeys. *Primates*, **50**, 95–104.

Visalberghi, E., Addessi, E., Spagnoletti, N., *et al.* (2009b). Selection of effective stone tools by wild capuchin monkeys. *Current Biology*, **19**, 213–217.

Waga, I., Dacier, A., Pinha, P. & Tavares, M. (2006). Spontaneous tool use by wild capuchin monkeys (*Cebus libidinosus*) in the Cerrado. *Folia Primatologica*, **77**, 337–344.

Whiten, A. & McGrew, W. C. (2001). Is this the first portrayal of tool use by a chimp? *Nature*, **409**, 12.

Whiten, A., Goodall, J., McGrew, W. C., *et al.* (1999). Cultures in chimpanzees. *Nature*, **399**, 682–685.

Wright, B. W., Wright, K. A., Chalk, J., *et al.* (2009). Fallback foraging as a way of life: using dietary toughness to compare the fallback signal among capuchins and implications for interpreting morphological variation. *American Journal of Physical Anthropology*, **140**, 687–699.

Wynn, T. & McGrew, W. C. (1989). An ape's view of the Oldowan. *Man*, **24**, 383–398.

Yamakoshi, G. (2004). Evolution of complex feeding techniques in primates: is this the origin of great ape intelligence? In A. Russon & D. R. Begun (eds.) *The Evolution of Thought: Evolutionary Origins of Great Ape Intelligence* (pp. 140–171). Cambridge: Cambridge University Press.

Part IV

Archaeological perspectives

11 From pounding to knapping: How chimpanzees can help us to model hominin lithics

Susana Carvalho

Leverhulme Centre for Human Evolutionary Studies, University of Cambridge
CIAS – Centre of Research in Anthropology and Health, University of Coimbra

Tetsuro Matsuzawa

Primate Research Institute, Kyoto University

William C. McGrew

Department of Archaeology & Anthropology, University of Cambridge

Introduction

Are pounding tools the missing link for the origins of technology?

All populations of modern humans (*Homo sapiens*) have repertoires of tool use and tool making (e.g., Ambrose, 2001). All well-studied extant populations of chimpanzees (*Pan troglodytes*) have repertoires of tool use and tool making (e.g., McGrew, 1992, 2010; Whiten *et al.*, 1999; Sanz *et al.*, 2010). No other living primate has lengthier inventories of tools than humans and chimpanzees (see van Schaik *et al.*, 2003; Ottoni & Izar, 2008; Gumert *et al.*, 2009 for inventories of tools in *Pongo pygmaeus*, *Cebus* spp. or *Macaca fascicularis*). During the Early Pleistocene, populations of hominins living in East and South Africa most likely had wider tool-use repertoires than the ones we have currently identified, which are only the non-organic, that is, non-perishable ones. Thus, archaeologists are mainly left with stone tools and fossilized bones to reconstruct the tool-use repertoire of the earliest hominins (e.g., Semaw *et al.*, 2003; Plummer, 2004). In West Africa, for at least 4300 years, chimpanzees have been using *stone tools*, namely hammers and anvils, to access the kernels of hard-shelled nuts (Mercader *et al.*, 2007).

Since studies began at Olduvai Gorge, Tanzania, and Koobi Fora, Kenya, stone tools have played a crucial role in inferring the evolution of early hominin behavior, beginning with research on the earliest stone tool industries (Leakey, 1971; Isaac, 1984). Leakey's typological approach to the Oldowan was based on a continuous sequence of artifacts corresponding to the succession of cultures identified through the Olduvai stratigraphy. Metrics and types were applied to categorize lithics, to provide relative

Tool Use in Animals: Cognition and Ecology, eds. Crickette M. Sanz, Josep Call and Christophe Boesch. Published by Cambridge University Press.

timelines and to make comparisons across industries/assemblages (mainly the Oldowan from Olduvai and KBS from Koobi Fora). In the 1980s lithic studies progressed, as artifacts needed to be understood and described in contexts, as they represent only one part of the hominin activity budget (Isaac, 1981). Isaac inspired a new research direction: integration of lithic studies into a paleoecological approach. This interdisciplinary approach aimed to get closer to site-formation processes and functions, and to understand ranging patterns in hominins (Isaac & Harris, 1997). Isaac (1984, 1986) also proposed an alternative terminology to describe the earliest stone assemblages. All artifacts were placed into three main categories (flaked, detached, pounded) with the aim of finding a simple alternative to terminologies based on post hoc, function-inferred labels (Leakey, 1971). Carbonell *et al.* (1983) refined this non-functional approach by categorizing artifacts according to their stage of production. The limitations of inferring behavior from archaeological data were soon recognized (Toth, 1985): choppers had been given their name because they were thought to be the end-product of a core reduction process, yielding the tool (Leakey, 1971). However, experimental studies showed that the real tools were instead the pieces of stone detached from the core (Toth, 1985). Thus, what Leakey had termed "debitage" (waste) was later seen to be the goal. The difficulties inherent in recognizing the intentions of early tool makers are highlighted in this dilemma: What was intended in this process – to produce the core tool, the flake, or both?

Since then, much has been achieved in the field of Oldowan lithic studies, especially in reconstructing an overall scenario of the early tool makers, but also in redefining the extension of the Oldowan in space (out of Africa) and time (earlier than 1.5 Ma) (Braun & Hovers, 2009; also see Hovers & Braun, 2009 for a complete review). Braun and Hovers (2009) argue that studies of the Oldowan have had cycles of progress and stagnation, but that after the 1990s a new dynamic period started. Recent research has extended knowledge about early hominin adaptations and ranging patterns, raw material selection and quality, technical and cognitive skills necessary for stone knapping, raw material transport and variability of assemblages (Plummer, 2004; Delagnes & Roche, 2005; Mora & de la Torre, 2005; Roux & Bril, 2005; Braun *et al.*, 2008, 2009, 2010; Stout *et al.*, 2008, 2010; Harmand, 2009).

The need to implement primate archaeology as a discipline is not new, but early efforts were few and non-empirical (e.g., Napier, 1962). Research that explicitly and specifically connects archaeology and primatology, especially on the origins of stone technology, dates to the 1980s (Wynn & McGrew, 1989), but has been developing more recently with multiple scientific approaches (McGrew, 1992, 2004; Joulian, 1996; Mercader *et al.*, 2002, 2007; Davidson & McGrew, 2005). Primatologists pioneered the application of archaeology to chimpanzee lithic technology in natural conditions by monitoring spatial movements within activity areas (nut-cracking sites) or by indirectly recording the transport of hammer stones (Boesch & Boesch, 1984; Sakura & Matsuzawa, 1991). Simultaneously, archaeologists carried out a ground-breaking study to see if apes in captivity could learn how to make chipped tools in order to compare the results with Oldowan morphotypes (Toth & Schick, 1993; Schick *et al.*, 1999). Kanzi, a bonobo (*Pan paniscus*) successfully made stone tools, even if these were ultimately differentiable from the Oldowan artifacts. The Kanzi study showed that the bonobo, a non-technical species in the wild, had the necessary cognitive flexibility and motor coordination to solve a problem that requires the making and using of a stone tool. This research also suggests that during learning, innovation and idiosyncrasy may have played crucial roles in the emergence and variability of technology. Kanzi was shown how to knap by human demonstrators, but he developed his own technique of throwing stones at the substrate, in order to produce sharp-edged implements. However, this and other captive studies are constrained by the impossibility of accounting for the role that environmental variables or other natural processes may have played in triggering the emergence and complexification of technology.

More recently, the first archaeological study at an archaic non-human primate site was done by Mercader *et al.* (2002, 2007) in the Taï forest. Excavations at three nut-cracking sites revealed an LSA (Later Stone Age) for chimpanzee stone tool use. Radiometric dating by ^{14}C indicated 4300-year antiquity, and analysis of starch residues confirmed the presence of five nut species on stone tools; all of these nuts are currently consumed by Taï chimpanzees (Boesch & Boesch, 1983). However, cultural variation in nut cracking across populations questions the degree to which we can generalize Mercader's findings to chimpanzees as a species.

To study how the observed differences across populations of chimpanzee nut cracking relate to archaeological findings, we applied archaeological theory and methods at Bossou, along with behavioral observations of the apes. In this chapter we start by giving an overview comparison of population variation in nut cracking, based on ethological data. Then we compare the findings of Mercader and colleagues at Taï with our results at Bossou, with the goal of improving our interpretation of archaeological findings and our understanding of hominin evolution.

Nut cracking by wild chimpanzees: Bossou, Taï and other sites in West Africa

Nut cracking by wild chimpanzees has long been known (Savage & Wyman, 1844) but has been reported only in West African populations. Since 1976 systematic research at Bossou, Guinea has continuously studied nut cracking in both natural and experimental conditions, the latter in the so-called "outdoor laboratory" in Bossou forest (Matsuzawa, 1994). This long-term project has contributed important insights for the understanding of tool use in wild chimpanzees, especially for the ontogeny of behavior, social learning and transmission of traditions (Biro *et al.*, 2003, 2006, 2010). The hierarchical structure of behavior in tool use is shown in the long-term "master–apprenticeship" process that chimpanzees undergo in order to proficiently use stone tools to crack open nuts (Matsuzawa *et al.*, 2001; but see also Matsuzawa *et al.*, 2011 for a full review of nut-cracking studies at Bossou). In 2006 archaeological methods were added, using a hybrid approach that focused on behavioral observations (Carvalho *et al.*, 2008; Matsuzawa *et al.*, 2011). The Bossou apes offer a unique opportunity for investigating technological evolution, as they are the only population customarily to use portable stones as hammers and anvils (Sugiyama & Koman, 1979; Sugiyama, 1997; see Figure 11.1).

Figure 11.1 An example of two movable stone tools (anvils and hammer) selected by the chimpanzees of Bossou to crack open nuts during one experimental session in 2009.

The physical properties of objects constrain the emergence of technological innovation: embedded outcrops do not allow the use of insertable wedges; boulders are unlikely to fracture and so to produce by-products to be reused and transported; log or fallen tree-trunk anvils do not yield serendipitous flakes that could be employed in different tasks, etc. Such limited affordances may curtail the rate of innovation in populations of wild chimpanzees using stone tools (see van Schaik *et al.*, 2006 for rates of invention in populations of orangutans).

On the other hand, many of the data from Taï, the other chimpanzee field site where nut cracking has been a long-term study, came from chimpanzees who were not fully habituated to observers (Boesch & Boesch, 1983, 1984, 1990. See Figure 11.2 for a summary of diversity in nut cracking across sites. This made difficult the accurate behavioral recording of known individuals, especially for direct transport, learning processes, reuse of tools or variation in site use.

The following summarizes the nut-cracking behavioral variety recorded from nine West African chimpanzee populations (see Figure 11.2, based on the original diagram in Whiten *et al.*, 1999).

Guinea: Bossou and Diecké

At Bossou, chimpanzees (*Pan troglodytes verus*) crack open oil-palm nuts (*Elaeis guineensis*) year round, but consumption peaks when other wild fruits are scarce (Yamakoshi, 1998). This contrasts with Taï, where nuts are a staple food of chimpanzees, and 12–15% of their feeding time is spent cracking nuts (Yamakoshi, 2001). Bossou's apes use movable stones as hammers and anvils and consume only one species of nut (Sugiyama & Koman, 1979). They do not practice customary arboreal nut cracking, although it has been seen twice (by immatures when the terrestrial nut-cracking site on the ground was crowded with adults; Matsuzawa *et al.*, 2011). Occasionally, the chimpanzees observed use one or two wedge-stones to stabilize and to level the anvil's working surface (Matsuzawa, 1994; Carvalho *et al.*, 2008). The available

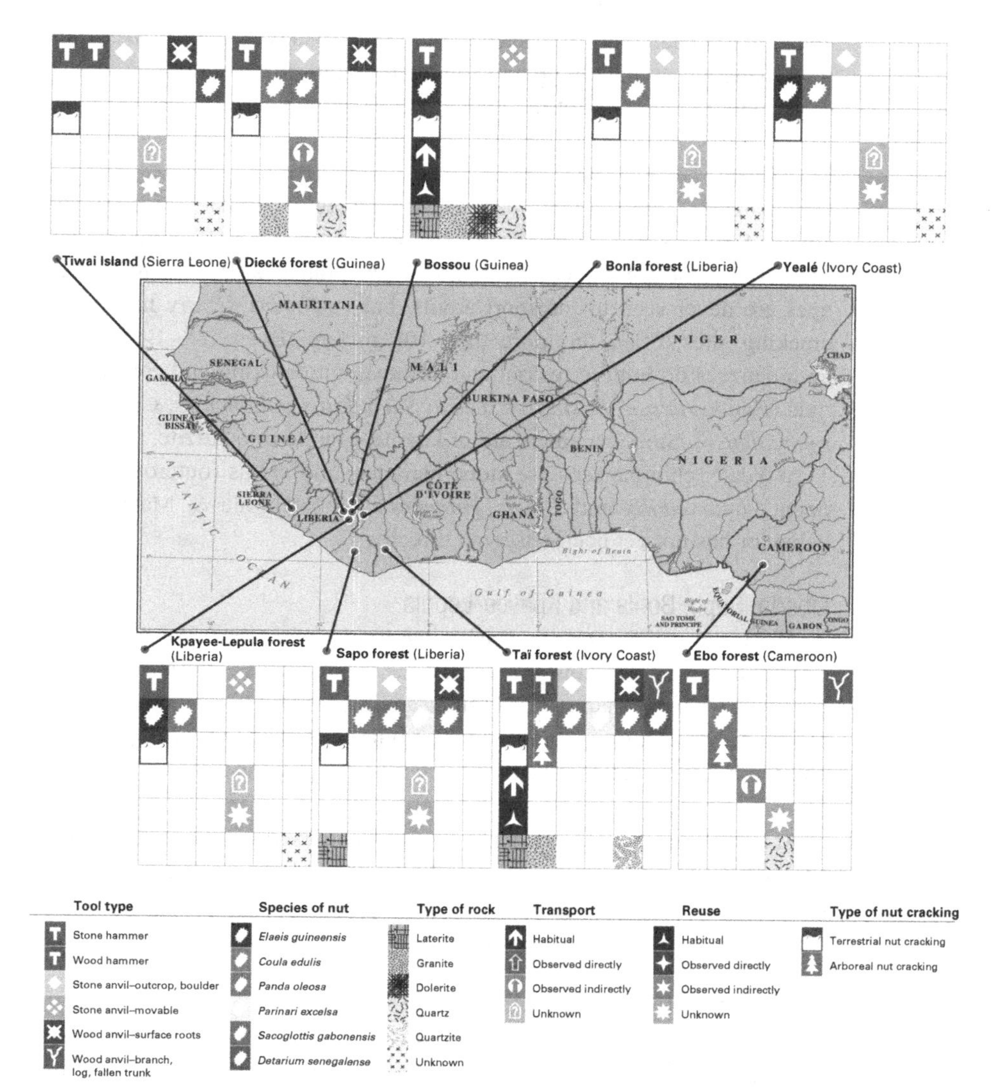

Figure 11.2 Diversity in nut cracking across the nine chimpanzee populations where this extractive technology was directly or indirectly recorded (design based on Whiten *et al.*, 1999): 1 – Tiwai Island, Sierra Leone (Whitesides, 1985); 2 – Diecké forest, Guinea (Carvalho *et al.*, 2007); 3 – Bossou forest, Guinea (Sugiyama & Koman, 1979; Carvalho *et al.*, 2008; Matsuzawa et al., 2011); 4 – Bonla forest, Liberia (Ohashi, 2011); 5 – Yealé (Nimba Mountains), Ivory Coast (Humle & Matsuzawa, 2001); 6 – Kpayee-Lepula forest, Liberia (Ohashi, 2011); 7 – Sapo forest, Liberia (Anderson *et al.*, 1983); 8 – Taï forest, Ivory Coast (Struhsaker & Hunkeler, 1971; Boesch, 1978); 9 – Ebo forest, Cameroon (Morgan & Abwe, 2006). From Carvalho and McGrew, 2012.

raw materials are laterite, granite, dolerite and quartz. Transport and reuse of tools is seen to be customary.

At Diecké, chimpanzees use movable stone hammers; stone outcrops, and on occasion root surfaces, serve as anvils (Carvalho *et al.*, 2007; Carvalho, 2011a). No arboreal nut cracking has been seen. Two species of nut are cracked: *Panda oleosa* and *Coula edulis*. Raw materials are mainly granite and quartz. Transport and reuse of tools was indirectly confirmed through monitoring known nut-cracking sites.

Ivory Coast: Taï and Yealé

Chimpanzees (*Pan troglodytes verus*) use movable hammers (wood or stone) and fixed anvils (Struhsaker & Hunkeler, 1971; Boesch, 1978). Of the anvils, 97% are surface roots, but some are embedded stones or tree branches (Boesch & Boesch, 1983). The apes are never seen to transport anvils, but customarily carry hammers between nut-cracking sites, often for more than 100 meters (Boesch & Boesch, 1984). At Taï, chimpanzees do both terrestrial and arboreal nut cracking and consume five species of nuts (*Panda oleosa*, *Parinari excelsa*, *Sacoglottis gabonensis*, *Coula edulis*, *Detarium senegalense*). Stone hammers are of laterite, granite or quartzite.

At Yealé, chimpanzees use stone hammers and anvils (outcrop or boulder) to crack open *Coula edulis* and *Elaeis guineensis* nuts (Humle & Matsuzawa, 2001). This behavior has not yet been observed directly.

Liberia: Sapo, Bonla and Kpayee-Lepula

At Sapo, forest-living chimpanzees (*Pan troglodytes verus*) use stone hammers to crack open four species of nuts (*Panda oleosa*, *Coula edulis*, *Parinari excelsa*, *Sacoglottis gabonensis*), using mainly (71%) stone anvils but also (29%) root anvils (Anderson *et al.*, 1983). As at Diecké, most of the stone anvils are embedded rocks. Nut-cracking sites are close to nut trees. Laterite hammers are reported, but the transport and reuse of the tools has not been seen.

At Bonla, chimpanzees use stone hammers and fixed stone anvils (outcrop or boulder) to get the kernels of *Coula edulis* nuts (Ohashi, 2011). Recent surveys reported remnant signs of terrestrial nut cracking. The Kpayee-Lepula forest revealed patterns of nut cracking identical to Bonla forest, but chimpanzees there also consume *Elaeis guineensis* nuts.

Sierra Leone: Tiwai

Chimpanzees at Tiwai (*Pan troglodytes verus*) seem to use both stone and root surfaces as fixed anvils, and stones and wooden clubs as hammers to open *Detarium senegalense* nuts (Whitesides, 1985). Only terrestrial nut cracking is mentioned. Nut-cracking sites are near nut trees and transport is suggested to be absent, but all data are indirect.

Cameroon: Ebo

At Ebo forest, the discovery of nut cracking in Cameroon challenges the *ecological frontier* model (i.e., absence of nut cracking east of the N'Zo-Sassandra River), based on riverine barriers to the cultural diffusion of nut-cracking behavior. Chimpanzees (*Pan troglodytes ellioti*, formerly *P. t. vellerosus*) were seen using stone hammers to crack open the nuts of *Coula edulis* (Morgan & Abwe, 2006). Only arboreal nut cracking with quartz hammers was reported, at about 5–8 m above the ground. *Pan troglodytes ellioti* is the least studied and most endangered subspecies of chimpanzee (Kormos *et al.*, 2003).

Wrangham (2006) asserted that the existence of a "culture-zone" is now uncertain, but populations that once lived between West and Central Africa may have gone extinct, severing the chain of transmission (Wrangham, 2006).

Methodology

Why chimpanzee archaeology needs to compare hammers with hammers

Previous research (Mercader *et al.*, 2002) comparing chimpanzee and Oldowan stone assemblages suggested parallels between the end-products of chimpanzee percussion and of hominin flaking activities. The earliest Oldowan knappers, at 2.6 Ma, produced flakes using a variety of strategies to reduce cores (Semaw *et al.*, 2003; Delagnes & Roche, 2005). Wild chimpanzees have not been seen to flake stone tools, and we lack evidence from chimpanzee assemblages that signal goal-directed reduction sequences. Recently, de la Torre and Mora (2009: 49) raised methodological concerns, asserting that several studies try to compare the incomparable: percussion processes versus flake making. The latter process "always entails one more step" through the production of an object, from a previous one, to be used afterwards (de la Torre and Mora, 2009: 49). We agree in part with this view. Chimpanzee percussion and hominin flaking technologies allow only limited direct comparison. If the aim is to gain new insights into the origins of technology, the logical comparison is of similar technological processes, such as the percussive technology of chimpanzees and hominins. On the other hand, little is known about the first steps that led to stone flaking. One explanatory hypothesis (Mora & de la Torre, 2005; Carvalho *et al.*, 2009) proposes that continuous reuse of the same tool composites (hammers and anvils) leads to an accumulation of unintentional detachments of stone pieces during such battering. This hypothesis has emerged from analyses of percussion tools from both hominins and chimpanzees, confirming the need to further understand unintentional reduction processes. Furthermore, wild chimpanzees have been seen to detach flakes from anvils that were then employed as hammers, thus showing that their percussive technology can entail one or more steps, if we include the use of wedges to stabilize stone anvils (Carvalho *et al.*, 2008). Hence, instead of focusing current research on comparing the forms of end-products or by-products, assuming that these may be the best indicators of differential primate cognitive complexity and capacity for motor coordination, it may be equally important to comprehend how and why different contexts can lead to typological and technological variability in simple percussive technologies, like the ones currently used by wild chimpanzees. Opportunities for innovation, raw material constraints, environmentally imposed variation, different strategies to exploit available resources, or number of individuals accessing the resources are some of the variables that may have triggered the emergence of technology in human and non-human primates.

Chimpanzee archaeology combining archaeological methods with direct and indirect observations of individuals and their assemblages has been ongoing in Guinea since 2006 (Carvalho *et al.*, 2008, 2009), with the objective of understanding the natural behavioral patterns and contexts that generate the artifacts. Experimental studies conducted in the forests of Bossou (that is, experimentation in the natural context of free-living individuals) are based on the principle that the lithic technique may be simple, but the strategies of selection, transport and exploitation of raw materials by several individuals can be complex (Toth [1985] drew similar conclusions for the Oldowan). Direct observation of tool functions and processes of tool use and tool making adds other advantages: Primatologists define categories of percussive stone tools by naming the type of food consumed combined with a precise functional description – nut crack (Sugiyama, 1997), baobab smash (Marchant & McGrew, 2005), *Treculia* cleave (Koops *et al.*, 2009) – an approach that cannot be repeated with the archaeological record. Primatologists can record how tool production is organized, the social contexts in which it occurs (see Chapter 8) and how individuals interact with their ecological contexts.

Results and discussion

African tropical rainforests are thought to be difficult ecotypes by most archaeologists (Bailey *et al.*, 1989), but they contain important data to help understand primate technology in eastern Africa (Mercader, 2003). However, most previous studies have not capitalized on the main advantage of combining archaeology with behavioral primatology: being able to observe directly and simultaneously behavior and its products, in order to compare these complementary data. This combination of indirect and direct data enables cross-validation and bridges the knowledge gaps that hinder our attempts to infer past acts (Carvalho, 2011b).

Mercader *et al.* (2002, 2007) showed that classical archaeological methods and techniques can be applied to sites and assemblages created by non-human primates. Furthermore, they raised the issue that different primates may leave different signatures in the archaeological record, and that it *may* be possible to recognize and define these signatures, that is, discriminate between systematic flaking versus thrusting percussion. However, if the Taï archaeological data (Mercader *et al.*, 2007) are re-examined in light of recent data collected at Bossou using direct and indirect measures of chimpanzee stone-tool use in different ecological settings (Carvalho *et al.*, 2007, 2008, 2009; Biro *et al.*, 2010; Carvalho, 2011b), then some of Mercader *et al.*'s (2007) conclusions are debatable:

(1) Most artifacts (the granitoid specimens) at Taï previously recognized as products of thrusting percussion by chimpanzees are of raw materials that are poor for knapping. Objects resulting from systematic flaking by humans are made of quartz or quartzite stones – that is, rocks that fracture conchoidally. The nature and quality of the lithic raw material may affect the identification of different types of percussion.
(2) Mercader *et al.* (2007: 3046) wrote that review of hammer metrics from early hominin sites indicates that the hominin hand accommodates hammer stones up to 12 cm (maximum length), but usually about 8 cm. Hominin hammers weigh less than 400 g and even anvils, the largest component of this bimodal technology, on average, weigh up to only 1000 g, but often much less. Depending on the nature of the food item, chimpanzees use large hammer stones that do not fit readily in their hands

(Boesch & Boesch, 1983; Carvalho *et al.*, 2008). Instead, the size of their hammers depends strongly on the nature of the food item being processed.

(3) Mercader's comparative measures for tools included only Taï forest data, while excluding findings from long-term research at Bossou. Bossou's chimpanzees use tools with very similar characteristics to *hominin* hammers, which differ from the values provided for the so-called *chimpanzee hammers* (see Table 11.1).

(4) Chimpanzee nut cracking is more than a *bimodal technology* of hammer and anvil. When the raw material permits, this technology combines the simultaneous use of three or four stones to construct the same nut cracker. Thus, Bossou's tool composites (McGrew, 2010) for nut cracking are more complex than Taï's.

(5) "This marked difference (between measures for hominins and chimpanzees) probably is related to the larger size of the chimpanzee hand, its morphology" (Mercader *et al.*, 2007: 3046): Two data sets on nut cracking, from Bossou and Diecké, thus different sites with different nut species but with same-sized hands, yielded distinct values for the tools used to crack open oil-palm nuts at Bossou and panda nuts at Diecké. This marked difference arose from the types of raw material available and from the nature of the food being processed (Carvalho *et al.*, 2008), not from hand size.

(6) When portable tools are available and used, as at Bossou, the products generated by bashing technologies are reused with new functions, and thus are "recycled." Thus, among collected artifacts, a "bashing product" may be an unrecognized wedge, for instance, so pounding technologies with non-portable elements, as at Taï, may require different analyses.

(7) "Stone pieces unintentionally generated by thrusting percussion and stone pieces intentionally produced by systematic flaking can be differentiated" (Mercader *et al.*, 2007: 3046). At Bossou, stone pieces unintentionally generated by thrusting percussion are reused, thereby changing their original function. To identify the function of such chimpanzee tools, it may not be enough to recognize types of percussion. It may be necessary to combine use-wear analysis, refitting and direct observation.

(8) Use-wear patterns from both human and non-human nut cracking also include small pitting and depressions (hollow concavities), both now (Carvalho *et al.*, 2007) and in prehistory (Goren-Inbar *et al.*, 2002), so presence/absence data are not sufficiently discriminatory.

(9) Any claim that chimpanzees select raw materials (such as granitoids) requires data on the relative availability of raw materials in the areas of use and access. If granitoid rocks are the most commonly found, then their frequent use may be random, not preferential.

(10) Environmental constraints and local traditions can create technological and typological differences between different nut-cracking tools. Generalizing across sizes and morphotypes of hominin and chimpanzee percussive technologies may gloss over the existence of technological and typological variation within and between assemblages.

Past research (Carvalho *et al.*, 2007, 2008; Carvalho, 2011b) shows that by monitoring chimpanzee nut-cracking sites in both the absence and presence of individuals, one can analyze the spatial distribution of artifacts on a daily basis and over an extended period

Table 11.1 Mean values for hammers and anvils at Bossou sites in later study (Carvalho *et al.*, 2008) and in early study (Sakura & Matsuzawa, 1991).

Site		Length (cm)		Width (cm)		Height (cm)		Weight (kg)	
		Hammer	Anvil	Hammer	Anvil	Hammer	Anvil	Hammer	Anvil
Bossou lab	Outdoor	11.91 ± 1.90	17.20 ± 3.16	7.68 ± 1.49	13.87 ± 1.14	5.53 ± 1.48	9.00 ± 2.45	0.68 ± 0.21	3.2 ± 1.4
Bossou SA13	Moblim	8.41 ± 1.51	11.75 ± 2.89	13.92 ± 5.35	16.24 ± 3.88	6.62 ± 1.99	6.30 ± 2.49	1.38 ± 0.79	2.1 ± 0.78
Bossou (1991)	Forest	12.0 ± 2.65	16.1 ± 6.47	7.7 ± 1.86	11.1 ± 4.38	5.2 ± 1.08	7.3 ± 2.97	0.7 ± 0.33	2.2 ± 2.21

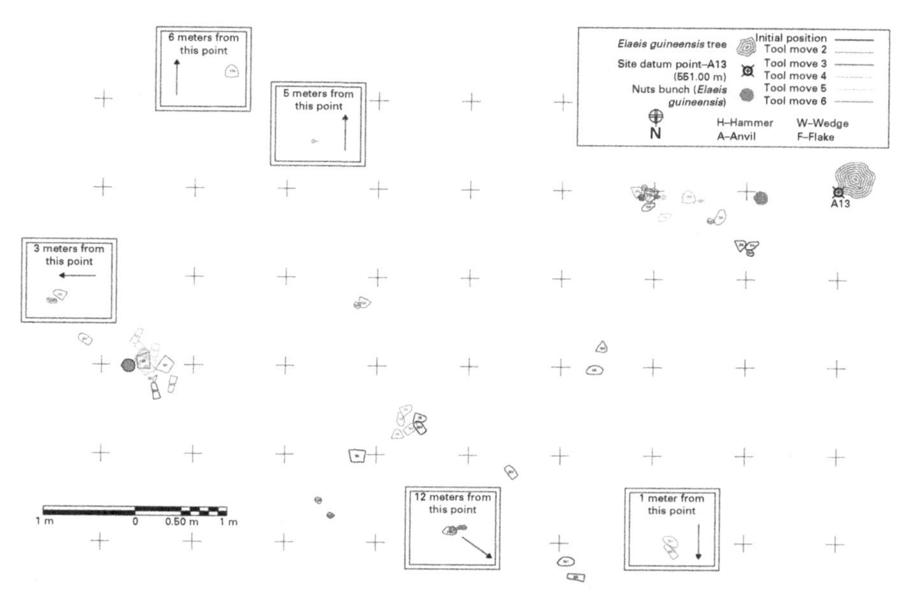

Figure 11.3 Spatial distribution of tools at a Bossou nut-cracking site, during six visits in 2006.

(Figure 11.3). This precise temporal scale is impossible to achieve in the archaeological record. The analysis of daily records, both in experimental conditions and at natural nut-cracking sites, reveals repeated patterns of exploitation of the available resources (nuts and stones). Measurements of tool characteristics, combined with direct observation of individuals practicing nut cracking, suggest that chimpanzees: discriminate tool function based on tool dimensions; reuse fractured tools if they fall within the mean values for hammers and anvils; and discard tools when they stop being functional. The variation in these stone tools appears to relate to the need to process different nut species, and so chimpanzees select different stone tools to pound variants of the same food type (nuts) in different environmental contexts.

Transport of tools is customary within nut-cracking sites and during stone-tool use, which leads to the question: Why is transport being done during percussive sessions when apparently it is not necessary? By directly observing nut cracking we confirmed that detachment of stone pieces occurs incidentally during the pounding processes. To further understand if chimpanzees systematically reuse particular tool composites (which could help explain the unintentional detachment of pieces during percussion processes), we devised an experiment. Results show that chimpanzees have both individual- and population-level preferences for particular tool composites (same hammer combined with same anvil) that are repeatedly reused (Carvalho *et al.*, 2009).

We also described systematic sequences of behavior during nut cracking, applying the concept of *chaîne operatóire* (Carvalho *et al.*, 2008). These sequences were flexible in some actions, but were stable in their main categories (e.g., selection of raw material, transport, discard). This allows description of the full behavioral process that starts with acquisition of raw materials and ends with the abandonment of tools, so leading to the formation of a true archaeological site. *Chaîne operatóire* applied to chimpanzee nut cracking does *not* equal analysis of lithic reduction. The *chaîne operatóire* framework was originally introduced in ethnography to describe behavioral sequences in human groups (Mauss, 1967), then borrowed by archaeology to infer behavioral sequences from reduction strategies (Pelegrin *et al.*, 1988). In other words, reconstitution of the technical processes of stone-tool making was used as an inferential measure of behavioral and cognitive complexity in early hominins. In the chimpanzee archaeology approach, the aim was not to study reduction strategies during stone-tool use, but instead to understand the complexity and flexibility of the nut-cracking behavior, and to describe how a chimpanzee nut-cracking site is transformed into an archaeological site. We applied *chaîne operatóire* side by side with standard ethological methods to measure behavior in the wild (Martin & Bateson, 2007).

Conclusion

Archaeological methods and concepts can be applied to enhance our primatological knowledge of chimpanzee material culture. This bridging of the disciplines should be symbiotic and help archaeology in modeling the evolutionary origins of technology in human and non-human primates.

Continuing research

Current research aims to expand preliminary studies, as many questions remain unanswered, such as: (1) the role and influence of raw material types and raw material availability in chimpanzee stone industry; and (2) definition of a chimpanzee signature in the archaeological record, both to help identify future archaeological sites and potentially to draw typological-technological comparisons across the percussive technologies of different species of primates. The latest data collected in 2008–2009 included: (1) field experimentation with stone tools and introduced new raw materials (basalt from Kenya, flint from Portugal); (2) monitoring and increasing the number of sites sampled in Bossou forest, including a direct record of sites being used and reused, with mapping of tool movements; (3) description of the technological assemblages; (4) measurement of raw material availability in three different Guinean forests using transect lines and quadrat sampling; (5) excavation of a chimpanzee nut-cracking site; and (6) application of archaeological techniques (radiometric dating, residue and use-wear analysis, 3D scanning) to a sample of chimpanzee stone assemblages. All of these methods and approaches are already in use to varying extents in mainstream human archaeology, but extending them to ape archaeology is an increasingly exciting and productive prospect.

Acknowledgments

We thank the Direction National de la Recherche Scientifique et Technique, République de Guinée for permission to conduct field work at Bossou. The research was supported by

Grants-in-Aid for scientific research from the Ministry of Education, Science, Sports, and Culture of Japan: MEXT-20002001 and 24000001, JSPS-HOPE, JSPS-gCOE-Biodiversity (A06), and F-06-61 of the Ministry of Environment, Japan to Tetsuro Matsuzawa. We thank B. Zogbila, H. Camara, J. Doré, P. Goumy, M. Doré, P. Cherif, J. M. Kolié, J. Malamu, L. Tokpa, A. Kbokmo, Cé Koti, O. Mamy and F. Mamy for essential field support. SC thanks C. Sanz, C. Boesch and J. Call for the invitation to participate in the 2009 conference. We thank K. Hockings and V. Carvalho for helpful comments on the manuscript. We are extremely grateful to V. Batista, P. Fuentez and P. Gonçalves for research support. SC was supported by Cambridge European Trust (RIB00107), Fundação para a Ciência e Tecnologia (SFRH/BD/36169/2007), and Wenner-Gren Foundation for Anthropological Research; WCM was supported by National Science Foundation, Researching Hominid Origins Initiative, grant to T.D. White and F.C. Howell.

References

Ambrose, S. (2001). Palaeolithic technology and human evolution. *Science*, **291**, 1748–1753.

Anderson, J. R., Williamson, E. A. & Carter, J. (1983). Chimpanzees of Sapo Forest, Liberia: density, nests, tools and meat-eating. *Primates* **4**, 594–601.

Bailey, R. C., Head, G., Jenike, M., *et al.* (1989). Hunting and gathering in tropical rain forest: is it possible? *American Anthropologist*, **91**, 59–82.

Biro, D., Inoue-Nakamura, N., Tonooka, R., *et al.* (2003). Cultural innovation and transmission of tool use in wild chimpanzees: evidence from field experiments. *Animal Cognition*, **6**, 213–223.

Biro, D., Sousa, C. & Matsuzawa, T. (2006). Ontogeny and cultural propagation of tool use by wild chimpanzees at Bossou, Guinea: case studies in nut cracking and leaf folding. In T. Matsuzawa, M. Tomonaga & M. Tanaka (eds.) *Cognitive Development in Chimpanzees* (pp. 476–508). Tokyo: Springer.

Biro, D., Carvalho, S. & Matsuzawa, T. (2010). Tools, traditions and technologies: interdisciplinary approaches to chimpanzee nut-cracking. In E. V. Lonsdorf, S. R. Ross & T. Matsuzawa (eds.) *The Mind of the Chimpanzee: Ecological and Experimental Perspectives* (pp. 128–141). Chicago, IL: University of Chicago Press.

Boesch, C. (1978). Nouvelles observations sur les chimpanzés de la forêt de Taï (Côte d'Ivoire). *Terre et Vie*, **32**, 195–201.

Boesch, C. & Boesch, H. (1983). Optimization of nut-cracking with natural hammers by wild chimpanzees. *Behaviour*, **83**, 265–286.

Boesch, C. & Boesch, H. (1984). Mental maps in wild chimpanzees: an analysis of hammer transports for nut-cracking. *Primates*, **25**, 160–170.

Boesch, C. & Boesch, H. (1990). Tool use and tool making in wild chimpanzees. *Folia Primatologica*, **54**, 86–99.

Braun, D. R. & Hovers, E. (2009). Current issues in Oldowan research. In E. Hovers & D. R. Braun (eds.) *Interdisciplinary Approaches to the Oldowan* (pp. 1–14). Dordrecht: Springer.

Braun, D. R., Plummer, T., Ditchfield, P. W., *et al.* (2008). Oldowan behavior and raw material transport: perspectives from the Kanjera Formation. *Journal of Archaeological Science*, **35**, 2329–2345.

Braun, D. R., Plummer, T., Ferraro, J. V., *et al.* (2009). Raw material quality and Oldowan hominin toolstone preferences: evidence from Kanjera South, Kenya. *Journal of Archaeological Science*, **36**, 1605–1614.

Braun, D. R., Harris, J. W. K., Levinc, N. E., *et al.* (2010). Early hominin diet included diverse terrestrial and aquatic animals 1.95 Ma in East Turkana, Kenya. *Proceedings of the National Academy of Sciences USA*, **107**(22), 10002–10007.

Carbonell, E., Guilbaud, M. & Mora, R. (1983). Utilizacíon de la lógica analítica para el estudio de tecno-complejos a cantos tallados. *Cahier Noir*, **1**, 3–64.

Carvalho, S. (2011a). Dieckė Forest, Guinea: delving into chimpanzee behavior using stone tool surveys. In T. Matsuzawa, T. Humle & Y. Sugiyama (eds.) *The Chimpanzees of Bossou and Nimba* (pp. 301–312). Tokyo: Springer,.

Carvalho, S. (2011b). Extensive surveys of chimpanzee stone tools: from the telescope to the magnifying glass. In T. Matsuzawa, T. Humle & Y. Sugiyama (eds.) *The Chimpanzees of Bossou and Nimba* (pp. 145–155). Tokyo: Springer.

Carvalho, S. & McGrew, W. C. (2012). The origins of the Oldowan: why chimpanzees (*Pan troglodytes*) are still good models for technological evolution in Africa. In M. Domínguez-Rodrigo (ed.) *Stone Tools and Fossil Bones: Debates in the Archaeology of Human Origins* (pp. 201–221). Cambridge University Press.

Carvalho, S., Sousa, C. & Matsuzawa, T. (2007). New nut-cracking sites in Dieckė forest, Guinea: an overview of the etho-archaeological surveys. *Pan African News*, **14**, 11–13.

Carvalho, S., Cunha, E., Sousa, C. & Matsuzawa, T. (2008). *Chaînes opératoires* and resource exploitation strategies in chimpanzee nut-cracking (*Pan troglodytes*). *Journal of Human Evolution*, **55**, 148–163.

Carvalho, S., Biro, D., McGrew, W. C. & Matsuzawa, T. (2009). Tool-composite reuse in wild chimpanzees (*Pan troglodytes*): archaeologically invisible steps in the technological evolution of early hominins? *Animal Cognition*, **12**, 103–114.

Davidson, I. & McGrew, W. C. (2005). Stone tools and the uniqueness of human culture. *Journal of the Royal Anthropological Institute*, **11**, 793–817.

de la Torre, I. & Mora, R. (2009). Remarks on the current theoretical and methodological approaches to the study of early technological strategies in East Africa. In E. Hovers & D. R. Braun (eds.) *Interdisciplinary Approaches to the Oldowan* (pp. 15–24). Dordrecht: Springer.

Delagnes, A. & Roche, H. (2005). Late Pliocene hominid knapping skills: the case of Lokalalei 2C, West Turkana, Kenya. *Journal of Human Evolution*, **48**, 435–472.

Goren-Inbar, N., Sharon, G., Melamed, Y., *et al.* (2002). Nuts, nutcracking, and pitted stones at Gesher Benot Ya'aqov, Israel. *Proceedings of the National Academy of Sciences USA*, **99**, 2455–2460.

Gumert, M. D., Kluck, M. & Malaivijitnond, S. (2009). The physical characteristics and usage patterns of stone axe and pounding hammers used by long-tailed macaques in the Andaman Sea region of Thailand. *American Journal of Primatology*, **71**, 594–608.

Harmand, S. (2009). Variability in raw material selectivity at the late Pliocene sites of Lokalalei, West Turkana, Kenya. In E. Hovers & D. R. Braun (eds.) *Interdisciplinary Approaches to the Oldowan* (pp. 85–97). Dordrecht: Springer,.

Hovers E. & Braun D. R. (eds.) (2009). *Interdisciplinary Approaches to the Oldowan*. Dordrecht: Springer.

Humle, T. & Matsuzawa, T. (2001). Behavioural diversity among the wild chimpanzee populations of Bossou and neighbouring areas, Guinea and Cote d'Ivoire, West Africa: a preliminary report. *Folia Primatologica*, **72**, 57–68.

Isaac, G. L. (1981). Archaeological tests of alternative models of early hominid behaviour: excavation and experiments. *Philosophical Transactions of the Royal Society of London B*, **1057**, 177–188.

Isaac, G. L. (1984). The archaeology of human origins: studies of the Lower Pleistocene in East Africa. In F. Wendorf & A. Close (eds.) *Advances in World Archaeology* (pp. 1–87). New York: Academic Press.

Isaac, G. L. (1986). Foundation stones: early artifacts as indicators of activities and abilities. In G. N. Bailey & P. Callow (eds.) *Stone Age Prehistory* (pp. 221–241). Cambridge: Cambridge University Press.

Isaac, G. L. & Harris, J. W. K. (1997). The stone artifact assemblages: a comparative study. In G. L. Isaac (ed.) *Koobi Fora Research Project: Plio-Pleistocene Archaeology* (pp. 262–362). Oxford: Clarendon Press.

Joulian, F. (1996). Comparing chimpanzee and early hominid techniques: some contributions to cultural and cognitive questions. In P. Mellars & K. Gibson (eds.) *Modelling the Early Human Mind* (pp. 173–189). Cambridge: McDonald Institute Monographs.

Koops, K., McGrew, W. C. & Matsuzawa, T. (2009). Do chimpanzees (*Pan troglodytes*) use cleavers and anvils to fracture *Treculia africana* fruits? Preliminary data on a new form of percussive technology. *Primates*, **51**, 175–178.

Kormos, R., Humle, T., Brugiere, D., *et al.* (2003). The Republic of Guinea. In R. Kormos, C. Boesch, M. I. Bakarr, *et al.* (eds.) *Status Survey and Conservation Action Plan for West African Chimpanzees* (pp. 63–76). Gland, Switzerland and Cambridge: IUCN/SSC Primate Specialist Group.

Leakey, M. D. (1971). Olduvai Gorge. Vol. **3**, *Excavations in Beds I and II, 1960–1963.* Cambridge: Cambridge University Press.

Marchant, L. F. & McGrew, W. C. (2005). Percussive technology: chimpanzee baobab smashing and the evolutionary modeling of hominin knapping. In V. Roux & B. Bril (eds.) *Stone Knapping: The Necessary Conditions for a Uniquely Hominin Behaviour* (pp. 341–350). Cambridge: McDonald Institute for Archaeological Research.

Martin, P. & Bateson, P. (2007). *Measuring Behaviour: An Introductory Guide*. 3rd edn. Cambridge: Cambridge University Press.

Matsuzawa, T. (1994). Field experiments on the use of stone tools by chimpanzees in the wild. In R. W. Wrangham, W. C. McGrew, F. B. M. de Waal & P. G. Heltne (eds.) *Chimpanzee Cultures* (pp. 351–370). Cambridge, MA: Harvard University Press.

Matsuzawa, T., Biro, D., Humle, T., *et al.* (2001). Emergence of culture in wild chimpanzees: education by master-apprenticeship. In T. Matsuzawa (ed.) *Primate Origins of Human Cognition and Behavior* (pp. 557–574). Tokyo: Springer.

Matsuzawa, T., Humle, T. & Sugiyama, Y. (eds.) (2011). *The Chimpanzees of Bossou and Nimba.* Tokyo: Springer.

Mauss, M. (1967). *Manuel d' Etnographie*. Paris: Editions Payot.

McGrew, W. C. (1992). *Chimpanzee Material Culture: Implications for Human Evolution.* Cambridge: Cambridge University Press.

McGrew, W. C. (2004). *The Cultured Chimpanzee: Reflections on Cultural Primatology.* Cambridge: Cambridge University Press.

McGrew, W. C. (2010). Chimpanzee technology. *Science*, **328**, 579–580.

Mercader, J. (ed.) (2003). *Under the Canopy: The Archaeology of Tropical Rain Forests*. New Brunswick, NJ: Rutgers University Press.

Mercader, J., Panger, M. & Boesch, C. (2002). Excavation of a chimpanzee stone tool site in the African rainforest. *Science*, **296**, 1452–1455.

Mercader, J., Barton, H., Gillespie, J., *et al.* (2007). 4,300-year-old chimpanzee sites and the origins of percussive stone technology. *Proceedings of the National Academy of Sciences USA*, **104**, 3043–3048.

Mora, R. & de la Torre, I. (2005). Percussion tools in Olduvai Beds I and II (Tanzania): implications for early human activities. *Journal of Anthropological Archaeology*, **24**, 179–192.

Morgan, B. J. & Abwe, E. E. (2006). Chimpanzees use stone hammers in Cameroon. *Current Biology*, **16**, R632–R633.

Napier, J. R. (1962). The evolution of the hand. *Scientific American*, **207**(6), 56–62.

Ohashi, G. (2011). From Bossou to Liberian forest. In T. Matsuzawa, T. Humle & Y. Sugiyama (eds.) *The Chimpanzees of Bossou and Nimba* (pp. 313–321). Tokyo: Springer.

Ottoni, E. B. and Izar, P. (2008). Capuchin monkey tool use: overview and implications. *Evolutionary Anthropology*, **17**, 171–178.

Pelegrin, J., Karlin, C. & Bodu, P. (1988). Chaînes opératoires: Un outil pour le pre′historien. In J. Tixier (ed.) *Technologie Pre′historique*. Paris: CNRS.

Plummer, T. (2004). Flaked stones and old bones: biological and cultural evolution at the dawn of technology. *Yearbook of Physical Anthropology*, **47**, 118–164.

Roux, V. & Bril, B. (eds.) (2005). *Stone Knapping: The Necessary Conditions for a Uniquely Hominin Behaviour*. Cambridge: McDonald Institute for Archaeological Research.

Sakura, O. & Matsuzawa, T. (1991). Flexibility of wild chimpanzee nut-cracking behavior using stone hammers and anvils: an experimental analysis. *Ethology*, **87**, 237–248.

Sanz, C. M., Schöning, C. & Morgan, D. B. (2010). Chimpanzees prey on army ants with specialized tool set. *American Journal of Primatology*, **72**, 17–24.

Savage, T. S. & Wyman, J. (1844). Observations on the external characters and habits of the *Troglodytes niger*, Geoff. and on its organization. *Boston Journal of Natural History*, **4**, 362–386.

Schick, K. D., Toth, N., Garufi, G., *et al.* (1999). Continuing investigations into the stone tool-making and tool-using capabilities of bonobo (*Pan paniscus*). *Journal of Archaeological Science*, **26**, 821–832.

Semaw, S., Rogers, M. J., Quade, J., *et al.* (2003). 2.6-million-year-old stone tools and associated bones from OGS-6 and OGS-7, Gona, Afar, Ethiopia. *Journal of Human Evolution*, **45**, 169–177.

Stout, D., Toth, N., Schick, K. D. & Chaminade, T. (2008). Neural correlates of early Stone Age toolmaking: technology, language and cognition in human evolution. *Philosophical Transactions of the Royal Society of London B*, **363**, 1939–1949.

Stout, D., Semaw, S., Rogers, M. J., *et al.* (2010). Technological variation in the earliest Oldowan from Gona, Afar, Ethiopia. *Journal of Human Evolution*, **58**, 474–491.

Struhsaker, T. T. & Hunkeler, P. (1971). Evidence of tool-using by chimpanzees of the Ivory Coast. *Folia Primatologica*, **15**, 212–219.

Sugiyama, Y. (1997). Social tradition and the use of tool-composites by wild chimpanzees. *Evolutionary Anthropology*, **6**, 23–27.

Sugiyama, Y. & Koman, J. (1979). Tool using and making behavior in wild chimpanzees at Bossou, Guinea. *Primates*, **20**, 323–339.

Toth, N. (1985). The Oldowan reassessed: a closer look at early stone artifacts. *Journal of Archaeological Sciences*, **12**, 101–120.

Toth, N. & Schick, K. (1993). Early stone industries and inferences regarding language and cognition. In K. R. Gibson & T. Ingold (eds.) *Tools, Language and Cognition in Human Evolution* (pp. 346–362). Cambridge: Cambridge University Press.

van Schaik, C. P., Ancrenaz, M., Borgen, G., *et al.* (2003). Orangutan cultures and the evolution of material culture. *Science*, **299**, 102–105.

van Schaik, C. P., van Noordwijk, M. A. & Wich, S. A. (2006). Innovation in wild Bornean orangutans (*Pongo pygmaeus wurmbii*). *Behaviour*, **143**, 839–876.

Whiten, A., Goodall, J., McGrew, W. C., *et al.* (1999). Cultures in chimpanzees. *Nature*, **399**, 682–685.

Whitesides, G. H. (1985). Nut-cracking by wild chimpanzees in Sierra Leone, West Africa. *Primates*, **26**, 91–94.

Wrangham, R. W. (2006). Chimpanzees: the culture-zone concept becomes untidy. *Current Biology*, **16**, R634–635.

Wynn, T. G. & McGrew, W. C. (1989). An ape's view of the Oldowan. *Man*, **24**, 383–398.

Yamakoshi, G. (1998). Dietary responses to fruit scarcity of wild chimpanzees at Bossou, Guinea: possible implications for ecological importance of tool use. *American Journal of Physical Anthropology*, **106**, 283–295.

Yamakoshi, G. (2001). Ecology of tool use in wild chimpanzees: towards reconstruction of early hominid evolution. In T. Matsuzawa (ed.) *Primate Origins of Human Cognition and Behavior* (pp. 537–556). Tokyo: Springer.

12 Early hominin social learning strategies underlying the use and production of bone and stone tools

Matthew V. Caruana

Bernard Price Institute for Palaeontological Research, School of Geosciences and Institute for Human Evolution, University of the Witwatersrand

Francesco d'Errico

University of Bordeaux, UMR-CNRS PACEA, Equipe Préhistoire, Paléoenvironnement, Patrimoine
Department of Archaeology, History, Cultural Studies and Religion, University of Bergen

Lucinda Backwell

Bernard Price Institute for Palaeontological Research, School of Geosciences and Institute for Human Evolution, University of the Witwatersrand

Introduction

Current trends in research toward the integration of primatological and archaeological models have provided significant insight into the emergence of tool use from a multidisciplinary perspective (e.g., Wynn & McGrew 1989; van Schaik *et al.*, 1999; Backwell & d'Errico, 2001, 2008, 2009; d'Errico *et al.*, 2001; Mercader *et al.*, 2002, 2007; van Schaik & Pradhan, 2003; Marzke, 2006; Lockwood *et al.*, 2007; Sanz & Morgan, 2007; Carvalho *et al.*, 2008, 2009; Gowlett, 2009; Haslam *et al.*, 2009; Hernandez-Aguilar, 2009; Uomini, 2009; Visalberghi *et al.*, 2009; Whiten *et al.*, 2009a; Chapter 11). Recently, this has culminated in the new "primate archaeology" subdiscipline (Haslam *et al.*, 2009), which has effectively modeled the advantages of incorporating comparative primatological research within the study of early hominin technologies. While this approach advances a unique perspective concerning the evolution of tool use and production, the predominantly ethological focus of primate archaeology has not fully benefited from exploring neuro-cognitive mechanisms in non-human primates and modern humans that might pertain to tool use in the deep past. Cognition remains a critical element in archaeological and paleoanthropological theories regarding the nature of early hominin technologies (e.g., Toth, 1985; Semaw, 2000; Delagnes & Roche, 2005; Stout *et al.*, 2008; Whiten *et al.*, 2009a). Thus, examining the cognitive capacities underlying tool use within the Order Primates is a critical pursuit toward understanding the social and cultural contexts of tool-mediated behavior, and the evolution of technology (van Schaik *et al.*, 1999; van Schaik & Pradhan, 2003; see also Chapters 2, 3 and 10). This chapter presents and explores various primatological perspectives concerning

Tool Use in Animals: Cognition and Ecology, eds. Crickette M. Sanz, Josep Call and Christophe Boesch.

tool use, and cognitive approaches regarding the emergence of technology within the hominin lineage. The infusion of cognitive perspectives within the primate archaeology framework is imperative for defining the biological, sociocultural and ecological contexts of tool use and production, thus enhancing its interpretive potential.

A pivotal issue within primatological research that lends itself toward the enrichment of archaeology and paleoanthropology has been the examination of social learning among non-human primates carried out by Tomasello, Byrne, Call, Whiten and colleagues (e.g., Tomasello *et al.*, 1987, 1993; Whiten & Ham, 1992; Tomasello, 1996; Byrne & Russon, 1998; Whiten, 1998, 2000; Byrne, 1999, 2003; Whiten *et al.*, 2004, 2005, 2007, 2009a, 2009b; Call *et al.*, 2005; Horner & Whiten, 2005; Hopper *et al.*, 2008; Tennie *et al.*, 2009) over the last two decades. In particular, the focus upon *emulation* versus *imitation* has engendered an intriguing debate concerning the nature of cultural transmission and diffusion of tool use (for a review see Whiten *et al.*, 2009b). As Whiten *et al.* (2009a) have recognized, this research has critical implications for cognitive and behavioral mechanisms underlying the emergence of technology among Plio-Pleistocene hominins. While they have provided a detailed analysis correlating the transmission of percussive technologies among non-human primate and early hominin populations, a vital question that arises from their research is the implication of comparing alternative tool types within this framework, such as purported bone tool collections from Early Pleistocene sites in southern and eastern Africa (Backwell & d'Errico, 2001, 2004; d'Errico & Backwell, 2003, 2009).

Backwell and d'Errico's (2001, 2003, 2004, 2005, 2008; d'Errico *et al.*, 2001; d'Errico & Backwell, 2003, 2009) interdisciplinary analyses of bone tools from Sterkfontein, Swartkrans and Drimolen in South Africa have demonstrated a uniformity of use and shaping techniques that primarily centered upon termite-mound foraging in the Early Pleistocene. The contrasts of this evidence with lithic technologies may be useful in distinguishing particular neuro-cognitive mechanisms that were central in the manifestation of social learning strategies supporting the cultural transmission of tool use within the deep past. Moreover, Backwell and d'Errico's (2004) examination of bone tools from Olduvai Gorge in East Africa appears to provide further information upon the significance of socioecology in the use of these tools.

The aim of this research is to further elucidate the interrelation of biological, behavioral, sociocultural and ecological contexts of early hominin tool use, and the implications concerning social learning strategies subserving the transmission of various Plio-Pleistocene technologies. In order to achieve this goal we will first summarize recent advances in the categorization of learning strategies, which will be followed by a comparative survey of cultural transmission mechanisms among non-human primate and modern human populations. This will provide a theoretical framework for investigating the possible social contexts of learning strategies underlying early hominin tool use. Next, contrasting Plio-Pleistocene tool types will be examined to highlight cognitive and behavioral capacities involved in their production and use, which will ultimately implicate the likelihood of particular social learning strategies that were critical to the transmission of early hominin technologies.

A categorical summary of learning strategies

Before an analysis of social learning among early hominins can be explored, it is necessary to outline the associated behavioral categories documented among living primates (non-human and human). This will potentially provide an analogous framework to outline possible cultural transmission mechanisms underlying tool use within Plio-Pleistocene hominin populations. To begin, it is important to first briefly consider vital learning strategies supporting the transmission of behavior that range outside of the "social learning" category. For instance, one of the most critical is *individual learning*, or the acquisition of novel behavior through the idiosyncratic processes of individual agents without direct external influences (Galef, 1988; Heyes, 1994; Whiten, 2000). While this strategy might not entail social interaction, it is the pivotal behavioral factor by which all innovations of tool use find their roots. Even within the bounds of socially structured learning environments, the manner in which individuals acquire tool-mediated behaviors through personal experimentation, play or error can redefine technological complexes through social diffusion processes. Ultimately, it is through such circumstances that individual learning can lead to the evolutionary threshold of cultural change. However, detecting the behavior of an individual within bone and stone assemblages that have accumulated over great expanses of time (millions of years in some cases) is extremely difficult, and most often unfeasible. As a result of this limitation, this discussion will restrict its focus to the possible social environments scaffolding learning strategies within the archaeological and paleoanthropological record.

Social learning strategies

Social learning can be generally defined as the processes by which an individual acquires novel behaviors through the observation of, or interaction with, other agents within a social setting (Galef, 1988; Heyes, 1994). However, this does not always involve the direct observation and mimicry of a conspecific's actions. In fact, Whiten and colleagues (Whiten & Ham, 1992; Whiten, 2000) have differentiated *social learning* from *social influence* strategies, which is important in defining the context of cultural transmission processes. For example, social influence includes strategies such as *exposure* in which agents come into contact with learning environments via their associations with other individuals (Whiten & Ham, 1992; Whiten, 2000). Furthermore, *social support* describes when an agent's motivation to learn and engage in novel behaviors is stimulated through the presence of associated individuals. While these scenarios may arise within socially structured environments, neither involves the active copying of conspecifics' actions, and instead are passively facilitated through the influence of other social agents.

Historically, social learning strategies have received substantial attention in a number of scientific disciplines, which has propagated a large body of literature detailing various behavioral and environmental contexts (for a review, see Galef, 1988). For the sake of brevity, four categories of social learning behaviors that have been widely discussed throughout the primatological literature will be outlined here (see e.g., Whiten & Ham, 1992; Whiten, 2000; Byrne, 2003). The first category is *stimulus enhancement*, where an agent's chances of interacting with a particular object or environmental condition are enhanced through observing a conspecific's interactions with it (Whiten, 2000; Byrne, 2003). Next is *response facilitation*, where an agent learns to respond to particular stimuli through observing another individual's reaction to it. While these behaviors remain distinct phenomena, Byrne (2003) has suggested that they may be interrelated through the involvement of neural priming mechanisms. In this sense, the neural substrates of a behavioral response to a given situation are stimulated within an agent through the observation of similar actions in conspecifics, which then increases the likelihood of that response (cf. Rizzolatti *et al.*, 1988; Jeannerod *et al.*, 1995). Thus priming behavioral interactions with objects or environmental conditions results in stimulus enhancement, and conversely, priming particular action responses that already exist within an agent's behavioral repertoire results in response facilitation. This priming effect is a critical aspect of the behavioral and cognitive assessment concerning the transmission of social knowledge among early hominin populations, which will be discussed in detail below.

The last two categories of social learning strategies have provoked significant debate within primatological circles as to which is more prevalent within non-human primate learning environments. These are *emulation* and *imitation*, which have been a point of contention due to the interpretive challenges in distinguishing them within experimental settings (see, e.g., Tomasello *et al.*, 1987, 2003; Whiten & Ham, 1992; Tomasello, 1996; Byrne & Russon, 1998; Whiten, 1998, 2000; Byrne, 1999, 2003; Whiten *et al.*, 2004, 2005, 2009a, 2009b; Call *et al.*, 2005; Horner & Whiten, 2005; Hopper *et al.*, 2008; Tennie *et al.*, 2009). By definition, *imitation* involves the processes in which an agent observes and copies the bodily actions of another individual; *emulation* entails the observation and re-enactment of the manner in which an object can be used to achieve a desired result, i.e., the object's effect upon the environment (Whiten, 2000; Call *et al.*, 2005; Whiten *et al.*, 2009b). This contrast separating imitation (bodily action-oriented) and emulation (object action-oriented) remains a contentious issue in the primatological literature (for discussion, see Whiten *et al.*, 2009b), and has important implications in defining the social context of early hominin learning environments.

Emulation

Furthermore, the phenomenon of emulation is divided into subcategories that are useful to consider, such as end-state emulation, affordance learning and object movement re-enactment (Whiten *et al.*, 2009b). *End-state emulation* involves the observation of a modeled behavioral goal, although learning to accomplish such ends are achieved through individualistic means without directly copying the actions observed. For instance, non-human primates presented with a demonstration of how to procure food with a tool may learn to manipulate the tool for the same ends, although the exact actions of the demonstrator are not replicated during this process; instead subjects copy the results of the manner in which tools can be manipulated to gain access to food sources (see Tomasello *et al.*, 1987, for example).

Affordance learning entails the perception of an object's physical properties in association with behavioral strategies that relate to achieving a particular goal (Tomasello, 1998). The concept of "affordance" was first described by Gibson (1979), who suggested that the perception of an object's physical properties determines its potential for manipulation; this in turn *affords* the means for behaviors toward achieving goals. For example, Gibson (1979) states that the perception of a hammer affords whole-hand grasping with respect to its handle shape that can be used in a variety of action contexts, such as hitting, raking or levering. The premise of this concept defines tool use as resulting from the integration of object perception and the potential for action it provides, which then affords means for tool-mediated behaviors.

The last emulative learning strategy, *object movement re-enactment* (OMR), is more difficult to classify as emulative versus imitative in nature because of its defining characteristic as the observation of tool–body interactions. OMR entails the observation of object movements relating to the manner in which objects affect the environment, whereby these particular actions are then re-enacted upon manipulating similar objects (see Whiten & Ham, 1992; Custance *et al.*, 1999; Caldwell & Whiten, 2002; Whiten *et al.*, 2004). While OMR is categorized as an emulative strategy due to its theoretical focus upon objects, Caldwell and Whiten (2002) suggest that distinguishing the copying of bodily versus object actions remains a matter of debate (cf. Whiten *et al.*, 2004, 2009b). It has been proposed that OMR might be better defined as an imitative strategy because of the difficulties in differentiating the focal point of observation, which has further been argued as inseparable (Whiten, 2000; Stoinski *et al.*, 2001; Caldwell & Whiten, 2002; Whiten *et al.*, 2004). However, OMR has also been theorized as including aspects of both emulation and imitation, and thus functions along a continuum. At one end, focusing upon bodily actions in tool use would tend to be more imitative in nature; at the other end, focusing upon the movements of the tool itself would qualify as emulative (Whiten *et al.*, 2009b: 2419). The interrelation of OMR and emulation may be a significant factor in defining the social contexts of learning environments among early hominin populations, as discussed below.

Imitation

The concept of imitation has also been at the center of debate in the comparative primatological literature regarding its definition and determination of its behavioral manifestation in experimental settings (e.g., Tomasello *et al.*, 1987; Byrne & Russon, 1998; Whiten, 1998, 2000; Byrne, 1999, 2003; Whiten *et al.*, 2004, 2005, 2009a, 2009b; Call *et al.*, 2005). To summarize, there has been some dispute as to the nature of imitative strategies that non-human primates display in terms of whether or not it can be considered as *true* imitation, or simply mechanized copying behavior (see Byrne & Russon, 1998; Byrne, 1999, 2003; Whiten *et al.*, 2000, 2004). While true imitation describes the copying of observed actions to achieve an intended result, alternative models such as "program-level imitation" propose that observed responses toward objects or environmental stimuli can be facilitated through neural priming, which elicit particular behaviors (see Byrne & Russon, 1998; cf. Byrne, 1999, 2003). In fact, Byrne (1999, 2003, 2005) has argued that non-human great apes do not learn by true imitation and instead parse observed behaviors into action units that prime similar responses, and are then acquired through individual, trial-and-error learning. In this sense, the process of program-level imitation describes the observation of action sequences that reveal hierarchical organized structures of behavior from which information about such actions is extracted. Byrne (1999, 2003) has concluded that in extracting statistical information about behaviors through observation, non-human great apes are able to mimic them at the program level – i.e., how actions are hierarchically structured in terms of exploiting goals.

However, Whiten (see Whiten, 2000 for review) has challenged the concept of program-level imitation in several ways. In terms of definition, he states that the particular elements of behavior analyzed within the program-level imitation model (sequences of stages, subroutine structure and bimanual coordination) are predominantly independent of one another (Whiten, 2002; Whiten *et al.*, 2006). Whiten (2002) suggests that this model creates confusion in strictly defining imitation as copying a virtually complete sequence of actions, whereas actions themselves are known to involve a myriad of elements (e.g., sequence, tempo, orientation, causal and intentional links) that do not always translate into the imitator's behavioral repertoire. Furthermore, Whiten *et al.* (2004: 45) report that while chimpanzees are able to copy the hierarchical organization of observed behaviors in experimental settings, they fail to replicate their sequential order, which challenges the validity of program-level imitation.

Despite such disagreements, both Byrne (2003) and Whiten *et al.* (2009b) posit that non-human primates indeed engage in some sort of imitative learning behavior, although the nature of such capacities remains a matter of contention. However, Whiten and colleagues (e.g., Whiten & Ham, 1992; Custance *et al.*, 1999; Caldwell & Whiten, 2002; Whiten *et al.*, 2004, 2005, 2007, 2009a, 2009b; Horner & Whiten, 2005; Hopper *et al.*, 2008) have provided substantial evidence that non-human primates, especially with respect to chimpanzees, are quite capable of true imitation in social learning settings. Importantly, they have yielded significant experimental data detailing both significant similarities and variances in comparative social learning strategies among non-human primates and modern humans. The differentiation of preference to engage in various learning mechanisms discussed below has critical implications for the comparison of bone-tool versus stone-tool use.

Social learning in chimpanzees and humans

Chimpanzees and humans engage in a variety of social learning strategies, which may provide a potential comparative framework for analyzing early hominin cognition and behavior. As such, the manner in which chimpanzees and human children engage in social learning can provide some insight into the use and transmission of Plio-Pleistocene tool types, and further elucidate the nature of cognitive demands that might have accompanied the behavioral correlates of these processes (see Whiten *et al.*, 2009a for discussion).

To begin, the recent emphasis upon social learning in primatology has largely resulted from the seminal experiments conducted by Tomasello *et al.* (1987) in which young chimpanzees were presented with a model of food procurement actions using a tool. This revealed that chimps did not copy the particular actions of the model in acquiring the food source, and instead used tools to procure food by idiosyncratic means. They interpreted these results as evidence that chimps observed what actions the tool afforded, and thus highlighted chimps as *emulators* and not imitators. This hypothesis has since been supported by further research, including that by Call *et al.* (2005), who studied both chimps and children in an experiment that involved opening a tube by emulative versus imitative strategies. In the first experiment, subjects were exposed to the open tube (end-state emulation condition), and in another they were provided with a full demonstration of how to open the tube (imitation condition). Their findings suggested that chimps were more likely to "copy the result upon seeing the result than upon seeing the action" (Call *et al.*, 2005: 157), whereas children were more likely to reproduce the demonstrator's actions. In accordance with Tomasello *et al.*'s (1987) original conclusion, this evidence further suggested that humans actively imitated tool-mediated actions, while chimps primarily emulated tool use in social learning environments.

However, Whiten and colleagues (e.g., Whiten & Ham, 1992; Custance *et al.*, 1999; Whiten *et al.*, 1999, 2004, 2005, 2007, 2009a, 2009b; Caldwell & Whiten, 2002; Horner & Whiten, 2005; Hopper *et al.*, 2008) have argued that few studies have directly tested for emulation in chimps, and instead rely upon the term as a fall-back explanation when action copying (i.e., imitation) is not observed in experimental settings. They have alternatively used diffusion-chain experiments to demonstrate that chimpanzees possess the capacity for imitation within social learning conditions (for a review, see Whiten *et al.*, 2007, 2009b). Two high-ranking females from different chimpanzee groups were taught opposing techniques (i.e., lifting versus poking) to manipulate an artificial foraging mechanism, and later were re-introduced into their respective groups (Whiten *et al.*, 2005). A high fidelity of copying either the lifting or poking technique was recorded within these groups according to how their females were taught to open the device. This result demonstrated a preference for imitating actions during the transmission of manipulation techniques among chimps without any contact from human experimenters during the *chimp-to-chimp* diffusion process (Whiten *et al.*, 2005). Likewise, further diffusion experiments (Whiten *et al.*, 2007, 2009b) have confirmed a high fidelity of copying the same techniques to manipulate the foraging device between interrelated chimp groups (but see Tennie *et al.*, 2009 for debate). However, questions of whether mechanisms such as OMR could scaffold diffusion-chain learning in these experiments implies that emulation cannot be firmly ruled out of this scenario.

In summary, the research discussed above presents conflicting results in terms of predominant behavioral means of social learning among chimp populations, which highlights both emulative and imitative strategies in different experimental circumstances. Horner and Whiten (2005) have recently suggested that chimps are quite capable of employing both emulative and imitative strategies, depending upon the context of tool-mediated actions they observe. They found that selection for either emulation or imitation was dependent upon the perceptual availability of causal information clearly relating actions with their results. For example, when actions were visually associated with their direct results, chimps primarily engaged in emulative behavioral strategies; conversely, when the causal relations of actions and results were not easily perceived, they tended to engage in imitative behaviors. These findings were attained through an experiment using two artificial foraging devices, one constructed from transparent materials and the other from opaque materials to demonstrate tool-mediated actions necessary to access a reward for both chimps and children. Results confirmed that chimps interacting with the opaque device displayed a more complete copy of the demonstrator's actions, including those that were irrelevant to gain access to the reward. Conversely, when these chimps interacted with the transparent device, they switched to emulative strategies, which suggests that behavioral strategies in learning situations are highly dependent upon perception–action mechanisms at the neuro-cognitive level.

By comparison, children in this experiment reproduced the demonstrator's actions regardless to the availability of causal information, which led Horner and Whiten (2005) to conclude that an increased attentional capacity in children to focus upon actions without consideration for their relevancy to the task at hand was responsible for this condition. Similar developmental psychology experiments involving the demonstration of relevant and irrelevant actions relating to tool use have provided similar results, which Whiten *et al.* (2009b: 2424) describe as a capacity in children to over-copy observed actions. However, the ubiquity of this phenomenon is currently under investigation, as several studies have shown that children at various ages predominantly rely upon either emulative or imitative learning strategies. For instance, children under the age of three to six months old have been shown to replicate object affordances in experiments, while 24-month-olds have been suggested as displaying overimitative behaviors (Call *et al.*, 2005; Jones, 2009; Tennie *et al.*, 2009; Whiten *et al.*, 2009b). McGuigan *et al.* (2007) have argued that maturation has been shown to produce an increased reliance upon imitation in social learning experiments. When presented with a demonstration to manipulate a reward device, children at three years of age tended to employ emulative strategies when insufficient information concerning the replication of actions by a demonstrator was presented (in this case a video of just hands operating the device). Children at five years of age tended to copy actions at a high rate of fidelity, even if relevant information to do so was degraded, which appears to relate to the cognitive sophistication of five-year-olds to "fill in the gaps" of actions they observed. However, Lyons *et al.* (2007, 2011) have recently found that the capacity to overimitate is a surprisingly rigid process in children between three and five years who directly observe actions. When observing an adult demonstrator manipulate a novel object, children at this age seem to automatically encode the causal relationships between the observed actions and the object, which aids in their understanding of how objects function in a given situation. While these findings may differ, the understanding is that overimitation is a uniquely human trait that aids children in quickly and proficiently acquiring causal relationships between actions and objects, according to sociocultural demands.

Furthermore, Whiten and colleagues (see Whiten *et al.*, 2006; Flynn & Whiten, 2008) have demonstrated that children imitate the hierarchical structure of complex tasks (i.e., the sequential relation of actions toward achieving a behavioral goal), rather than the action styles (i.e., the particular motions involved in such actions) they observe. This suggests that children tend to concentrate upon the causal relations of actions and intended goals in social learning situations, which highlights the cognitive demands of attention toward *causality* as a critical component of imitation among humans. This further attests to the significance of bodily action as a pivotal focal point in social learning strategy throughout the evolution of the hominin lineage (McGuigan *et al.*, 2007), with imitation proposed as an increasingly important learning mechanism in relation to the evolution of stone-tool industries (e.g., Steele, 1996; Shennan & Steele, 1999).

The significance of this comparative research is to first highlight the fact that both human and non-human primates possess a "portfolio" of social learning strategies, which they regularly employ depending upon the context of observed tool–body interactions (Whiten, 2009b: 2417). Second, the causal association of tool-use actions with intended results ultimately dictates the learning strategy employed within social settings. For instance, when causality is readily perceptible (i.e., transparent condition), then emulative strategies are predominant, and conversely when the perception of actions to their intended results is not available (i.e., opaque condition), then imitative strategies will likely be employed to overcome the lack of causal information. Third, with respect to modern human learning repertoires, this dichotomy clearly breaks down as children increasingly rely upon imitative strategies with age (McGuigan *et al.*, 2007).

In sum, these contrasts have important implications for interpreting the development of hominin technologies, which is further highlighted in the comparison of bone- versus stone-tool modification and use. In the following sections, a selection of early tool traditions found throughout Plio-Pleistocene sites in Africa will be compared in terms of cognitive and behavioral substrates in relation to social learning strategies. Of particular interest to this discussion are the purported bone tools from the Sterkfontein Valley in South Africa and Olduvai Gorge in East Africa (Backwell & d'Errico, 2001, 2003, 2004, 2005, 2008; d'Errico *et al.*, 2001; d'Errico & Backwell, 2003, 2009), which will be compared with stone tool artifacts from Lokalalei Lake 2C, Kenya (Delagnes & Roche, 2005) in order to highlight potential insights into variances of cultural transmission relating to the archaeological timing of these traditions. In addition, a synthesis of the technology seen at Olduvai Gorge will also be presented to highlight the importance of ecological contexts surrounding tool use in interpreting the cognitive and behavioral repertoires of various hominin species, which is a critical aspect in determining the emergence of technology.

The Sterkfontein Valley bone tools

Purported bone tools from the Early Pleistocene Sterkfontein Valley were first identified by Brain and Shipman (1993) from Swartkrans (Members 1–3; ca. 1.8–1.0 Mya) (Figure 12.1) using scanning electron microscopy to identify use-wear patterns, which were compared with experimental tools to demonstrate their function as digging implements used in the extraction of underground storage organs. Recently, Backwell and d'Errico (2001, 2003, 2004, 2005, 2008; d'Errico *et al.*, 2001; d'Errico & Backwell,

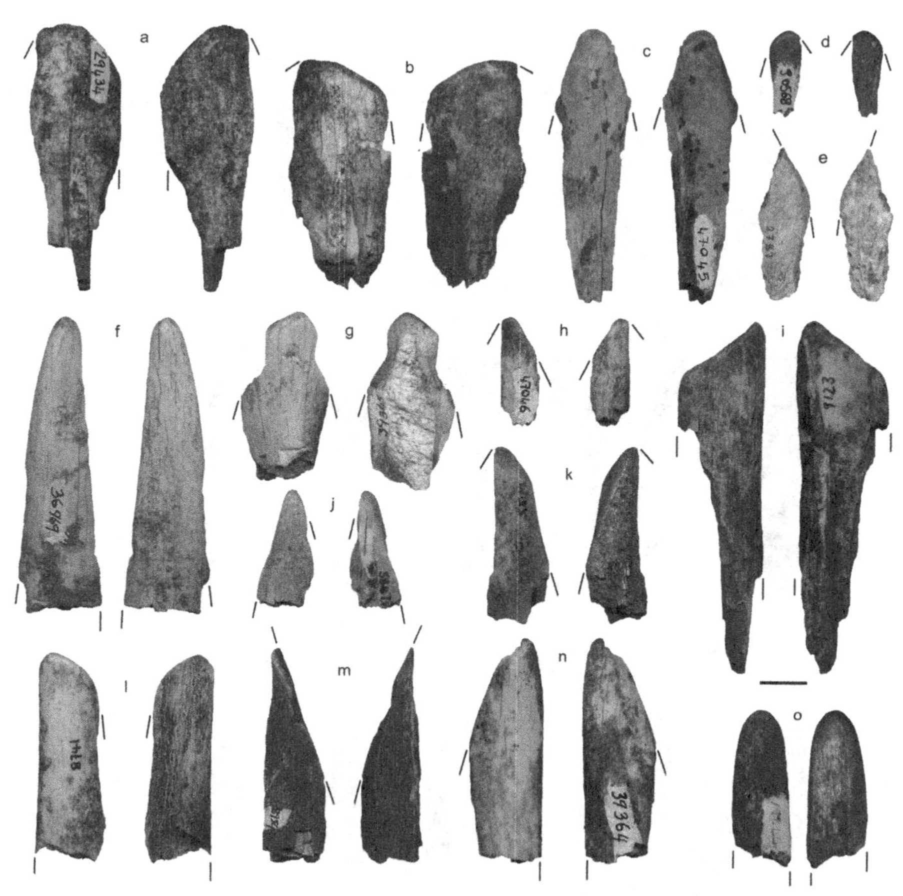

Figure 12.1 Bone tools from Swartkrans. (a) SKX 29434; (b) SKX b; (c) SKX 47045; (d) SKX 30568; (e) SKX 2787; (f) SKX 36969; (g) SKX 39365; (h) SKX 47046; (i) SKX 9123; (j) SKX 5847; (k) SKX SEM; (l) SKX 8741; (m) SKX 8954; (n) SKX 39364; (o) SKX 19845. Scale = 1 cm.

2003, 2009) have expanded upon this research in corroborating the anthropogenic use hypothesis through further interdisciplinary macro- and microscopic analyses. Subsequently, their research has increased the Swartkrans collection, and described purported bone tools from two nearby localities, Sterkfontein (Member 5; ca. 2.0–1.7 Mya) and Drimolen (2.0–1.5 Mya) (Figures 12.2–12.3) (for a review, see d'Errico & Backwell, 2009). Furthermore, they have demonstrated a homogeneity of features among these bone tool assemblages in terms of element, shape, size, weathering and surface modifications, which remain consistent throughout the respective Members of Swartkrans (Backwell & d'Errico, 2003).

The co-occurrence of early *Homo* and *Paranthropus* has complicated the attribution of bone tools to a distinct hominin species. However, evidence at Swartkrans indicates the likelihood of *Paranthropus robustus* as the predominant user of these tools, where the Member 3 deposit has yielded the highest concentration of bone tools and are devoid of

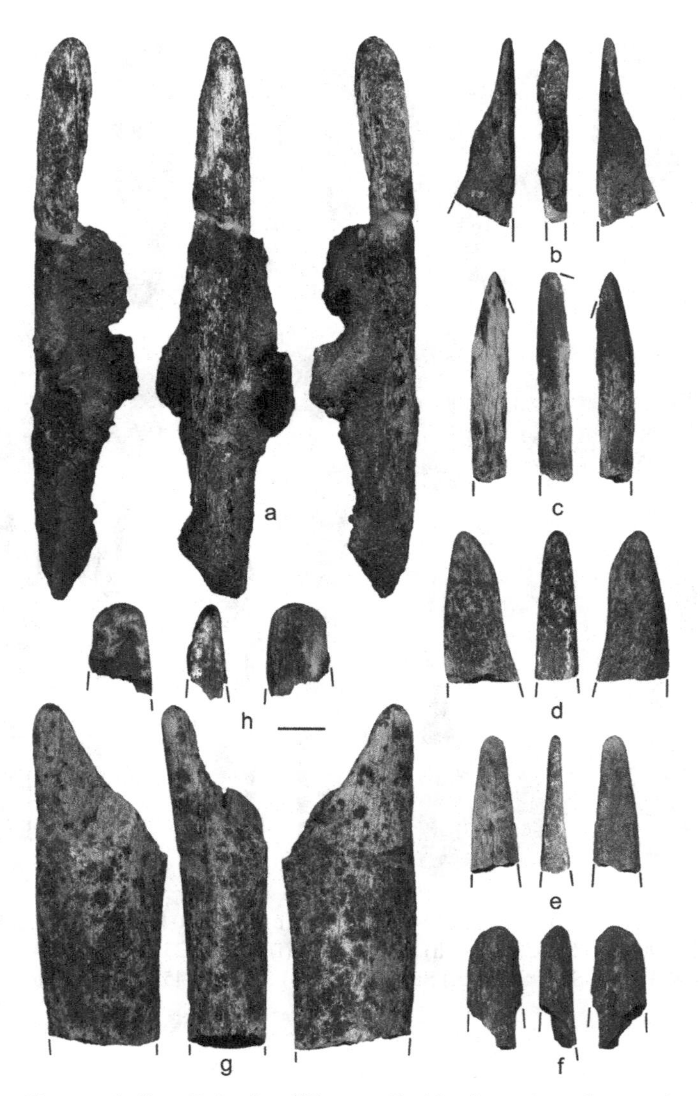

Figure 12.2 Bone tools from Drimolen. When applicable, the periosteal aspect is on the left and the medullary aspect is on the right. Lines indicate breaks. Scale = 1 cm.

early *Homo* remains (Brain & Shipman, 1993; Backwell & d'Errico, 2001; d'Errico & Backwell, 2003). Moreover, this argument is supported by evidence from Drimolen, which has predominantly yielded bone tool artifacts in association with *P. robustus* remains (Backwell & d'Errico, 2008). Thus, it is plausible that non-*Homo* species favored using bone tools over stone in some cases, such as at Drimolen, which is almost devoid of stone tools (Backwell & d'Errico, 2008).

While the debate surrounding the attribution of bone tools to a particular hominin taxon might remain a contentious issue, the nature of their use is less controversial. Brain and Shipman (1993) first hypothesized that the Swartkrans bone tool collections were

Figure 12.3 Bone tools from Drimolen. The delineated area in (d) marks a weathered surface altered by porcupine gnawing. Lines indicate breaks. Scale = 1 cm.

used to extract tubers from the ground, and additionally to work soft skins. More recently Backwell and d'Errico (2001; d'Errico & Backwell, 2009) have demonstrated that the subparallel striations found upon the surfaces of these bone tools closely matched experimental tools used to forage in termite mounds (Figure 12.4). Recently, Lesnik (2011) has corroborated Backwell and d'Errico's results through confocal microscopy and sensitive fractal analysis software, which found that the use-wear patterns on the Swartkrans bone tools resemble those resulting from experimental tools used to perforate and dig in termite mounds. Thus, it seems evident that southern African hominin populations engaged in a subsistence strategy of bone-tool use that centered upon termite

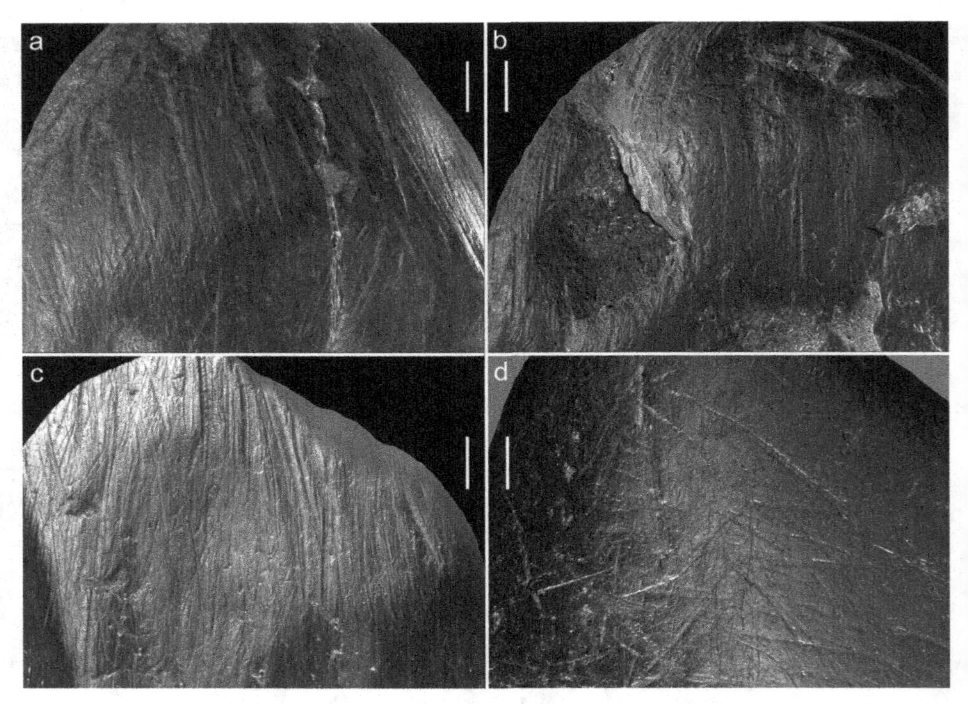

Figure 12.4 Transparent resin replicas of wear patterns on Swartkrans, Drimolen and experimental bone tool tips photographed in transmitted light. (a) Bone tool from Swartkrans Member 2 (SKX 1142). (b) Bone tool (DN 414) from the early hominin site of Drimolen showing rounded tip covered with longitudinal striations and flake scars smoothed by use-wear. (c) Experimental bone tool used to dig in a termite mound. (d) Bone tool used in Brain's experiment to dig up *Scilla marginata* bulbs. Note the similarity in the orientation and width of the striations in (a), (b) and (c). Scale = 1 mm.

foraging activities (Backwell & d'Errico, 2001; d'Errico & Backwell, 2003), and possibly other activities, including fruit processing (Backwell & d'Errico, 2008; d'Errico & Backwell, 2009).

While the Sterkfontein Valley bone tools appear to be of a relatively simple type, Backwell and d'Errico (2008; d'Errico & Backwell, 2009) have demonstrated that their use was indeed conditioned by morpho-functional constraints. As mentioned above, the examination of Early Pleistocene bone tools has confirmed homogeneous physical characteristics among this collection that are the result of selection in relation to their use. For example, weathered long bone fragments and horn cores from medium to large bovids were more robust than the remaining bone accumulations from these assemblages, and likely selected due to their morphology and resistance to the hard, compact outer crusts of termite mounds (Backwell & d'Errico, 2001). In addition, it has been demonstrated that the shape of these bone tool types were more suitable for perforating and controlling the mass of termite mounds during foraging tasks when compared to stone tools (Lesnik & Thackeray, 2007).

Moreover, experimental research has shown that actions of tool use involved in foraging events, described as "controlled, repeated motions parallel to the main axis of the tool" (Backwell & d'Errico, 2001: 1359), indicate a technique employed by early

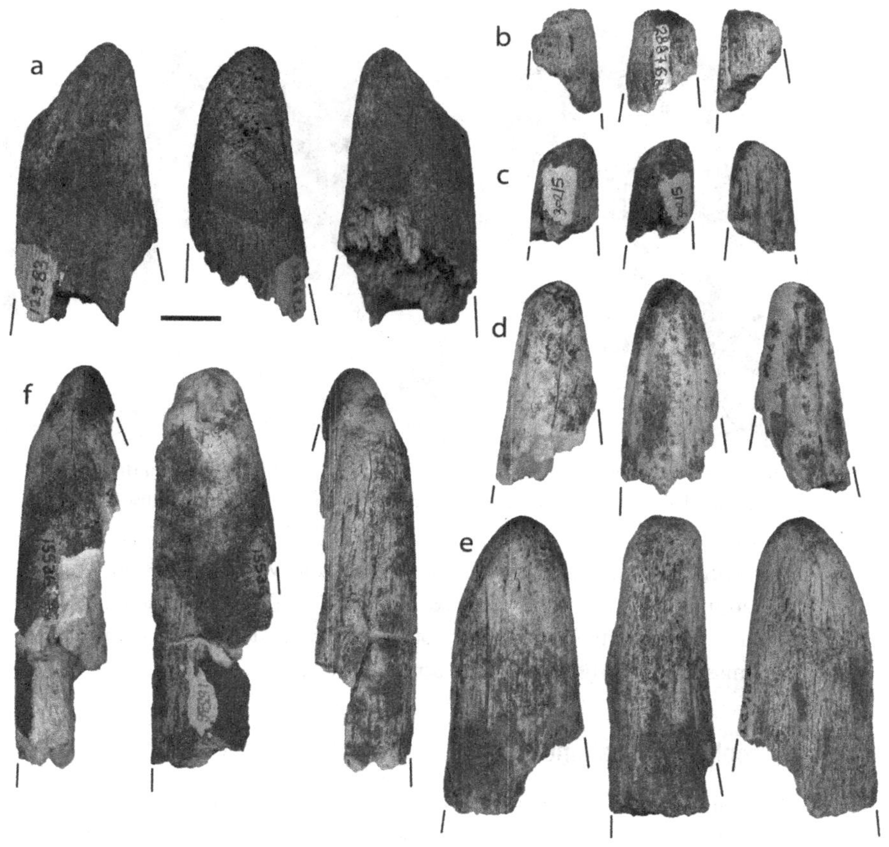

Figure 12.5 Bone tools from Swartkrans bearing possible traces of grinding. (a) SKX 12383; (b) SKX 28876B; (c) SKX 30215; (d) SKX 7068; (e) SKX 28437; (f) SKX 15536. Scale = 1 cm.

hominin tool users. In addition, analyses of horn cores from Swartkrans and Drimolen revealed oblique tip facets covered by striations oriented perpendicularly in relation to the longitudinal axis of these tools (Figures 12.5–12.6). This indicates that they were intentionally re-shaped through grinding motions to sharpen their tips after prolonged use (d'Errico & Backwell, 2003; Backwell & d'Errico, 2008), which suggests that an optimal shape for bone tools was preferred based upon their efficiency in termite-mound foraging. As such, the nature of the relationship between bone tools and their intended use as subsistence implements implies an intimate knowledge of bone as a raw material, which was selected based upon its suitable properties for shaping and use in termite-mound foraging (d'Errico & Backwell, 2003: 1572). Lastly, Backwell and d'Errico's (2001; d'Errico & Backwell, 2009) use and examination of experimental bone tools confirms a commonality of neurophysiological constraints governing tool-use actions by Early Pleistocene hominins in South Africa. Therefore, a neurophysiological analysis of the technical constraints underlying bone-tool use can contribute a useful perspective to model interrelated neuro-cognitive substrates, which might have supported Early

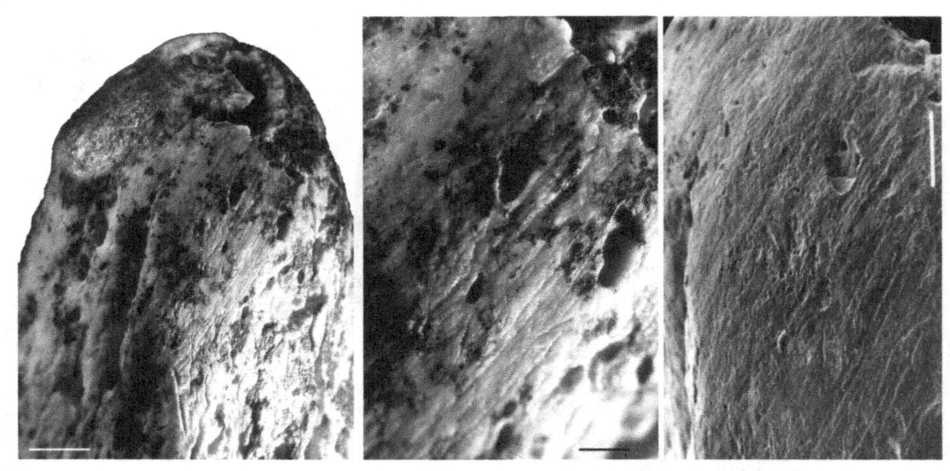

Figure 12.6 Distal portion of a horn core from Swartkrans (SKX 7068) showing a distinct facet (left) with clear parallel striations (center and right). Scale: left = 2 mm; center and right = 1 mm.

Pleistocene foraging activities in the Sterkfontein Valley. The following section will thus expand upon the possible cognitive and behavioral repertoire subserving bone-tool use, which will be further linked to social learning mechanisms underlying the cultural transmission of Early Pleistocene bone-tool use.

Neurophysiological implications

The neurophysiological perspective of Early Pleistocene bone-tool use is derived from the integration of visual and tactile perception involved in termite-mound foraging activities, which may indicate possible neuro-cognitive correlates underlying this subsistence strategy within the Sterkfontein Valley. As discussed above, Backwell and d'Errico (2001; d'Errico & Backwell, 2003) have demonstrated homogeneity in shape, morphology, element and weathering among these tools, which implies the selection of bone types as a visually driven process. The proclivity for locating suitable bone types (i.e., long bone fragments and horn cores), based upon physical characteristics (i.e., state of weathering, morphology, robustness, etc.), was directly related to their functionality as tools (i.e., termite-mound foraging implements). This strongly indicates a clear *perception–action* association of bone types with the intended goals of use driving their selection.

Furthermore, experimental research has revealed that a developed power-grip was necessary for the use of bone tools as foraging implements, and possibly a precision-grip type for shaping actions (cf. Backwell & d'Errico, 2001; Lesnik & Thackeray, 2007). The standardization and uniformity of striation patterns resulting from these activities suggests a high level of motor control confirmed through 2D and 3D roughness analysis (Figure 12.7) (d'Errico & Backwell, 2009). The striation patterns resulting from controlled motions in experimental bone-tool use were comparable to the archaeological collections, and significantly differed from experimental tools that implemented alternative motions. Additionally, Susman (1988, 1998) has suggested that *Paranthropus*,

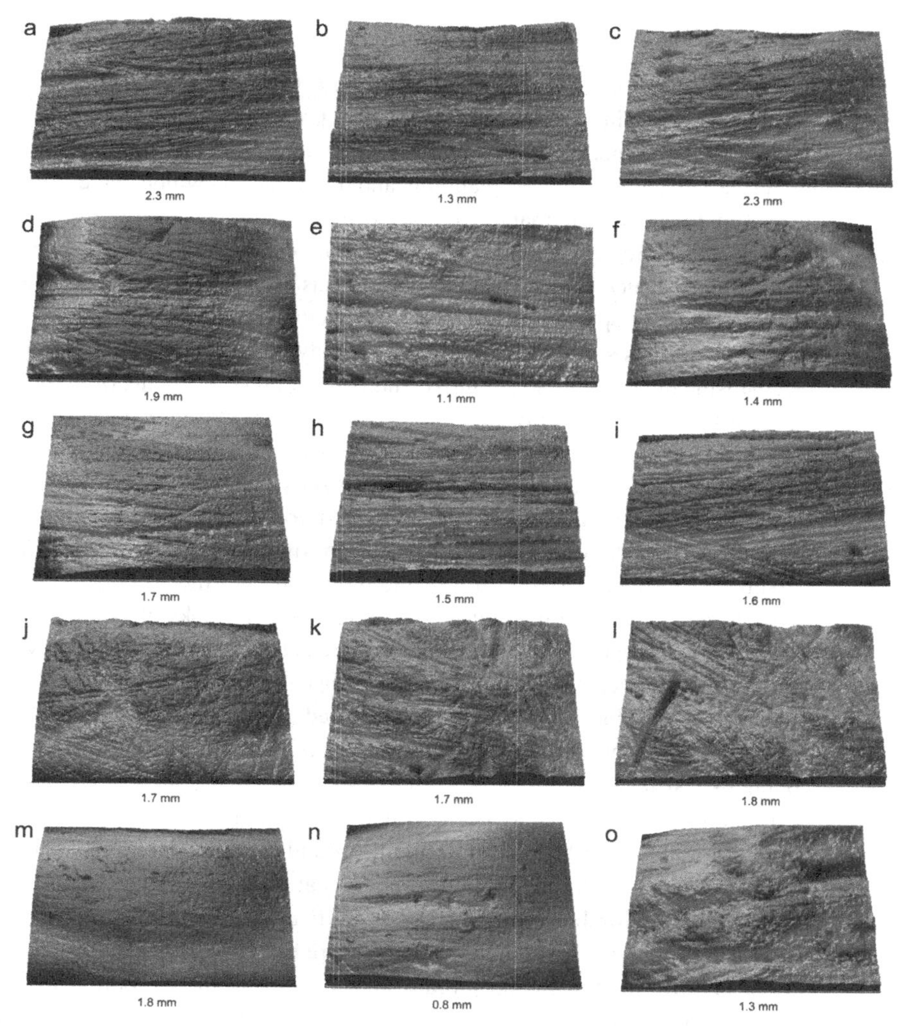

Figure 12.7 Microtopography of use-wear on the tips of bone tools from the early hominin sites of Swartkrans (a) SKX 38830; (b) SKX 35196; (c) SKX 20081; and Drimolen (d) DN 414; (e) DN 412; (f) DN 409; experimental bone tools used to dig in termite mounds (g–i), to dig out tubers from the ground (j: Nad; k: Lucinda; l: Francesco) and to process marula fruits – (m) BEM 391a; (n) BEM 391b; (o) BEM 393c. Note the similarities between the wear patterns on archaeological specimens and pieces used in termite foraging experiments (a–i) compared to the randomly oriented striations produced by subterranean digging activities (j–l), and the fine striations resulting from defleshing marula fruits (m–o). Vertical scales = 10 μm.

associated with both the Swartkrans and Drimolen bone tool collections, was capable of human-like gripping action based upon their hand morphologies (cf. Marzke & Wullstein, 1996; Marzke, 1997, 2006). This evidence indicates that Early Pleistocene hominins possessed a well-developed repertoire for controlled actions that was likely governed by elaborate neurophysiological functioning.

Neuro-cognitive substrates

In terms of cognition, Backwell and d'Errico (2003: 1572) have concluded that Early Pleistocene hominins understood the physical properties of bone as a raw material type relating to its efficiency of use in subsistence activities. This implies the existence of cognitive capacities such as planning and forethought, which have commonly been attributed to stone-tool manufacturing techniques (e.g., Semaw, 2000; Delagnes & Roche, 2005). However, while behavioral implications of tool use might suggest the involvement of such cognitive abilities, there has been little mention of neural correlates that might have supported such functions. Nonetheless, the neurophysiological evidence for technical rules subserving bone-tool use may indeed reveal possible integral neural mechanisms underlying the cognitive and behavioral constituents of these activities.

For instance, the selection of bone types according to their physical characteristics suggests an association of object affordances based upon manipulability and effectiveness during their implementation in foraging activities. The morphologies of bone tool types are conducive to gripping ease, in which power and precision grips can be easily utilized, and more importantly correlate to their efficiency in foraging activities (Lesnik & Thackeray, 2007). In fact, the robust and weathered bone types are efficient for perforating the hard outer crusts of termite mounds (Backwell & d'Errico, 2001), and furthermore, bone tools have been demonstrated as more efficient in removing soil mass during foraging events when compared to stone tools (Lesnik & Thackeray, 2007). These aspects indicate that bone tools were selected upon the basis of their efficiency as foraging implements, and thus the affordances of their physical properties to effectively conduct subsistence activities.

In relation to neurological mechanisms, this strongly implies the involvement of the visuomotor neural system – including mirror and canonical neurons – which functions to transform the visual perception of actions and object characteristics into potential motor schemes for behavior (e.g., Rizzolatti *et al.*, 1988; Jeannerod *et al.*, 1995; Gallese *et al.*, 1996). Visuomotor neurons were first discovered in the ventral premotor cortex (vPMC) of the monkey brain area F5 during the execution of hand actions (specifically gripping) directed at objects intended for use. The single neuron-firing patterns that were recorded during the execution of these actions were also identically recorded during their observation in conspecifics (Rizzolatti *et al.*, 1988; Jeannerod *et al.*, 1995; Gallese *et al.*, 1996). Hence, this prompted the term "mirror" neurons, which refers to their activation during the observation of transitive (i.e., tool-mediated) actions *as if* they were being executed (i.e., motor simulation) (Rizzolatti *et al.*, 1988, 1996, 2001, 2002; Gallese *et al.*, 1996; Craighero *et al.*, 1998; Blakemore & Decety, 2001; Rizzolatti & Luppino, 2001; Umiltà *et al.*, 2001; Rizzolatti & Craighero, 2004; Ferrari *et al.*, 2005; Iacoboni *et al.*, 2005).

In addition, further testing led to the identification of a second class of visuomotor neurons in F5, termed canonical neurons, which activated during both the execution of goal-oriented hand movements (e.g., power and precision grip configurations) toward objects, and identically during the observation of objects with physical characteristics that require these particular grips. For instance, the observation of a hammer activates canonical neurons governing the whole-hand gripping action necessary to manipulate the hammer, without any actual physical action. Likewise, the observation of a small stone activates canonical neurons relating to precision gripping (Jeannerod *et al.*, 1995; Grèzes & Decety, 2002; Garbarini & Adenzato, 2004). Hence, canonical neurons have been labeled as "affordance neurons" (Garbarini & Adenzato, 2004: 102; cf. Grove & Coward, 2008) in the sense that they activate according to the perception of an object's characteristics in terms of the opportunities for tool–body actions they provide (i.e., hand-grip types in relation to their use).

In accordance with this neurological evidence, it is possible that the location, selectivity and use of bone tools to forage in termite mounds within the Early Pleistocene Sterkfontein Valley was governed by visuomotor neural functioning. For example, the visual perception and selection of bone types may have been substrated by canonical neural activity, which potentiated appropriate grip configurations for their implementation in subsistence activities. This is corroborated by evidence presented above for the direct relation of bone tool characteristics and the efficiency of their use in termite-mound foraging. In effect, visuomotor neurons may have integrated the perception of bone types (i.e., robust, weathered long bone fragments and horn cores) with the appropriate motor representations for their manipulation (i.e., hand gripping), to effectuate subsistence goals (i.e., termite-mound foraging).

The possible involvement of visuomotor neurons has important implications relating to the cognitive substrates of bone tool-mediated behavior within the Early Pleistocene hominin brain. In fact, Rizzolatti and colleagues (Rizzolatti *et al.*, 1996, 2001, 2002; Gallese, 2000, 2003, 2009; Blakemore & Decety, 2001; Rizzolatti & Luppino, 2001; Rizzolatti & Craighero, 2004; Iacoboni *et al.*, 2005) have posited that the simulative (*as-if*) activation of visuomotor neurons underpins action-understanding, or the ability to understand the "meaning" of observed object-mediated actions via neurally simulating their execution (see Gallese, 2000, 2003, 2009; Blakemore & Decety, 2001; Rizzolatti & Luppino, 2001; Jeannerod, 2006). This is a critical aspect for the implication of cognitive capacities, such as forethought and planning, involved in the use of bone tools in the Early Pleistocene. Essentially, the ability to associate the physical characteristics of bone types with the motor representation of actions intended toward their use provides a mechanism for understanding the meaning of such actions involved in termite-mound foraging, which is a critical aspect in the planning and sequencing of successive actions to effectively achieve subsistence goals (Hommel *et al.*, 2001; Romo *et al.*, 2004).

For example, a similar scenario has recently been described by Visalberghi *et al.* (2009), who demonstrated that wild capuchin monkeys discriminate and select stones for nut cracking based upon their physical affordances. They found that capuchins, when presented with a range of stones similar in size and appearance, although varying in weight, selected appropriately heavy stones based upon visual-tactile assessments of their shape and dimension (see Chapter 11). If stones tended to be too light, capuchins resorted to manually manipulating the remaining stones to assess their suitability for use. This supports the notion that object selection for intended tool use in capuchins is contingent upon the affordances of their physical properties, which is facilitated through the integration of visual and somatic perception (Visalberghi *et al.*, 2009). Importantly, the visual perception of stones was always accompanied by tactile manipulation, which suggests that the accurate assessment of object use involves the perception of visuomotor information. This resulted in the selection of stones that are most effective for nut cracking by capuchins, and as such the planning of future nut-cracking events was hinged upon the perception–action assessment of object characteristics, in relation to their implementation in subsistence activities (Visalberghi *et al.*, 2009).

Thus, determining the involvement of the visuomotor neural system is a critical aspect in discerning the possible cognitive and behavioral mechanisms underlying tool use among early hominin populations. Furthermore, it is reasonable to consider that visuomotor neurons are common to the Family Hominidae, including extinct hominin forms, for several reasons. First, recent neuroscientific testing has confirmed F5 visuomotor functioning as homologous to area BA44 (Broca's area) within the vPMC of the human brain (Gallese, 2000, 2003, 2006, 2009; Blakemore & Decety, 2001; Rizzolatti & Luppino, 2001; Grèzes & Decety, 2002; Rizzolatti & Craighero, 2004; Iacoboni *et al.*, 2005; Rizzolatti, 2005). This corroborates the existence of visuomotor neural mechanisms within the early hominin brain based upon the commonalities of motor constraints demonstrated in Backwell and d'Errico's (2001) experimental use of bone tools. Furthermore, the ubiquity of visuomotor functioning throughout non-human primate species strongly argues for Hominidae homologies; however, testing for such mechanisms within non-human great apes remains in the preliminary stages (Whiten *et al.*, 2004). This is problematic in the sense that the coalescence of primatological and paleoanthropological research has implicated the modeling of great ape behavior as a suitable analogue for early hominin behavior.

Notwithstanding, Cantalupo and Hopkins (2001) have shown that asymmetries of F5 in great apes and BA44 in humans are uncannily similar, which implies a common evolutionary origin of the vPMC and the visuomotor neural system (cf. Rizzolatti *et al.*, 2001). Moreover, a number of authors have demonstrated behavioral data that implies visuomotor neural functioning in addition to neurological implications. For example, young chimpanzees have been found to perform identically to children in familiarization tests, which showed their capability for attributing goals to actions (see Premack & Woodruff, 1978; Uller & Nichols, 2000). Also, chimps understand the psychological states of others based upon the outcomes of their actions (Tomasello *et al.*, 2003), and likewise understand what conspecifics know based upon what they observe (Hare *et al.*, 2000, 2001).

On this basis, it is reasonable to assume that visuomotor neuron functioning is common throughout Hominidae, and that a homologous functional mechanism existed within the neural repertoire of early hominin tool users (cf. Stout & Chaminade, 2007, 2009). The involvement of the visuomotor neural system is a key aspect of social learning in supporting the translation of visual perceived actions into motor representations for behavior, which are then engendered to achieve intended goals relating to tool use. This perception–action mechanism is critical in the organizing and structuring of action sequences, which suggests that bone-tool use emerged from the integration of neurological functioning, cognition and behavioral capacities as a unified "body+tool system" (Stout *et al.*, 2008: 1946; cf. Stout, 2002, 2006; Stout & Chaminade, 2007). Furthermore, neuro-cognitive functioning ultimately plays a vital role in social learning strategies surrounding the transmission of tool use, as the nature of observation toward objects or the actions behind their use has important implications for the evolution of technology.

Social learning

From the perspective of social learning, the archaeological timing of termite-mound foraging, lasting from approximately 1.8 to 1 Mya (d'Errico *et al.*, 2001; Backwell & d'Errico, 2008), strongly suggests that these activities incorporated social opportunities for learning in the transmission of bone tool use. Critically, the visuomotor neural mediation of object affordance argued above implicates learning strategies that were emulative in nature, which likely included affordance learning and possibly OMR. The argument for affordance learning is corroborated by evidence for the relation of bone types and their intended use. Hence, the observational learning processes involved in the acquisition of bone tool use were likely centered upon the manner in which the physical properties of these tools afforded intended subsistence activities.

However, the consistency of motion demonstrated in Backwell and d'Errico's (2008) analyses suggests the possibility that the bodily actions behind bone-tool use were imitated to some degree. This implicates OMR as another potential social learning strategy, which does not directly involve the copying of bodily actions, only re-enacting the movements of the tools themselves (see Custance *et al.*, 1999 for discussion). However, it has been previously stated that OMR functions along an emulative/imitative continuum, which may have correlated with possible strategy switching within Early Pleistocene social learning environments. Nonetheless, the perception–action characteristics of termite-mound foraging behaviors are strongly indicative of visuomotor functioning, suggesting the predominance of automatic imitation resulting from motor simulation (Heyes, 2009). As such, the focal point of OMR strategies relating to bone-tool use were likely centered upon the action styles, or the details of the tool's movement embedded within foraging behaviors (see Flynn & Whiten, 2008 for discussion). In this sense, it is likely that imitation was centered upon the movement of the tool rather than the bodily actions of the user, although this hypothesis remains difficult to test in light of the complications in making such distinctions.

Nonetheless, it is more likely that the intention behind bone tool use for termite-mound foraging was translated via the action–understanding properties of motor simulation, which thus did not require the imitation of bodily actions in social learning environments. Therefore this evidence provides support for the predominance of emulative learning in the transmission of Early Pleistocene bone tool use, which was likely contingent upon the manner in which the physical properties of bone afforded termite-mound foraging, rather than the imitation of social actors demonstrating their use. Hence, it is more likely that the emulative nature of acquiring bone-tool use was driven by the observation of proficient tool users within the context of individual learning processes, in which young individuals related tool function with the outcome of subsistence rewards. This suggests the possibility that a combination of both social influence and social learning strategies may have subserved the transmission of Early Pleistocene bone tool culture.

For instance, Backwell and d'Errico (2008: 2893) have suggested that termite-mound foraging at Drimolen was a "female aggregation" activity, which was based on the female aggregation practices present in chimpanzees and gorillas, and the fact that both are proposed as models for early hominin cultural and social behavior (d'Errico *et al.*, 2001; Lockwood *et al.*, 2007). They hypothesize that if *Paranthropus robustus* was the user of the bone tools, the foraging activity in which they were used was conducted mainly by females. Backwell and d'Errico (2008) have implied that in this scenario social learning is transferred from mother to infant. Although the role of mothers has been implicated as critical in the facilitation of tool-use transmission among non-human primates (e.g., Boesch, 1991; Lonsdorf, 2006; Chapter 8), it has also been demonstrated that young chimpanzee and capuchin monkey individuals often seek out opportunities for observing tool use based upon the efficiency and skill of users (e.g., Biro *et al.*, 2003; Ottoni *et al.*, 2005). Thus, at Swartkrans, in addition to observing female relatives, young hominins may have "sought out the best" (Ottoni *et al.*, 2005: 215) tool user for observation in learning to facilitate termite-mound foraging activities.

Furthermore, it is possible that neural mechanisms, including the visuomotor system, facilitated the recognition of these bone-tool-mediated actions as *meaningful* acts among observing young hominins. This may have been similar to the manner in which non-human primates and children are capable of understanding the intended goals of others' actions, such as objects grasped for intended uses (for example, see Umiltà *et al.*, 2001; Sommerville & Woodward, 2005; Sommerville *et al.*, 2005; Rochat *et al.*, 2008). As young hominins continually observed bone tool use in termite-mound foraging activities, the familiarity of action–goal associations with these objects potentiated suitable motor representations for similar tool-mediated behaviors. This priming effect may implicate social learning strategies such as *stimulus enhancement* or *response facilitation*. Additionally, social influence strategies including *exposure* and *social support* may have provided access to the tools and ecological contexts of foraging activities, in which the presence of experienced tool users may have motivated young hominins to observe and acquire these activities through individual learning.

Such a scenario can be compared to young chimpanzees (Biro *et al.*, 2003; Lonsdorf, 2006) that routinely observe their mothers' subsistence behaviors such as termite-mound fishing and nut cracking. In the case of termite fishing, young chimps from the Gombe National Park, Tanzania, spent significant periods of time observing their mothers' tool-mediated behaviors during fishing activities, including the use of sticks to penetrate mounds for termites (Lonsdorf, 2006). As such, the regular observation of this practice greatly improved the acquisition of similar behaviors in young chimps; however, the transmission of termite-mound fishing did not involve active demonstrations by mothers. Instead, this process was facilitated through social influence, in which young chimps acquired termite-fishing behaviors through individual learning processes (Lonsdorf, 2006). Furthermore, both chimps and capuchin monkeys have been shown to seek out expert nut crackers for observing the use of a hammer stone and anvil to open the hard outer casings of various nut species (see Biro *et al.*, 2003; Ottoni *et al.*, 2005 for discussion). Similarly, the regular observation of nut-cracking events also increases the rate of acquiring these skills among juvenile individuals (Biro *et al.*, 2003; Ottoni *et al.*, 2005). As in the case of termite fishing, these social learning situations were facilitated through the tolerance of older, passive tool users in the presence of juvenile observers, in which nut-cracking behavior is observed; acquired skills are then developed through individual learning.

Nonetheless, as mentioned above, the Early Pleistocene bone tool use likely involved social learning strategies; however, such mechanisms were possibly contingent upon social and ecological contexts. This implies that the transmission of bone-tool use was predominantly effectuated through the exposure of young hominin individuals to tool-using subsistence activities who emulated the results of observed foraging behaviors. In this sense, learning to use and shape bone tools was likely dependent upon the neural-simulation nature of visuomotor priming, which potentiated innate motor schemes replicating observed tool–body interactions. As such, the hypothesized social learning mechanisms underpinning Early Pleistocene bone tool use within the Sterkfontein Valley present an intriguing contrast with those possibly involved in stone tool knapping. Next, an example of knapping skill will be presented to examine the differences in cognitive and social mechanisms underlying the transmission of technology in the hominin lineage.

Lokalalei 2C stone tools

In comparison to Early Pleistocene bone tool use and shaping known from South Africa, the percussive technologies of the Late Pliocene present a divergent perspective regarding cognitive and behavioral capacities, and the social learning mechanisms subserving cultural transmission. This provides a useful contrast in examining the origin and timing of these archaeological behavioral adaptations, in which stone tools persist throughout the duration of technological evolution in various forms, while bone tool use dies out in the Early Pleistocene and re-emerges in later time periods and archaeological contexts. Among other reasons, this disparity in timing could possibly reflect differences in behavioral, cognitive and social processes involved in the cultural transmission of these technologies. To investigate the probability of this hypothesis, a Late Pliocene site, Lokalalei 2C (LA2C) (2.34 Mya) in West Turkana, Kenya, yielding some of the earliest and consistent evidence for percussive technologies will be examined.

Delagnes and Roche's (2005) analysis of LA2C artifacts presents a portrait of Late Pliocene hominin knapping skill, which was previously hypothesized as constituting a "pre-Oldowan" phase, due to the perceived poor quality of tool characteristics and manufacturing techniques (e.g., see Roche, 1989). However, excavations at Gona, Ethiopia (Semaw *et al.*, 1997, 2003; Semaw, 2000, 2006) and LA2C (Roche *et al.*, 1999; Delagnes & Roche, 2005) revealed that Late Pliocene knapping indeed challenged these conceptions. In fact, several lines of evidence that have emerged from LA2C clearly present the existence of technical rules governing stone knapping, which have been reconstructed from core refitting analysis (Figure 12.8) (see Delagnes & Roche, 2005 for discussion). First, raw material selectivity suggests that predominantly fine-grained lava cobbles were preferred and transported to the site for knapping. This has been directly related to the knapping suitability of selected stones as cores mainly consist of angular

Figure 12.8 Core refit group 16 from Lokalalei Lake 2C. Three core subgroups (a, b, c) are shown, all part of a large porphyritic fine-grained phonolite cobble that was first split (indicated by lines) and subsequently flaked. The initial series (IS) of removals are thin flakes struck from the longest opposing edges of the original cobble, and relatively non-invasive due to the possibility that they belonged to a "core preparation" phase. The first series (bI) of removals is increasingly invasive, consisting of small flakes. Initial point of *débitage* in group c (cI) produced a suitable knapping plane from which a series of flakes were removed. Black arrows indicate direction of refitted flakes, white arrows indicate direction of missing flakes (from Delagnes & Roche, 2005: 446).

cobbles (<90°), which were carefully flaked to maintain workable surfaces (Delagnes & Roche, 2005: 467). Furthermore, larger stones (>15 cm in maximum dimension) were split into several pieces off-site, in which suitable fractured cobbles were presumably selected and transported for later knapping events (Delagnes & Roche, 2005: 462).

Second, Delagnes and Roche (2005: 449) have proposed that the knapping process at LA2C demonstrates a system of "organized *débitage*" concerning flaking practices. Specifically, core reduction sequences reveal that platforms were successively flaked according to their naturally suitable surfaces (i.e., longest available with <90° angles). The contiguous removal of flakes revealed that cores were heavily reduced, and yielded a maximum of 71 flakes in some cases (an average of 18 flakes per core) (Delagnes & Roche, 2005: 466). Also, there is evidence of retouch upon flake surfaces, suggesting a preference for edge shape in relation to utility (also see Semaw, 2000; Semaw *et al.*, 2003 for example). Third, hammerstones tended to be larger and heavier than cores, which implies that knappers distinguished and selected hammerstones according to their weight, and possible shape, for percussive striking (Delagnes & Roche, 2005: 462). In addition, the impact zones upon hammerstones are highly localized, suggesting that they were used in multiple knapping events with a high degree of dexterous precision in directing blows during the flaking process.

The attribution of LA2C artifacts to particular hominin taxa also lacks firm evidence, considering the occurrence of *Australopithecus*, *Paranthropus* and *Homo* at East African Pliocene sites. While *A. aethiopicus* remains have been found in contemporaneous deposits in West Turkana, Prat *et al.* (2005) have found an early *Homo* molar in association with Lokalalei 1. This implicates *Homo* in the manufacturing of LA2C artifacts; however, the tools at Gona exhibit comparable knapping skills and have been tentatively attributed to *A. garhi* (Semaw, 2000; Semaw *et al.*, 2003), which suggests that pre-*Homo* species were probably capable of producing the LA2C assemblage.

In any case, the significant aspect of LA2C knapping skills derives from implications toward the cognitive capacities of early hominin tool makers (Delagnes & Roche, 2005; also see Semaw, 2000; Semaw *et al.*, 2003). Delagnes and Roche (2005: 467) attribute "planning" and "foresight" as evident cognitive features of manufacturing processes demonstrated within their core refitting analysis. For instance, the initial breaking of large, unworked cobbles during raw material selection implies anticipatory capacities in obtaining core blanks with suitable flaking surfaces for later knapping events (Delagnes & Roche, 2005: 463). Additionally, the strict technical constraints surrounding knapping activities suggest a high degree of resource management, in conjunction with an ability to repair knapping mistakes, which indicates a propensity for sequencing complex actions.

Lastly, the authors (Delagnes & Roche, 2005: 466) mention the importance of manual dexterity demonstrated in the LA2C assemblage, such as localized impact zones on hammerstones; flaking blows being restricted to suitable edges in a successive manner; and a continuous firm grip in effecting knapping actions. Their analysis compartmentalizes the technical skills and cognitive capacities demonstrated at LA2C; nonetheless, it is proposed here that this separation is likely a matter of interpretation rather than a reality of knapping processes (cf. Stout, 2002). In extending some of the cognitive arguments presented by Delagnes and Roche, the interdependence of technical skill supported by neurophysiological functioning and cognition is considered here as an integrated system (Stout, 2002, 2006; Stout & Chaminade, 2007, 2009; Stout *et al.*, 2008), which has critical implications for social learning and cultural transmission amongst Late Pliocene hominin populations.

Neurophysiological implications

As argued above, the neurophysiological implications for tool use are indicative of cognitive substrates that function as a cognitive-behavioral system (see Stout, 2002, 2006). As such, the justification of planning and forethought capacities extends beyond the examples of tools themselves as suggested in Delagnes and Roche's (2005) core refitting analysis. In fact, the actions behind stone knapping may offer a more abundant source of evidence for pivotal cognitive capacities, as they are ultimately the direct link between hominin cognition and the artifacts examined in their research.

To begin, stone tool knapping is composed of individual actions aimed at achieving intended subgoals, which are then organized together into a hierarchical structure of sequences. The subgoals involved in the knapping process relate to the various stages of tool production, including raw material selectivity, knapping activities and finally tool use. For instance, the subgoals of raw material procurement discussed above involved the selection of cobbles with physical characteristics suitable for knapping, and additionally fracturing larger stones to create suitable cores for later tool-making activities. Furthermore, the subgoals of knapping events involved successive percussion blows to detach usable flakes for cutting, in which some of these flakes were further shaped through retouch to achieve the desired edge morphology. Finally, the production of knapped flakes relates to their use in subsistence activities, which will be further discussed below. Thus, the structuring of stone tool knapping and use required explicit organization and sequencing of subgoals within the overall process of tool production.

From a neurophysiological perspective, the actions relating to these subgoals were likely regulated by neural substrates that are similar to those involved in bone tool use, such as the visuomotor neural system. For example, raw material selectivity may have been based upon the visual affordance of cobbles exhibiting physical characteristics suitable for flaking (e.g., fine-grained materials with naturally acute angles, etc.). Furthermore, the importance of hand-gripping has been demonstrated as a key component of knapping activities. In fact, Stout and Chaminade (2009: 90) have posited that "the fingertips of the left hand, positioned directly under the point of impact, can serve as a proprioceptive guide to percussion . . . and exert pressure that may help guide flake detachments" in modern knapping experiments of Oldowan stone tools. This indicates that the integration of perception–action elements of percussive striking determines the successful detachment of usable stone flakes, which strongly suggests the involvement of visuomotor neural functioning in Late Pliocene hominin knapping activities. In this case, visuomotor neurons may have potentiated appropriate grips upon the observation of raw materials, prepared cores and hammer stones that were necessary for their effective manipulation during tool production (Stout & Chaminade, 2007). This argument has been substantiated by Stout and Chaminade (2007), who have confirmed the activation of the vPMC and other mirror neuron areas through PET scanning human subjects participating in Oldowan tool knapping activities.

Neuro-cognitive substrates

In terms of cognition, the integration of perceptual–action associations involved in the precision of stone flaking suggests that knapping actions were highly organized and imposed cognitive demands that would include planning and forethought, albeit at a basic level. For example, the organization of skilled knapping behaviors is contingent upon integrating the perceptual-motor coordination of actions and object manipulation; organizing precision movements to maintain knapping proficiency; planning successive actions to maximize the efficiency of knapping sequences; and tracking the progression of actions toward the intended subgoals of tool production (cf. Stout, 2002; Stout & Chaminade, 2007). This indicates that the integral cognitive capacities subserving the knapping process were not necessarily those that involved anticipatory faculties, but rather those centered on the control and organization of action sequences.

Thus, the capacity for planning future events may have been governed through the execution of controlled actions, which Roux and David (2006) have recently demonstrated in examining modern knapping activities. They argue that planning in stone knapping does not result from mental templates or preconceived notions of tool shape, but through the dynamic interactions of the knapper's body in relation to the stone being knapped. They propose that "[c]ourses of action are the actualization of methods through sub-goals which organize differently the courses of action from one piece to the other" (Roux & David, 2006: 101). In other words, planning of future actions is highly dependent upon the manner in which knappers effectuate preceding actions. Forethought and planning within this scheme can be considered as elements of a larger integrated system of neuro-cognitive networks involved in the generation of technical skill, and cognitive control in organizing and regulating action sequences (Stout, 2002, 2006).

While the action-understanding properties of visuomotor neurons may have been a critical aspect of acquiring stone knapping behaviors, the Late Pliocene hominin capacity for organizing, regulating and structuring actions suggests a more intensified cognitive repertoire when compared to Early Pleistocene bone tool use. This implies a more robust neuro-cognitive network involving increased demands upon attentional and action control to support Oldowan tool manufacturing. Thus, the perception–action perspective of early hominin tool use cannot sufficiently account for the knapping skills of the Late Pliocene hominins. In fact, Stout (2010: 10) has further argued that a full range of Oldowan knapping skill involves additional demands relating to the inhibition of automatic imitative behaviors and self-regulation. He suggests that these capacities are facilitated through activation in the prefrontal cortex (PFC), although minimally, nonetheless they implicate an intensified neuro-cognitive demand surrounding the executive control of actions. This hypothesis has recently been corroborated by recent neuroscientific research, which has demonstrated the importance of the PFC in cognitive capacities surrounding action planning and organization (e.g., Bechara *et al.*, 1999; Pochon *et al.*, 2001; Wagner *et al.*, 2001; Brass & von Cramon, 2002; Johnson-Frey *et al.*, 2005). While it is likely that the cognitive basis for bone versus stone tool use may have involved similar neural networks, including the vPMC and/or PFC, it is argued here that the demands upon such substrates were increased with respect to stone knapping activities in comparison to the use and shaping of Early Pleistocene bone tools.

Social learning

The notion that stone tool knapping was neuro-cognitively more demanding than bone-tool use has important implications for social learning strategies subserving the transmission of knapping behaviors among Late Pliocene hominins. Unlike the transparency of intended bone-tool use regarding the direct causal association of actions and goals, the process of stone knapping is comparatively opaque. The intended goals of Late Pliocene stone knapping have recently been confirmed as producing sharp flakes for the processing of animal carcasses, corroborated by the discovery of stone tools associated *in situ* with faunal remains that bear cutmarks at Gona (see Semaw *et al.*, 2003; Domínguez-Rodrigo *et al.*, 2005; Semaw, 2006). Hence, the subgoals of raw material selectivity, core preparation, successive knapping of flakes and the processing of animal carcasses are further removed from their overall intentions of meat consumption. The action sequences at the onset of selecting suitable cobbles for knapping are less likely to be directly attributed to consuming meat than selecting a bone that can be immediately used in termite-mound foraging.

This contrast in causality between bone versus stone tool use indicates that the opacity of action goals involved in percussive technologies is likely correlated with a predominance of imitative strategies in social learning and cultural transmission (see Horner & Whiten, 2005; Whiten *et al.*, 2009a for discussion). While it must be acknowledged that all social learning among the Hominidae likely involves strategy switching to some degree, it is plausible that the causal opacity of stone tool knapping was primarily reliant upon the copying of bodily actions as opposed to object movements or the affordances of tools. The emphasis upon action in learning to knap stone is clearly supported by the complexity of activities involved in subgoals and their sequential structuring in the process of tool production. Furthermore, this inherent complexity likely demanded an actively imitative strategy to comprehend and learn, and thus cannot be fully explained by priming mechanisms.

The cultural transmission of stone tool knapping, from raw material procurement to meat consumption, probably required prolonged observation to acquire such skills (see Toth *et al.*, 1993; Schick *et al.*, 1999). In this sense, it is possible that the consistent chaining of actions underlying subgoals by individuals learning the stone knapping process was eventually organized into routines. As such, learning knapping routines likely required selective attentional faculties that were focused upon specific actions involved in achieving subgoals, which were then ultimately organized into a hierarchical structure of tool production. Hence, the focal point of learning to knap stone was not likely to emphasize the manner in which tools changed the environment, but the actions surrounding their manufacture and use. In sum, it is likely that the acquisition of Late Pliocene stone-tool manufacturing was supported by a more structured social context when compared to Early Pleistocene bone-tool use, in which novice individuals were continually habituated to the various actions and subgoals underlying the knapping process. Thus, the most effective means of social learning to acquire the complex motor details of action sequences was possibly to imitate them upon observation of experienced knappers.

Evidence for this hypothesis can be drawn from stone knapping experiments involving a male bonobo (*Pan paniscus*), Kanzi, who was taught to flake stone tools and use them as cutting implements in order to gain access into a reward box (Toth *et al.*, 1993; Schick *et al.*, 1999; Savage-Rumbaugh & Fields, 2006; Whiten *et al.*, 2009a). Kanzi was initially shown how to detach stone flakes from cores via the use of a hammerstone through demonstrations, in which he was able to master these basic skills after 18 months of continued practice (Toth *et al.*, 1993). This indicated that Kanzi learned to knap stone through imitating the observed actions of demonstrators modeling knapping procedures. Moreover, further observation, vocal encouragement and practice led Kanzi to slowly refine his knapping skills after approximately 120 hours (Whiten *et al.*, 2009a). Interestingly, Kanzi's time required for the acquisition of stone knapping significantly differed from that of modern human novices, who only required a few hours to achieve the same technical proficiency (Stout & Semaw, 2006). Thus, while the continued observation and practice of stone knapping played a significant role in the mastery of tool production for both Kanzi and modern humans, differences in cognitive-behavioral repertoires show marked differences in the acquisition of skill.

Comparatively speaking, Late Pliocene knapping skills fall between those observed in *Pan* and modern humans, and given the learning contexts in which Kanzi acquired stone-tool making, Stout and Semaw (2006: 316) have argued that knapping at Gona was indeed achieved through "effortful practice." Moreover, they state that the degree of skill necessary to produce early hominin stone tools was likely acquired through "habitual knapping behaviour associated with a relatively extended learning period" (Stout & Semaw, 2006: 318). This corroborates the argument for a structured social setting in which novice stone knappers were habituated into cultural routines, from which the process of stone tool production was transmitted through the imitation of observed actions, and continued practice.

At the neuro-cognitive level, it has been argued that the neural substrates of action observation and execution involved in stone knapping, predominantly governed by visuomotor functioning, are crucial for social cognitive capacities among the Hominidae (Stout & Chaminade, 2009; cf. Gallese *et al.*, 2004). As mentioned above, the motor simulation of observed actions has been demonstrated as subserving action-understanding in generating meaning, which is a critical aspect of reading the intentions of conspecifics' behaviors within social contexts (see Rizzolatti *et al.*, 2001; Gallese *et al.*, 2004; Gallese, 2009 for discussion). Thus, it is possible that the demands upon neuro-cognitive networks in the facilitation of acquiring and executing stone tool knapping also substantiated a highly structured social learning environment based upon understanding the meaning of intended actions involved in complex social behaviors necessary for survival.

Furthermore, the causal opacity of intentions behind the knapping process probably required imitative learning strategies, which were reinforced through socially mediated observation and practice. This also likely involved intensified neuro-cognitive activity as the ability to imitate actions and reproduce them in behavioral schemes relates to increasing cognitive demands (Heyes, 1993). For instance, the copying of causally opaque actions imposes escalated cognitive requirements, as the imitator is not only required to recreate observed action sequences without any guidance from external influences; they also must do so in a manner that is relevant to their own perspective, i.e., copy actions that fulfill social or biological requirements for survival (Heyes, 1993; cf. Whiten, 2000; Whiten *et al.*, 2004). Thus, the cultural transmission of stone tool production also plausibly involved increasing neuro-cognitive demands that facilitated the acquisition of complex motor skills through socially structured learning environments.

Tomasello and colleagues (Tomasello *et al.*, 1993; Tennie *et al.*, 2009) support this hypothesis in their work comparing modern human and non-human primate social learning in the generation of cultural complexes. They argue that the learning mechanisms typifying modern humans are focused more so upon the processes constituting cultural activities rather than their material outcomes (Tennie *et al.*, 2009: 2405–2406). This provides a significant advantage over non-human primate correlates in that the orientation toward process engenders a cooperative basis in the transmission of culture, such as active teaching and encouraging conformity to social norms. Tennie *et al.* (2009) suggest that building upon social practices in the formation of cultural knowledge creates a "ratchet up" effect, in which current generations build upon the innovations of the previous, which tends to sustain the transmission of innovative practices throughout cultural evolution. Within this framework, active teaching tends to motivate the replication of actions among learning individuals toward achieving a specific behavioral goal, which is driven by social conformity in fitting into a distinct cultural milieu. Thus, modern humans tend to overimitate observed actions in social learning environments due to this focus upon *process*, which allows for the rapid incorporation of innovative behaviors into modern human cultural repertoires. Conversely, non-human primates focus more upon the *outcomes* of such actions under the same circumstances (Tennie *et al.*, 2009: 2412), which subsequently does not sustain the behavioral details of technological innovations in the same manner. As such, the orientation toward social cooperation in modern humans tends to "transmit cultural items across generations much more faithfully" when compared to non-human primates (Tennie *et al.*, 2009: 2406).

In drawing a median between modern human and non-human primate cultural transmission, as suggested in the analysis of Kanzi's versus modern humans' acquisition of stone knapping, the social environments subserving the transmission of Late Pliocene stone knapping likely supported increasingly cooperative and structured opportunities for learning. This may have cultivated a hyper-sensitivity of knapping novices toward imitating actions that conformed to the norms of tool production processes. Moreover, the involvement of active demonstrations might have further encouraged learning opportunities, which were likely an effective means of transmitting the overall intentions of the knapping process. As such, while it is plausible that social influence strategies played a role in the acquisition and transmission of stone tool production and use, the evidence presented above might imply that the imitation of intended demonstrations might have been a significant strategy of social learning, which structured the cultural transmission of stone knapping in the Late Pliocene.

It can be argued that imitation may likely be a social learning particular to the genus *Homo*, suggesting that a neural hard-wiring might have taken place during the transition from australopithecine, which gave precedence to the neural and behavioral correlates that manifest imitative learning. However, the possible evidence for australopithecine knapping at Gona tentatively contradicts this hypothesis, favoring a socioecological interpretation for imitation among early hominins. This might suggest that while imitative behavior requires specific neural substrates, social relations promoting intense cooperation may be the integral aspect supporting its constitution as a learning mechanism. Despite the lack of direct evidence associating particular hominin taxa to the Late Pliocene assemblages discussed above, the possibility of australopithecine knappers seems evident, and thus it is more likely that tool-use strategies are dictated by biological, social and ecological contexts, which will be discussed in the following section.

The implications of context in examining tool use and social learning

In reviewing the evidence above, it is apparent that the complexity of tool use and social learning among the hominin lineage is not easily mapped onto a linear evolutionary scale. This is largely due to the contextual basis surrounding these phenomena as contingent upon biological, sociocultural and ecological constraints across taxonomic distinctions. As discussed earlier, it seems that non-human primates, early hominins and modern humans possess a ubiquitous latent capacity for tool-mediated behavior and a social tendency to transmit them throughout generational lines. However, there are important distinctions that also differentiate the Primate taxa in terms of these behavioral features, which are critical in detailing their evolutionary development (Chapter 10).

For example, while selectivity of raw materials and effective tools have been demonstrated as universal characteristics of the Primate order (e.g., Toth *et al.*, 1993; Schick *et al.*, 1999; Semaw, 2000; Delagnes & Roche, 2005; Visalberghi *et al.*, 2009), there are obvious differences in neurophysiological mechanisms governing tool use (see Toth *et al.*, 1993, 2006). Kanzi, for instance, can knap similar stone tools to those found in the Late Pliocene artifactual record of eastern Africa; however, his capacity for percussive force in the flaking process is less than that implicated in the tool-manufacturing practices of early hominins (Toth *et al.*, 1993, 2006). As a result, Kanzi's assemblage characteristically shows less *débitage* and higher core vs. flake percentages compared to the East African artifactual record. Interestingly, Toth *et al.* (2006) have suggested that in terms of percussive velocity, early hominins range between *Pan* and modern humans, which indicates an evolutionary continuum for the neurophysiology of arm-swing dynamics in stone-flaking actions. Additionally, the same may be said about the motion of grinding observed among the horn core implements of the Sterkfontein Valley bone tool assemblage. While the hand and wrist morphologies of *Paranthropus* are similar to those of modern humans, allowing for the execution of grinding motions that typify those of Early Pleistocene bone tools, grinding in this manner has never been observed in the wild among non-human primates. This is likely a consequence of the morphological constraints upon similar hand and wrist structures rather than a cognitive disparity, which is likely a consequence of environmental factors that do not necessitate such behaviors (i.e., the use of bone to forage in termite mounds).

While differences in hominin tool industries may be interpreted in terms of cognitive complexity, the neurophysiological basis of tool use is indeed a predominant factor in determining variances between tool types and tool-mediated behaviors (Backwell & d'Errico, 2004). Furthermore, genetic influence may play a key role in the manifestation of neurophysiological capacity across taxa; however, the ecological context of the manner in which these capacities phenotypically unfold ultimately dictates the constitution of individual and group behavior. Kanzi may possess the latent neuro-cognitive capacities for hominin-like tool-mediated behavior, although the biological, social and ecological contexts of *Pan paniscus* have constrained their tool using habits to organic materials that do not require complex manufacturing methods, such as knapping. In this light, the interpretation of tool use and social learning must consider how cognitive,

Figure 12.9 Intentionally shaped bone flakes from Olduvai Gorge. (a) MNKII 1133; (b) DKI 067-4259; (c) FCII 068-6679; (d) FLKII 45; (e) BKII 068-6666; (f) DKI 4200. Scale bar = 1 cm.

cultural and environmental factors influence the technological behaviors of species (see Chapters 2 and 9).

From an archaeological perspective, the importance of contextualizing tool use interpretation can be seen at Olduvai Gorge, which has yielded a collection of bone tools that are contemporaneous with the Sterkfontein Valley assemblage (Figure 12.9) (see Backwell & d'Errico, 2004 for review). Leakey (1971) first identified 125 purported bone tools from Olduvai Beds I and II (2.1–1.15 Mya), which showed distinct signs of intentional flaking in the same manner as "handaxes" typifying the early Acheulean industry (see specimen C in Figure 12.9). In the following decade, Shipman (1984, 1989) reviewed this collection using microscopic analysis and actualistic experiments to compare naturally occurring modification on bone surfaces, such as weathering, chewing and digestion. This led to the identification of 41 specimens resulting from anthropogenic use, 35 of which were flaked prior to their implementation (see Backwell & d'Errico, 2004 for review).

Subsequently, Backwell and d'Errico (2004) have verified Leakey and Shipman's hypotheses using microscopic and comparative analysis with experimental and taphonomic collections. While their results confirmed flaking techniques among Olduvai bone tools, it was also found that these artifacts were most likely exclusive to Bed II (1.7–1.15 Mya), which suggests that in fact they likely coincided with the appearance of *Homo erectus* and the Acheulean industry at Olduvai. This has interesting implications for the use of bone within this context as the Acheulean has been argued as a "technological revolution" in hominin lithic industries, which is distinguished from earlier tool kits by the imposition of "arbitrary design" as seen in the bifacial flaking techniques of the Acheulean handaxe (Wynn & McGrew, 1989: 394).

It has been argued that interpersonal relations supporting observational learning and imitation of actions were necessary to transmit the shared knowledge inherent in manufacturing the symmetrical forms typifying Acheulean tools (e.g., Petraglia *et al.*, 2005; Porr, 2005). Thus, it is also likely that "bone knapping" in the Early Pleistocene at Olduvai Gorge was supported by complex social networks based upon interpersonal cooperation in achieving subsistence means via carcass processing and meat consumption. Termite-mound foraging in southern Africa did not necessitate the demands for group cooperation to effectively carry out the use and shaping of bone tools. As such, the learning mechanisms subserving bone-tool use in the Sterkfontein Valley are focused upon individual learning strategies that were likely transmitted from adult to juvenile. By contrast, the bone knapping process was probably transmitted through intense intergroup interactions as implicated in the example of stone knapping. The emphasis upon group cooperation may have encouraged the proliferation of imitative learning strategies to rapidly and effectively transfer the necessary skills for proficient tool making. It can therefore be argued that the skills necessary for tool use develop within a sociocultural context that in turn directly correlates with ecological contexts of behavior.

Furthermore, the innovation of transferring stone knapping skills onto bone likely required the support of social networks that may have "ratcheted up" (Tennie *et al.*, 2009) the shared cultural knowledge of tool manufacturing to sustain the use of bone as a knappable raw material throughout this time period. While knapped bone tools among the Olduvai artifactual record are by no means predominant, their presence suggests that individual innovations, experiments or even accidents can redefine technical behavior within certain environmental contexts. In this respect the transference of knapping actions from stone to bone at Olduvai Gorge also implicates the importance of individual innovation for the diffusion of a bone-modifying technique from one raw material to another. It is likely that this transference was caused in some part through the observation of bone-fracturing mechanics within subsistence practices, where bone was probably broken to extract marrow. The regular fracturing of bone might have thus led to the understanding of its properties, which then resulted in the application of controlled knapping actions to predictively shape bone for tool use. In essence, the transference of knapping action from stone onto bone resulted from an innovation that likely required a complex understanding of fracturing mechanics borne out of observation and practice.

However, such innovations are not sustained through individual learning, they instead proliferate through their incorporation into the shared knowledge of cooperative social networks. Support for the sociocultural context of bone knapping is corroborated through the existence of technical rules supporting the manufacturing of the Olduvai bone tools, which suggests that this process was an extension of stone knapping techniques (Backwell & d'Errico, 2004). For instance, the Olduvai bone tool collection is composed of fresh bone at the time of their production, which is more easily utilized for knapping compared to weathered pieces. Furthermore, epiphyses were used as bone hammerstones (Figure 12.10), which suggests that the makers of these bone tools understood the properties of this raw material and used it in multiple tasks. Thus, both of these tool-manufacturing methods were likely structured through social networks that supported a common knowledge of the production process relating to either bone or stone tools.

The evidence presented above argues for similar cognitive repertoires when comparing geographically disparate but contemporaneous cases of bone tool use among early hominins (see Backwell & d'Errico, 2005). The manifestation of neurophysiological skill within the different hominin populations was no doubt to a large degree dictated by ecology. In turn, adaptations to local environments also shaped social organization and the mechanisms of social learning and cultural transmission employed for the exploitation of resources in a particular environment. This is a critical aspect in interpreting tool use within the deep past, as a number of different hominin genera including *Paranthropus*, *Australopithecus* and early *Homo* have been associated with both stone and bone tool technology. Although subtle nuances in cognition and behavior may have differed among these populations, it is likely that neurophysiology was a predominant factor in determining the development of technologies among hominin species. While biological constitution plays a significant role in the unraveling of neurophysiological capacities, it has been argued above that socioecological contexts may be the paramount constraint in shaping tool use, and ultimately the artifactual record. As Wynn and McGrew (1989: 384) have recognized, early hominin artifacts reveal the "minimum necessary competence" for their production and use, thus complicating cognitive and behavioral capacities as adequate means for comparing hominin technologies. Hence, it is likely that similarities in tool use throughout the hominin lineage result from socioecological contexts that engender similar learning environments and strategies. Ultimately, it is through the coalescence of biological, sociocultural and ecological contexts that tool use and social learning should be interpreted in terms of cognitive and behavioral complexity (Chapter 2). This provides a viable framework for assessing the nature of cultural transmission within species and across taxa, which is necessary to detail possible factors underlying such mechanisms within the early hominin lineage.

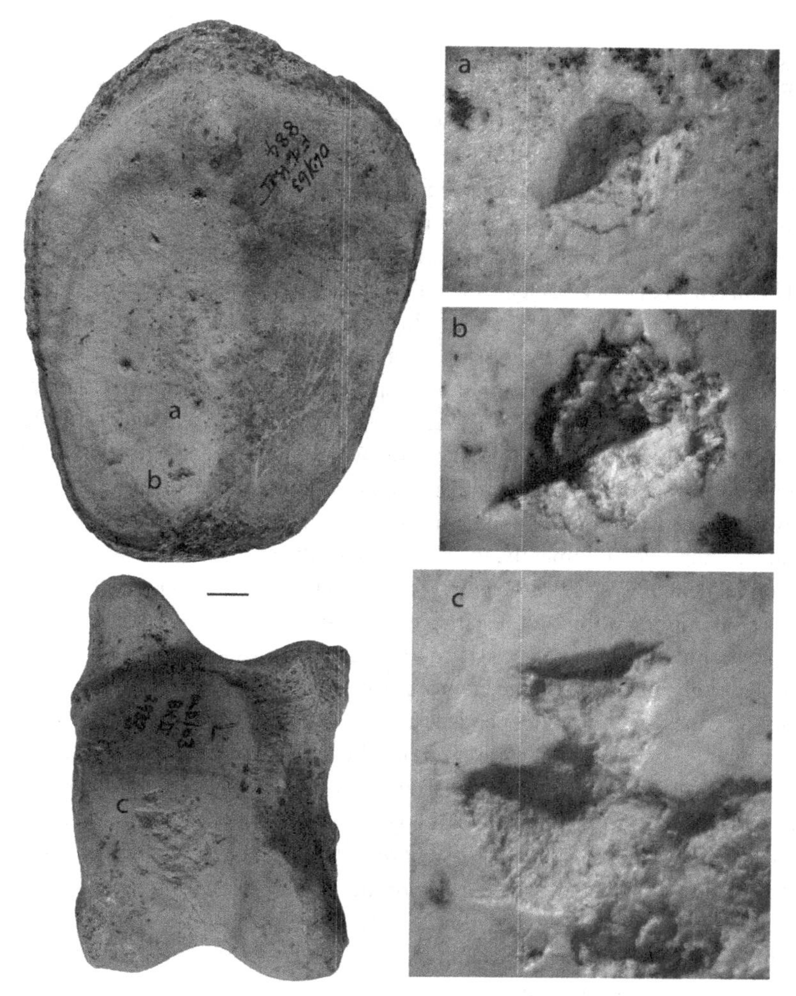

Figure 12.10 Elephant patella (FLKII 884) and astragalus (BKII 2933) from Olduvai Gorge used as hand-held hammers. (a–c) Close-up views of the puncture morphology.

Conclusions for the cultural transmission of tool use

We argue that different mechanisms have been implicated in the predominance of social learning strategies underlying the cultural transmission of Late Pliocene and Early Pleistocene tool use. While visual affordances predominantly structured gestures relating to the Early Pleistocene use of bone tools indicates emulative learning strategies, the intensity of action-related learning argues for imitative learning in the transmission of stone knapping. Importantly, this contrast in learning mechanisms further infers a variance of tool-use transmission that may relate to the archaeological timing of both bone and stone tools. As mentioned above, bone tool use within the emergence of hominin technologies is relatively sporadic when compared to percussive tool manufacturing, which might relate to differing socioecological contexts in eastern and southern Africa.

For instance, the uniform nature of bone tool shaping and use suggests a social basis for bone-tool behavior, which Backwell and d'Errico (2008) have posited in their female aggregation hypothesis. While termite-mound foraging may have been a socially structured activity, the predominant mechanisms supporting its transmission across generational lines was likely based upon the affordances of bone tools that were acquired through individual learning and experience. However, this neuro-cognitive factor should not be taken as evidence of a less socially cohesive structure among the Early Pleistocene hominin populations of the Sterkfontein Valley. It is more likely that the local abundance of termite mounds conditioned subsistence activities centered upon their exploitation, and hence the use of tool types that were specialized for the efficient management of such activities (see Backwell & d'Errico, 2001; d'Errico & Backwell, 2003; Lesnik & Thackeray, 2007). This ecological circumstance then dictated the social processes surrounding the use, shaping and transmission of bone tools and foraging practices, which may have primarily relied upon emulative learning strategies to facilitate the acquisition of these behaviors. Ultimately, the role of individual learning and practice were the most effective mechanisms in transmitting the affordances of bone tool shaping and use, which finds contemporary homologues in extant non-human primates, such as nut cracking and termite fishing in chimpanzees (Biro *et al.*, 2003; Lonsdorf, 2006).

Conversely, the East African environments exploited by hominin populations during the Late Pliocene were highly dependent upon animal carcass processing in terms of subsistence practices, which ultimately promoted knapping processes (de Heinzelin *et al.*, 1999; Semaw, 2000). However, while consistencies in the Oldowan industry have been demonstrated in terms of knapping skill and cognitive implications throughout East African Late Pliocene (see Semaw, 2000; Semaw *et al.*, 2003; Delagnes & Roche, 2005), it must be noted that tool-mediated behaviors were highly constrained by their local environments (e.g., de Heinzelin *et al.*, 1999; Potts *et al.*, 1999; Backwell & d'Errico, 2004). This has been demonstrated as the determining factor in the case of bone-tool use at Olduvai Gorge when other suitable raw materials for butchery were not available. Furthermore, the tendency for East African hominins to engage in animal butchery practices during the Late Pliocene was also predominantly shaped by local ecologies and environments. For instance, a number of Plio-Pleistocene paleoanthropological sites in East Africa, including Lokalalei, Bouri, Koobi Fora and Olduvai Gorge, are situated near lakeside zones (Cerling *et al.*, 1988; de Heinzelin *et al.*, 1999; Potts *et al.*, 1999). This may be directly related to the abundance of resources, and additionally influenced by the variety of animal species attracted to stable water sources. It is likely that such conditions would have yielded greater opportunities for scavenging by hominin populations based upon the increased number of mammal fauna also reliant upon similar resources. In turn, such ecological conditions may have provided relatively consistent access to carcasses, which promoted meat consumption within Late Pliocene hominins. This may explain the substantial number of hominin sites found near lakeside zones in East Africa, mentioned above, and their long expanses of occupation by various hominin taxa and species throughout the Plio-Pleistocene. Thus, the ecology of East African environments encouraged the development and later refinement of knapping practices, which were best suited for butchery practices. In contrast to bone tool use in the Sterkfontein Valley, the most efficient means to transmit processes of stone knapping was through imitative learning, which focused more upon the structuring of action sequences in the acquisition of such behaviors.

In accordance with this view, the emergence and disappearance of hominin technologies can also be traced to the socioecological constraints upon tool use (Backwell & d'Errico, 2004). For instance, the Early Pleistocene environment hosted an abundance of termite mounds within the Sterkfontein Valley, thus encouraging foraging practices in local hominin populations. However, bone-tool use disappears from the artifactual record at 1 Mya, which coincides with the extinction of *Paranthropus robustus* (Backwell & d'Errico, 2003). Thus, the highly specialized practice of using bone tools for termite-mound foraging ceased as contemporaneous and emerging hominin species were largely reliant upon alternative resources for subsistence. Additionally, the social mechanisms supporting the transmission of termite-mound foraging practices were not conducive toward faithful transmission outside this specialized context. The largely individualistic nature of learning strategies did not preserve a cultural knowledge of bone-tool use in the same manner as percussive tool making. On the other hand, the knapping process produced tools that were both highly versatile and applicable to changing ecological circumstances, as stone tools could be used to process most plant-based resources as well as animal carcasses. While Lesnik and Thackeray (2007) have demonstrated their inefficiency in termite-mound foraging when compared to bone tools, stone tools could accommodate for the changing nature of resource exploitation prompted by the variability in diet among different hominin taxa and species. Thus, the general utility of stone tools and the focus upon observing and learning action sequences in the cultural transmission of knapping activities led to the predominance of stone technologies throughout the emergence of the hominin lineage.

In conclusion, the integration of primatological and paleoanthropological data has been used here to model the variability of social learning strategies associated with different early hominin technologies. Elucidating the neuro-cognitive demands of bone and stone tool technologies from the Plio-Pleistocene period has implicated the predominance of social learning strategies that subserved the cultural transmission of tool use throughout early hominin populations. The largely *result-focused* nature of bone tool use in the Early Pleistocene Sterkfontein Valley relied upon visuomotor functioning, which engendered emulative learning strategies at the individual level during the transmission of these practices. Conversely, the *process-focused* nature of Late Pliocene stone knapping at LA2C primarily involved the organization and sequencing of actions, which required imitative learning strategies to effectively transmit tool making across generational lines.

Furthermore, it has been argued that the cognitive and social circumstances gleaned from these contrasting tool traditions were likely conditioned through varying ecological conditions of eastern and southern African environments. These constraints upon tool use have been suggested as a determining factor in the emergence, preservation and disappearance of Plio-Pleistocene hominin technologies. Such contextual approaches toward understanding tool use highlight the interrelatedness of biological, social and environmental contexts surrounding early hominin technologies, which has been suggested as a primary concern of the primate archaeology framework (Haslam *et al.*, 2009: 339). However, the interpretive potential of this subdiscipline cannot be realized without questioning the role of cognition in tool use among non-human primates and modern humans.

Nonetheless, while theoretical perspectives upon the contexts of tool use that indicate the predominance of social learning strategies can be constructed from existing paleoanthropological research, testing such hypotheses remains a critical challenge. However, the recent focus upon directly testing the acquisition of early hominin tool production and use among non-human primates (e.g., Toth *et al.*, 1993; Westergaard & Suomi, 1994; Schick *et al.*, 1999; Visalberghi *et al.*, 2009), in conjunction with comparative analyses of modern human versus non-human primate learning strategies (e.g., Tomasello *et al.*, 1993; Call *et al.*, 2005; Horner & Whiten, 2005; Whiten *et al.*, 2009a, 2009b), might serve as an effective methodology for verifying hypotheses concerning early hominin social learning. Nonetheless, further comparative experiments of this sort that wish to address learning mechanisms underlying the cultural transmission of Plio-Pleistocene tool use and production processes must target the acquisition and diffusion of such behaviors at both the individual and group level. Whiten and co-workers (2005; Whiten, 2005; Horner *et al.*, 2006) have provided a useful methodology for testing the transmission and diffusion of tool-using gestures across non-human great ape populations. If we are to test implications for early hominin technologies, they must directly involve tool use and gestures relating to Plio-Pleistocene tool types as set out by Toth, Schick and colleagues (see Toth *et al.*, 1993; Schick *et al.*, 1999). By such means, the continued integration of primatology and paleoanthropology can shed new light upon the mechanisms of learning responsible for the origin and evolution of tool use by hominins.

Acknowledgments

We thank the conference organizers Christophe Boesch, Josep Call and Crickette Sanz for inviting us to the "Understanding tool use" workshop held at the Max Planck Institute for Evolutionary Anthropology, and to contribute a chapter toward the resulting book. We are grateful to the conference coordinators, Claudia Nebel and Christina Kompo, for facilitating our stay in Leipzig, and thank our colleagues for the very stimulating and enjoyable time spent together. The research conducted in the framework of this chapter has been funded by the Palaeontological Scientific Trust (PAST); the National Research Foundation (NRF), South Africa; the University of the Witwatersrand Research Council; the Bernard Price Institute for Palaeontological Research and Institute for Human Evolution, University of the Witwatersrand; the Cultural Service of the French Embassy, South Africa; the OMLL/European Science Foundation Programme; the French Ministry for Research and Education; and the European Research Council (ERC Grant TRACSYMBOLS n°249587).

References

Backwell, L. R. & d'Errico, F. (2001). Evidence of termite foraging by Swartkrans early hominids. *Proceedings of the National Academy of Sciences USA*, **98**(4), 1358–1363.

Backwell, L. R. & d'Errico, F. (2003). Additional evidence on the early hominid bone tools from Swartkrans with reference to spatial distribution of lithic and organic artefacts. *South African Journal of Science*, **99**(May/June), 259–266.

Backwell, L. R. & d'Errico, F. (2004). The first use of bone tools: a reappraisal of the evidence from Olduvai Gorge, Tanzania. *Palaeontologia Africana*, **40**, 95–158.

Backwell, L. R. & d'Errico, F. (2005). The origins of bone tool technology and the identification of early hominid cultural traditions. In F. d'Errico & L. R. Backwell (eds.) *From Tools to Symbols: From Early Hominids to Modern Humans* (pp. 238–275). Johannesburg: Witwatersrand University Press.

Backwell, L. R. & d'Errico, F. (2008). Early hominid bone tools from Drimolen, South Africa. *Journal of Archaeological Science*, **35**, 2880–2894.

Backwell, L. R. & d'Errico, F. (2009). Additional evidence of early hominid bone tools from South Africa. *Palaeontologia Africana*, extended abstract, **44**, 91–94.

Bechara, A., Damasio, H., Damasio, A. R. & Lee, G. P. (1999). Different contributions of the human amygdala and ventromedial prefrontal cortex to decision-making. *Journal of Neuroscience*, **19**(13), 5473–5481.

Biro, D., Inoue-Nakamura, N., Tonooka, R., *et al.* (2003). Cultural innovation and transmission of tool use in wild chimpanzees: evidence from field experiments. *Animal Cognition*, **6**, 213–223.

Blakemore, S. J. & Decety, J. (2001). From the perception of action to the understanding of intension. *Nature Reviews Neuroscience*, **2**, 561–567.

Boesch, C. (1991). Teaching among wild chimpanzees. *Animal Behaviour*, **41**(3), 530–532.

Brain, C. K. & Shipman, P. (1993). The Swartkrans bone tools. In C. K. Brain (ed.) *Swartkrans: A Cave's Chronicle of Early Man* (pp. 195–215). Pretoria: Transvaal Museum.

Brass, M. & von Cramon, D. Y. (2002). The role of the frontal cortex in task preparation. *Cerebral Cortex*, **12**, 908–914.

Byrne, R. W. (1999). Imitation without intentionality. *Animal Cognition*, **2**, 63–72.

Byrne, R. W. (2003). Imitation as behaviour parsing. *Philosophical Transactions of the Royal Society of London B*, **358**, 529–536.

Byrne, R. W. (2005). Detecting, understanding, and explaining imitation by animals. In S. Hurely & N. Chater (eds.) *Perspectives on Imitation* (pp. 225–242). Vol. **1**. Cambridge, MA and London: MIT Press.

Byrne, R. W. & Russon, A. E. (1998). Learning by imitation. *Behavioural and Brain Sciences*, **21**, 667–721.

Caldwell, C. A. & Whiten, A. (2002). Evolutionary perspectives on imitation. *Animal Cognition*, **5**, 193–208.

Call, J., Carpenter, M. & Tomasello, M. (2005). Copying results and copying actions in the process of social learning: chimpanzees (*Pan troglodytes*) and human children (*Homo sapiens*). *Animal Cognition*, **8**, 151–163.

Cantalupo, C. & Hopkins, W. D. (2001). Asymmetric Broca's area in great apes. *Nature*, **414**, 505.

Carvalho, S., Cunha, E., Sousa, E. & Matsuzawa, T. (2008). *Chaînes opératoires* and resource-exploitation strategies in chimpanzee (*Pan troglodytes*) nut cracking. *Journal of Human Evolution*, **55**, 148–163.

Carvalho, S., Biro, D., McGrew, W. C. & Matsuzawa, T. (2009). Tool-composite reuse in wild chimpanzees (*Pan troglodytes*). *Animal Cognition*, **12**, S103–S114.

Cerling, T., Bowman, J. R. & O'Neil, J. R. (1988). An isotopic study of a fluvial-lacustrine sequence. *Palaeogeography, Palaeoclimatology, Palaeoecology*, **63**, 335–356.

Craighero, L., Fadiga, L., Rizzolatti, G. & Umiltà, C. (1998). Visuomotor priming. *Visual Cognition*, **5**(1/2), 109–125.

Custance, D. M., Whiten, A. & Fredman, T. (1999). Social learning of artificial fruit processing in capuchin monkeys (*Cebus apella*). *Journal of Comparative Psychology*, **113**, 13–23.

de Heinzelin, J., Clark, J. D., White, T., *et al.* (1999). Environment and behavior of 2.5 million-year-old Bouri hominids. *Science*, **284**, 625–629.

Delagnes, A. & Roche, H. (2005). Late Pliocene knapping skills. *Journal of Human Evolution*, **48**, 435–472.

d'Errico, F. & Backwell, L. R. (2003). Possible evidence of bone tool shaping by Swartkrans early hominids. *Journal of Archaeological Science*, **30**, 1559–1576.

d'Errico, F. & Backwell, L. R. (2009). Assessing the function of early hominin bone tools. *Journal of Archaeological Science*, **36**, 1764–1773.

d'Errico, F., Backwell, L. R. & Berger, L. R. (2001). Bone tool use in termite foraging by early hominids and its impact on our understanding of early hominid behaviour. *South African Journal of Science*, **97**(March/April), 71–75.

Domínguez-Rodrigo, M., Pickering, T. P., Semaw, S. & Rogers, M. J. (2005). Cutmarked bones from Pliocene archaeological sites at Gona, Afar, Ethiopia. *Journal of Human Evolution*, **48**, 109–121.

Ferrari, P. F., Rozzi, S. & Fogassi, L. (2005). Mirror neurons responding to observation of actions made with tools in monkey ventral premotor cortex. *Journal of Cognitive Neuroscience*, **17**(2), 212–226.

Flynn, E. & Whiten, A. (2008). Imitation of hierarchical structure versus component details of complex actions by 3- and 5-year-olds. *Journal of Experimental Child Psychology*, **101**, 228–240.

Galef, B. G., Jr. (1988). Imitation in animals. In T. Zentall & B. Galef (eds.) *Social Learning* (pp. 3–28). Hillsdale, NJ: Erlbaum.

Gallese, V. (2000). The inner sense of action. *Journal of Consciousness Studies*, **7**(10), 23–40.

Gallese, V. (2003). A neuroscientific grasp of concept. *Philosophical Transactions of the Royal Society of London B*, **358**, 1231–1240.

Gallese, V. (2006). Intentional attunement. *Brain Research*, **1079**, 15–24.

Gallese, V. (2009). Before and below "theory of mind." *Philosophical Transactions of the Royal Society of London B*, **362**, 659–669.

Gallese, V., Fadiga, L., Fogassi, L. & Rizzolatti, G. (1996). Action recognition in the premotor cortex. *Brain*, **199**, 593–609.

Gallese, V., Keysers, C. & Rizzolatti, G. (2004). A unifying view of the basis of social cognition. *Trends in Cognitive Sciences*, **8**(9), 396–403.

Garbarini, F. & Adenzato, M. (2004). At the root of social cognition. *Brain and Cognition*, **56**, 100–106.

Gibson, J. J. (1979). *The Ecological Approach to Visual Perception*. Hillsdale, NJ: Psychological Press.

Gowlett, J. A. J. (2009). Artefacts of apes, humans, and others. *Journal of Human Evolution*, **57**, 401–410.

Grèzes, J. & Decety, J. (2002). Does visual perception of objects afford action? *Neuropsychologia*, **40**, 212–222.

Grove, M. & Coward, F. (2008). From individual neurons to social brains. *Cambridge Archaeological Journal*, **18**(3), 387–400.

Hare, B., Call, J., Agnetta, B. & Tomasello, M. (2000). Chimpanzees know what conspecifics see and do not see. *Animal Behaviour*, **59**, 771–785.

Hare, B., Call, J. & Tomasello, M. (2001). Do chimpanzees know what conspecifics know? *Animal Behaviour*, **61**, 139–151.

Haslam, M., Hernandez-Aguilar, A., Ling, V., *et al.* (2009). Primate archaeology. *Nature*, **460**, 339–344.

Hernandez-Aguilar, R. A. (2009). Chimpanzee nest distribution and site reuse in a dry habitat. *Journal of Human Evolution*, **57**, 350–364.

Heyes, C. M. (1993). Imitation, culture and cognition. *Animal Behaviour*, **46**, 999–1010.

Heyes, C. M. (1994). Social learning in animals. *Biological Review*, **69**, 207–231.

Heyes, C. M. (2009). Evolution, development and intentional control of imitation. *Philosophical Transactions of the Royal Society of London B*, **364**, 2293–2298.

Hommel, B., Müsseler, J., Aschersleben, G. & Prinz, W. (2001). The theory of event coding (TEC). *Behavioral and Brain Sciences*, **24**, 849–937.

Hopper, L., Lambeth, S. P., Schapiro, S. J. & Whiten, A. (2008). Observational learning in chimpanzees and children through "ghost" conditions. *Philosophical Transactions of the Royal Society of London B*, **275**, 835–840.

Horner, V. & Whiten, A. (2005). Causal knowledge and imitation/emulation switching in chimpanzees (*Pan troglodytes*) and children (*Homo sapiens*). *Animal Cognition*, **8**, 164–181.

Horner, V., Whiten, A., Flynn, E. & de Waal, F. B. M. (2006). Faithful replication of foraging techniques along cultural transmission chains by chimpanzees and children. *Proceedings of the National Academy of Sciences USA*, **103**, 13878–13883.

Iacoboni, M., Molnar-Szakacs, I., Gallese, V., *et al.* (2005). Grasping the intentions of others with one's own mirror neuron system. *PLoS Biology*, **3**(3), 529–535.

Jeannerod, M. (2006). *Motor Cognition*. Oxford and New York: Oxford University Press.

Jeannerod, M., Arbib, M. A., Rizzolatti, G. & Sakata, H. (1995). Grasping objects. *TINS*, **18**(7), 314–320.

Johnson-Frey, S., Newman-Norlund, R. & Grafton, S. T. (2005). A distributed left hemisphere network active during planning of everyday tool use skills. *Cerebral Cortex*, **15**, 681–695.

Jones, S. S. (2009). The development of imitation in infancy. *Philosophical Transactions of the Royal Society of London B*, **364**, 2325–2335.

Leakey, M. D. (1971). *Olduvai Gorge*. Vol. **3**, *Excavations in Beds I and II*. Cambridge: Cambridge University Press.

Lesnik, J. (2011). Bone tool texture analysis and the role of termites in the diet of South African hominids. *Palaeoanthropology*, **2011**, 268–281.

Lesnik, J. & Thackeray, J. F. (2007). The efficiency of stone and bone tools for opening termite mounds. *South African Journal of Science*, **103**, 354–356.

Lockwood, C., Menter, C., Keyser, A. & Moggi-Cecchi, J. (2007). Extended male growth in a fossil hominid species. *Science*, **318**, 1443–1446.

Lonsdorf, E. V. (2006). What is the role of mothers in the acquisition of termite-fishing behaviors in wild chimpanzees (*Pan troglodytes schweinfurthii*)? *Animal Cognition* **9**, 36–46.

Lyons, D. E., Young, A. G. & Keil, F. C. (2007). The hidden structure of overimitation. *Proceedings of the National Academy of Sciences USA*, **104**, 19751–19756.

Lyons, D. E., Damrosch, D. H., Lin, J. K., Macris, D. M. & Keil, F. C. (2011). The scope and limits of overimitation in the transmission of artefact culture. *Philosophical Transactions of the Royal Society of London B*, **366**, 1158–1167.

Marzke, M. W. (1997). Precision grips, hand morphology and tools. *American Journal of Physical Anthropology*, **102**, 91–110.

Marzke, M. W. (2006). Who made stone tools? In V. Roux & B. Bril (eds.) *Stone Knapping* (pp. 243–256). Cambridge: McDonald Institute Monographs.

Marzke, M. W. & Wullstein, K. L. (1996). Chimpanzee and human grips. *International Journal of Primatology*, **17**(1), 117–139.

McGuigan, N., Whiten, A., Flynn, E. & Horner, V. (2007). Imitation of causally opaque versus causally transparent tool use by 3- and 5-year-old children. *Cognitive Development*, **22**, 353–364.

Mercader, J., Panger, M. & Boesch, C. (2002). Excavation of a chimpanzee stone tool site in the African rainforest. *Science*, **296**, 1452–1455.

Mercader, J., Barton, H., Gillespie, J., *et al.* (2007). 4,300-year-old chimpanzee sites and the origins of percussive stone technology. *Proceedings of the National Academy of Sciences USA*, **104**(9), 3043–3048.

Ottoni, E. B., Dogo de Resende, B. & Izar, P. (2005). Watching the best nutcrackers. *Animal Cognition*, **24**, 215–219.

Petraglia, M. D., Shipton, C. & Paddayya, K. (2005). Life and mind in the Acheulean. In G. Gamble & M. Porr (eds.) *The Hominid Individual in Context* (pp. 197–219). London and New York: Routledge.

Pochon, J., Levy, R., Poline, J., *et al.* (2001). The role of the dorsolateral prefrontal cortex in the preparation of forthcoming actions. *Cerebral Cortex*, **11**, 260–266.

Porr, M. (2005). The making of the biface and the making of the individual. In G. Gamble & M. Porr (eds.) *The Hominid Individual in Context* (pp. 68–80). London and New York: Routledge.

Potts, R., Behrensmeyer, A. K. & Ditchfield, P. (1999). Paleolandscape variation and early Pleistocene hominid activities: Members 1 and 7, Olorgesailie Formation, Kenya. *Journal of Human Evolution*, **37**, 747–788.

Prat, S., Brugal, J., Tiercelin, J., *et al.* (2005). First occurrence of early *Homo* in the Nachukui Formation (West Turkana, Kenya) at 2.3–2.4 Myr. *Journal of Human Evolution*, **49**, 230–240.

Premack, D. & Woodruff, G. (1978). Does the chimpanzee have a theory of mind? *Behavioral and Brain Sciences*, **1**, 516–526.

Rizzolatti, G. (2005). The mirror neuron system and its function in humans. *Anatomy and Embryology*, **210**, 419–421.

Rizzolatti, G. & Craighero, L. (2004). The mirror-neuron system. *Annual Review of Neuroscience*, **27**, 169–192.

Rizzolatti, G. & Luppino, G. (2001). The cortical motor system. *Neuron*, **31**, 889–901.

Rizzolatti, G., Camarda, R., Fogassi, L., *et al.* (1988). Functional organization of inferior area 6 in the macaque monkey. *Experimental Brain Research*, **71**, 491–507.

Rizzolatti, G., Fadiga, L., Gallese, V. & Fogassi, L. (1996). Premotor cortex in the recognition of motor actions. *Cognitive Brain Research*, **3**, 131–141.

Rizzolatti, G., Fogassi, L. & Gallese, V. (2001). Neuropsychological mechanisms underlying the understanding and imitation of action. *Nature Reviews Neuroscience*, **2**, 661–670.

Rizzolatti, G., Fogassi, L. & Gallese, V. (2002). Motor and cognitive functions of the ventral premotor cortex. *Current Opinion in Neurobiology*, **12**, 149–154.

Rochat, M. J., Serra, E., Fadiga, L. & Gallese, V. (2008). The evolution of social cognition. *Current Biology*, **18**, 227–232.

Roche, H. (1989). Technological evolution in early hominids. *Ossa*, **14**, 97–98.

Roche, H., Delagnes, A., Brugal, J., *et al.* (1999). Early hominid stone tool production and technical skill 2.34 Myr ago in West Turkana, Kenya. *Nature*, **399**, 57–60.

Romo, R., Hernández, A. & Zainos, A. (2004). Neuronal correlates of a perceptual decision in ventral premotor cortex. *Neuron*, **41**, 165–173.

Roux, V. & David, E. (2006). Planning abilities as a dynamic perceptual-motor skill. *Stone Knapping* (pp. 91–108). Cambridge: McDonald Institute Monographs.

Sanz, C. & Morgan, D. (2007). Chimpanzee tool technology in the Goualougo triangle. *Journal of Human Evolution*, **52**, 420–433.

Savage-Rumbaugh, S. & Fields, W. M. (2006). Rules and tools. In N. Toth & K. Schick (eds.) *The Oldowan* (pp. 223–241). Gosport, IN: Stone Age Institute Press.

Schick, K., Toth, N., Garufi, G., *et al.* (1999). Continuing investigations into the stone tool-making and tool-using capabilities of a bonobo (*Pan paniscus*). *Journal of Archaeological Science*, **26**, 821–832.

Semaw, S. (2000). The world's oldest stone artefacts from Gona, Ethiopia. *Journal of Archaeological Science*, **27**, 1197–1214.

Semaw, S. (2006). The oldest stone artefacts from Gona (2.6–2.5 Ma), Afar, Ethiopia. In N. Toth & K. Schick (eds.) *The Oldowan* (pp. 43–76). Gosport, IN: Stone Age Institute Press.

Semaw, S., Renne, P., Harris, J. W. K., *et al.* (1997). 2.5-million-year-old stone tools from Gona, Ethiopia. *Nature*, **385**, 333–336.

Semaw, S., Rogers, M. J., Quade, J., *et al.* (2003). 2.6-million-year-old stone tools and associated bones from OGS-6 and OGS-7, Gona, Afar, Ethiopia. *Journal of Human Evolution*, **45**, 169–177.

Shennan, S. J. & Steele, J. (1999). Cultural learning in hominids. In H. O. Box & K. Gibson (eds.) *Mammalian Social Learning*. Cambridge: Cambridge University Press.

Shipman, P. (1984). The earliest tools. *Anthroquest*, **29**, 9–10.

Shipman, P. (1989). Altered bones from Olduvai Gorge, Tanzania. In R. Bonnichsen & M. H. Sorg (eds.) *Bone Modification* (pp. 317–334). Orno, ME: Thompson-Shore, Inc.

Sommerville, J. A. & Woodward, A. L. (2005). Pulling out the intentional structure of actions. *Cognition*, **95**, 1–30.

Sommerville, J. A. Woodward, A. L. & Needham, A. (2005). Action experience alters 3-month-old infants' perception of others' actions. *Cognition*, **96**, B1–B11.

Steele, J. (1996). On the evolution of temperament and dominance style in hominid groups. In J. Steele & S. Shennan (eds.) *The Archaeology of Human Ancestry* (pp. 110–129). London and New York: Routledge.

Stoinski, T. S., Wrate, J. L., Ure, N. & Whiten, A. (2001). Imitative learning by captive western lowland gorillas (*Gorilla gorilla gorilla*) in simulated food-processing tasks. *Journal of Comparative Psychology*, **115**, 272–281.

Stout, D. (2002). Skill and cognition in stone tool production. *Current Anthropology*, **43**(5), 693–772.

Stout, D. (2006). Oldowan toolmaking and hominid brain evolution. In N. Toth & K. Schick (eds.) *The Oldowan* (pp. 267–306). Gosport, IN: Stone Age Institute Press.

Stout, D. (2010). The evolution of cognitive control. *Topics in Cognitive Science*, **2**(4), 614–630.

Stout, D. & Chaminade, T. (2007). The evolutionary neuroscience of tool making. *Neuropsychologia*, **45**, 1091–1100.

Stout, D. & Chaminade, T. (2009). Making tools and making sense. *Cambridge Archaeological Journal*, **19**(1), 85–96.

Stout, D. & Semaw, S. (2006). Knapping skills of the earliest Stone Age toolmakers. In N. Toth & K. Schick (eds.) *The Oldowan* (pp. 307–320). Gosport, IN: Stone Age Institute Press.

Stout, D., Toth, N., Schick, K. & Chaminade, T. (2008). Neural correlates of Early Stone Age toolmaking. *Philosophical Transactions of the Royal Society of London B*, **363**, 1939–1949.

Susman, R. L. (1988). Hand of *Paranthropus robustus* from Member 1, Swartkrans. *Science*, **240**, 781–784.

Susman, R. L. (1998). Fossil evidence for early hominid tool use. *Science*, **265**, 1570–1573.

Tennie, C., Call, J. & Tomasello, M. (2009). Ratcheting up the ratchet. *Philosophical Transactions of the Royal Society of London B*, **364**, 2405–2415.

Tomasello, M. (1996). Do apes ape? In C. M. Heyes & B. G. Galef (eds.) *Social Learning in Animals* (pp. 319–346). London: Academic Press.

Tomasello, M. (1998). Emulation learning and cultural learning. *Behavioral and Brain Sciences*, **21**, 703–704.

Tomasello, M., Davis-Dasilva, M., Camak, L. & Bard, K. (1987). Observational learning of tool use by young chimpazees. *Human Evolution*, **2**(2), 175–183.

Tomasello, M., Kruger, A. C. & Ratner, H. H. (1993). Cultural learning. *Behavioral and Brain Sciences*, **16**, 495–552.

Tomasello, M., Call, J. & Hare, B. (2003). Chimpanzees understand psychological states. *Trends in Cognitive Sciences*, **7**(4), 153–156.

Toth, N. (1985). The Oldowan reassessed. *Journal of Archaeological Science*, **12**, 101–120.

Toth, N., Schick, K. D., Savage-Rumbaugh, S., Sevcik, R. & Rumbaugh, D. (1993). *Pan* the tool-making. *Journal of Archaeological Science*, **20**, 81–91.

Toth, N., Schick, K. & Semaw, S. (2006). A comparative study of the stone tool-making skills of *Pan*, *Australopithecus*, and *Homo sapiens*. In N. Toth & K. Schick (eds.) *The Oldowan* (pp. 155–222). Gosport, IN: Stone Age Institute Press.

Uller, C. & Nichols, S. (2000). Goal attribution in chimpanzees. *Cognition*, **76**, B27–B34.

Umiltà, M. A., Kohler, E., Gallese, V., *et al.* (2001). I know what you are doing. *Neuron*, **31**, 155–165.

Uomini, N. T. (2009). The prehistory of handedness. *Journal of Human Evolution*, **57**, 411–419.

van Schaik, C. P. & Pradhan, G. R. (2003). A model for tool use traditions in primates. *Journal of Human Evolution*, **44**, 645–664.

van Schaik, C. P., Deaner, R. O. & Merrill, M. Y. (1999). The conditions for tool use in primates. *Journal of Human Evolution*, **36**, 719–741.

Visalberghi, E., Addessi, E., Truppa, V., *et al.* (2009). Selection of effective stone tools by wild bearded capuchin monkeys. *Current Biology*, **19**, 1–5.

Wagner, A. D., Maril, A., Bjork, R. A. & Schacter, D. L. (2001). Prefrontal contributions to executive control. *Neuroimage*, **14**, 1337–1347.

Westergaard, G. C. & Suomi, S. J. (1994). Hierarchical complexity of combinatorial manipulation in capuchin monkeys (*Cebus apella*). *American Journal of Primatology*, **32**(3), 171–176.

Whiten, A. (1998). Imitation of the sequential structure of actions by chimpanzees (*Pan troglodytes*). *Journal of Comparative Psychology*, **112**, 270–281.

Whiten, A. (2000). Primate culture and social learning. *Cognitive Science*, **24**(3), 477–508.

Whiten, A. (2002). The imitator's representations of the imitated. In A. Meltzoff & W. Prinz (eds.) *The Imitative Mind* (pp. 98–121). Cambridge: Cambridge University Press.

Whiten, A. (2005). The second inheritance system of chimpanzees and humans. *Nature*, **437**, 52–55.

Whiten, A. & Ham, R. (1992). On the nature and evolution of imitation in the animal kingdom. In P. J. B. Slater, J. S. Rosenblatt, C. Beer & M. Milinski (eds.) *Advances in the Study of Behaviour* (pp. 239–283). New York: Academic Press.

Whiten, A., Goodall, J., McGrew, W. C., *et al.* (1999). Cultures in chimpanzees. *Nature*, **399**, 682–685.

Whiten, A., Horner, V., Litchfield, C. A. & Marshall-Pescini, S. (2004). How do apes ape? *Learning and Behaviour*, **32**(1), 36–52.

Whiten, A., Horner, V. & de Waal, F. B. M. (2005). Conformity to cultural norms of tool use in chimpanzees. *Nature*, **437**, 737–740.

Whiten, A., Flynn, E., Brown, K. & Lee, T. (2006). Imitation of hierarchical action structure by young children. *Developmental Science*, **9**(6), 574–582.

Whiten, A., Spiteri, A., Horner, V., *et al.* (2007). Transmission of multiple traditions within and between chimpanzee groups. *Current Biology*, **17**, 1038–1043.

Whiten, A., Schick, K. & Toth, N. (2009a). The evolution and cultural transmission of percussive technology. *Journal of Human Evolution*, **57**, 420–435.

Whiten, A., McGuigan, N., Marshall-Pescini, S. & Hopper, L. M. (2009b). Emulation, imitation, over-imitation and the scope of culture for child and chimpanzee. *Philosophical Transactions of the Royal Society of London B*, **364**, 2417–2428.

Wynn, T. & McGrew, W. C. (1989). An ape's view of the Oldowan. *Man*, New Series, **24**, 383–398.

13 Perspectives on stone tools and cognition in the early Paleolithic record

Shannon P. McPherron
Max Planck Institute for Evolutionary Anthropology

Introduction

Cognitive archaeology is an unsettling area of study for many archaeologists, and it is unsurprising given the incredible challenges in taking the static remains of past people, most of them from a species separate from our own, and saying something about the processes of the mind that led to those remains. The attraction, however, remains great. As the chapters of this volume show, similarly challenging questions are being asked especially of extant primates, but also interestingly a variety of other living species, with experiments and field observations cleverly designed to probe into what these animals know about what they are doing. Archaeology has the potential to provide some insights into the evolutionary context of the modern cognitive condition, certainly for hominins who for 2.6 million years have left behind a record of their tool use, but also perhaps in a more limited fashion for non-human primates (Haslam *et al.*, 2009).

Of course, the modern human cognitive condition is easily experienced but very difficult to conceptualize, and these difficulties carry over into cognitive archaeology. Wynn (2009), while noting that the field is characterized by a diverse set of theoretical underpinnings, argues that evolutionary cognitive archaeology has been approached from three principle perspectives: the relationship between language and the mind; the relationship between the organization and context of actions and the mind; and representational theories of the mind based on cognitive psychology and neurosciences that look at how the brain actually works. If we ignore the methodological and theoretical debates within the cognitive sciences, as well as issues of communication between two different fields (i.e., archaeology and cognitive sciences), there are still significant hurdles. How does archaeology, a field that depends heavily on theoretical contributions from other fields, operationalize theoretical approaches in testable ways? This is especially the case for the Plio-Pleistocene archaeological record, which is limited in scope and dramatically different from data sets produced in the cognitive sciences. As part of this operationalization, often cognitive approaches to the archaeological record risk not

Tool Use in Animals: Cognition and Ecology, eds. Crickette M. Sanz, Josep Call and Christophe Boesch. Published by Cambridge University Press.

adequately taking into consideration some of the unique properties of this record and especially of stone tools.

Stone tools have the advantage of being durable and ubiquitous in the archaeological record, starting 2.6 million years ago (Semaw *et al.*, 1997). They are, however, challenging to interpret, particularly when it concerns concepts like intentionality and motivation; in other words, what were the desired end-products of the lithic reduction process and what decision process guided a particular solution to those needs? The two concepts are of course related in that studying the process should give some insight into goals. Thus in cognitive archaeology the principle methodological tool for operationalizing cognitive theory is to study the process of lithic reduction, rather than the final forms, using the so-called *châine opératoire* approach (de la Torre & Mora, 2009; Gowlett, 2009; Wynn, 2009). The term roughly translates as the study of operational sequences and has its roots in French ethnology; it was incorporated into studies of stone tool production in the 1960s (Leroi-Gourhan, 1993). Since then it has become the dominant approach to lithic studies in France and has greatly influenced lithic studies generally. While widely popular and influential, it is also not without its critics, particularly concerning the cognitive implications of the methodology (see discussion in Tostevin, 2011 and papers within).

In its application, the *châine opératoire* approach emphasizes the sequence of actions taken to produce desired end-products, in this case stone tools. This sequence starts from the initial selection of raw materials through the preparation of the core to the removal of the final flakes, which in later times are retouched into tools. The ideal method for reconstructing the sequence is refitting, putting flakes and cores back together, but this is difficult, time consuming and often not possible because long reduction sequences are not preserved in the limited samples archaeologists excavate. Additionally, because hominins moved raw materials around the landscape, reduction sequences also become spatially separated and fragmented. In the absence of refits, of particular importance in this approach are the flake scar patterns on the exterior surface of flakes and especially on the surfaces of cores whose orientation and relation to one another are "read" to determine how the core was prepared and manipulated to produce the desired end-products. Thus, *châine opératoire* reconstructions are typically accompanied by schematic drawings of cores showing the order and direction of the final flake removals and of flakes that might give insights into earlier stages of reduction. One weakness of this approach is that description and interpretation are often merged. Further, the method was initially explicitly non-quantitative as a reaction against what was perceived as over-quantification in rival schools of lithic analysis; however, some more recent studies have attempted to add some quantification (Soressi & Geneste, 2011; Tostevin, 2011).

For cognitive approaches to the archaeological record the *châine opératoire* approach is interesting and attractive because it flows directly and explicitly from a theoretical perspective that links actions to conscious decisions and intentions in the mind of the maker (Wynn, 2009; Tostevin, 2011). In this regard, of particular importance are moments in the sequence of actions when several possible alternative actions present themselves since then the action actually taken represents an intentional choice made by the maker. Inferring intentions, however, is a process filled with potential pitfalls as decisions are made during the reduction process in response to a variety of factors, including characteristics of the raw materials, technological constraints and functional requirements. Understanding the role of these factors and how they are interrelated is still quite challenging; even the basic first principals of how flake size and shape are related to strike force, strike direction, the hardness and texture of the material used to strike the core, raw material type, core surface morphology, platform (where the core is struck) morphology, etc. are still under investigation (e.g., Dibble & Rezek, 2009; Rezek *et al.*, 2011). Thus it is difficult to know which aspects of an artifact were constrained by intention and perhaps carry more cognitive importance and which were not.

The *châine opératoire* approach also cannot completely solve the problem of knowing which artifacts in an assemblage were desired end-products and which were by-products of production. There are numerous examples in lithic studies from the earliest Oldowan through to Holocene assemblages where, for instance, it is not clear whether certain artifact types are tools intended for use or cores to produce flakes intended for use (see McPherron, 2007b and papers within). In the absence of residues or microwear traces (both of which are fields with their own substantial methodological challenges), desired end-products are the ones selected for retouch (a re-working of the artifact's edge presumably to prepare it for a particular use or reuse). In the earliest Oldowan assemblages, however, there is virtually no retouch and in later times multiple studies have shown the difficulty of reading intention from retouched tools that are likely re-worked multiple times and discarded only after they are no longer useful (Dibble, 1995). Further, stone tool assemblages come from accumulations that represent uncertain periods of time. Time plus the repetition of basic technological behaviors by individual actors, including, importantly, decisions about when to discard artifacts into the archaeological record, can create unintended, emergent, assemblage-level patterning (cf. Pope & Roberts, 2005).

Problematic too is the explicitly non-empirical approach that as a result places elevated importance on the strengths and experience of the analysts, making it difficult to assess the representativeness of the described patterns in the overall assemblage, tends to channel variability into pathways prescribed by the analyst and makes it quite challenging to do comparisons between assemblages (Bar-Yosef & van Peer, 2009; Tostevin, 2003, 2011). Thus, in some lithic studies, including cognitive studies, a more quantitative attribute analysis approach to describing lithic reduction sequences is also applied and interpreted with a kind of residuals approach (Isaac, 1977) in which lithic variability is seen as structured by a series of factors beginning with basic flake mechanics and raw material properties and moving to higher levels of explanation, including knapping techniques, functional constraints and stylistic preferences. The key is to try to establish how each of these factors contributes to the observed variability in lithic assemblages, starting with the most basic levels of explanation. Attribute analysis when applied to *châine opératoire* is typically referred to as lithic reduction analysis (Shott, 2003). In effect, these are equivalents: both seek to understand the process by which blocks of stone are reduced into usable stone tools, but the former does away with the theoretical underpinnings of the latter and instead focuses on quantified description to test models that derive from a variety of theoretical perspectives. In Oldowan studies and to an extent in Acheulian studies, this theoretical perspective has tended to be based on ecological models and is sometimes seen as deterministic and, therefore, in conflict with cognitive approaches (de la Torre & Mora, 2009).

With these methodological considerations in mind, here some aspects of the early archaeological record which have been used to infer cognition are considered. The Oldowan record is limited and fragmentary, particularly before 2 Mya. The shift from a focus on artifacts to a focus more on process with an emphasis especially on raw material acquisition is providing a better data set from which we can start to try to ask questions about the cognitive abilities of these early hominins. In terms of cognition, however, the subsequent Acheulian and its defining artifact, the handaxe, have received substantially more attention and are generally viewed as representing a significant advancement in cognitive abilities. Here the focus is on the qualities of handaxes that have been used to argue for advanced cognitive abilities, with a critical look at the supporting data for these qualities. The primary point is that we still have a relatively poor idea of how these qualities vary and, importantly, what the cause of this variation might be.

Oldowan

Over 20 years have passed since Wynn and McGrew (1989) attempted a comparison between the kinds of behaviors that can be reconstructed from Oldowan archaeology and the kinds of behaviors observed among the higher primates. Their conclusion was that Oldowan behavior fell within the variability of extant primates. Since then both data sets have developed significantly with, in the case of the Oldowan, additional field work and the publication of key assemblages (see recent reviews by Plummer, 2004; Roche, 2005; Schick & Toth, 2006 [plus papers in Toth & Schick, 2006]; Braun & Hovers, 2009 [and papers in Hovers & Braun, 2009]; Roche *et al.*, 2009). The earliest stone tools remain those dated to soon after 2.6 Mya at Gona, a locality in the Awash valley of the eastern rift in Ethiopia. At the Bouri locality in the Middle Awash three bones with cutmarks have been reported from a deposit dating to approximately 2.5 Mya (de Heinzelin *et al.*, 1999). In this case stone tools were not found in immediate association. As summarized by Roche *et al.* (2009), between 2.4 and 2.3 Mya there are two sites reported from Hadar, an area immediately adjacent to Gona, an additional five sites in the Lower Omo and two more sites in the western portion of the Turkana Basin. Thus the database for the earliest Oldowan consists of a handful of sites, and this list has remained fairly stable for many years. Sometime after approximately 2.0–1.9 Mya the record starts to change, with far more archaeological occurrences in East Africa followed sometime after by archaeological occurrences (either Oldowan or Acheulian) in South Africa, North Africa, the Levant and southern Eurasia (see Whiten *et al.*, 2009 and citations within).

The stone tool assemblages at these early sites consist of flakes, cores, hammer stones and imported unworked stones (manuports). The technology of flake production consists of direct percussion (striking a held stone with a hard hammer stone or perhaps striking a held stone against a fixed anvil stone) and bipolar percussion (holding the stone with one hand on a hard anvil and striking it with a hard hammer stone held in the other hand)

(Roche *et al.*, 2009). There is limited evidence for retouch and, at least in the Oldowan prior to the Acheulian, no evidence for shaping (directed removals to form an object often referred to by the French term *façonnage*). Overall Oldowan stone tool technology is very simple. It is, in fact, about as simple as one can imagine a flake-stone technology to be, and certainly when looking only at the final forms of artifacts, in contrast to the later Acheulian, there is little to support an argument for the kinds of more complex cognitive abilities that may eventually lead to modern cognition (Wynn, 2002). However, some insights into the cognitive implications of initial stone tool production have been forwarded based on more recent studies of Oldowan flake production techniques and especially raw material selection criteria.

At a minimum, Oldowan stone tool production required that hominins recognize that certain stones in the environment were suitable for stone tool production (i.e., that they fracture conchoidally) and that a platform angle typically less than 90° was required in order to detach a flake through direct percussion. These two factors are related in the sense that a stone may have the mechanical properties for conchoidal fracture but not have a platform with a suitable angle for striking off a flake. A rounded cobble initially provides limited opportunities for flake removal; however, once a flake is removed this action changes the morphology of the cobble in such a way that additional areas with suitable platforms become available. In this light, selecting an angular block with a naturally occurring flat face can provide an extensive platform from which a series of flakes can be removed. Similarly, a rounded cobble can be first split in two as this will provide two blocks of material with flat faces along which suitable platforms can be found. Thus, care in raw material selection and the application of certain technological approaches can increase the productivity (number of flakes removed per core) of the material and result in flakes with particular and perhaps desired properties (e.g., a long sharp edge on a flake with a low overall mass might make a good cutting tool that can be efficiently transported) (Braun & Harris, 2003). In later time periods, for instance with Levallois technology in the Middle Stone Age, the intent to prepare a core in a particular way following a well-prescribed series of steps that result in particular kinds of flakes (Levallois flakes) is quite clear (Brantingham & Kuhn, 2001). The question is whether similar, though perhaps less involved, processes are visible in the earlier record.

In this regard the assemblage from Lokalalei 2C has become one of the more important data points for assessing the cognitive capacities of these early hominins. The site, located in northern Kenya on the west side of Lake Turkana and dated to about 2.34 Mya, was extensively excavated and produced an assemblage of 2614 lithics and 390 associated faunal remains (Delagnes & Roche, 2005). Importantly, a significant portion of the lithics could be refitted back together, meaning that the process of lithic reduction could be well documented from a *chaîne opératoire* perspective. What Delagnes and Roche (2005) argue is that two different approaches to the material, which they call "simple" and "organized," can be recognized. Of interest in the former are cores with only a very few flakes removed. They argue that these cores have the same morphology as "organized" cores that were extensively worked, but these were produced on a raw material of poor quality. As a result, they suggest that these cores were abandoned after being brought onto the site (a distance of 50 m), tested and found to be inferior to other similarly shaped nodules. Of the so-called "organized" cores, several could be refit with quite long reduction sequences. In one case 51 flakes were removed from a single nodule before it was discarded, and on average 18 flakes were removed from each core. To achieve this kind of productivity from the cores, according to Delagnes and Roche (2005), required more than careful selection for core morphology and raw material quality. They argue that additionally a purposeful strategy was employed wherein knappers alternated between multiple platforms exploiting a single surface. By alternating between these platforms a convex surface suitable for additional removals could be maintained. In other words, these hominins had an appreciation for how their own actions on the core would affect the subsequent core morphology and related platform angles, and they developed a strategy that maximized the productivity of the best nodules.

Similar levels of productivity have not been reported elsewhere for the early Oldowan, though similarly complex and skilled approaches to knapping have been argued for ca. 2.5 Mya assemblages from Gona (Stout *et al.*, 2010). Interestingly, at Lokalalei 1, which is roughly contemporaneous with Lokalalei 2C, the assemblage is characterized by far fewer flakes per core and by greater quantities of broken flakes, perhaps indicative of less control over the knapping process. Studies designed specifically to address whether the observed differences in flaking strategies at Lokalalei 1 and 2C were due to variation in raw material availability, both in terms of size and shape, failed to find any differences (Harmand, 2009). Roche *et al.* (2009) suggest that the differences seen in these two assemblages document variability in cognitive and motor skills, something that Harmand (2009) suggests could be linked to hominin species (robust *Australopithecines* versus early *Homo*). Harmand (2009), however, also considers the possibility that differential site use may be the cause, with Lokalalei 1 representing short-term or small-group occupation where a large number of flakes were not required versus a long-term or large-group occupation at Lokalalei 2C, with concomitant emphasis on increased flake production. In other words, perhaps the Lokalalei 1 hominins had the same abilities but not the same needs.

An early appreciation for differences in raw materials has been shown in several other studies as well. Stout *et al.* (2005), for instance, have shown that hominins at Gona preferentially selected or preferentially transported raw materials that were fine-grained and had fewer inclusions. This was demonstrated by comparing the distribution of rock types in the artifact assemblages with their prevalence in a randomly selected sample of workable stone in locally available outcrops. The motivation behind this selection bias is unclear, but Stout *et al.* (2005) suggest that it likely has to do with how the material fractures and its suitability to Oldowan flaking technology. Braun *et al.* (2009b), working with assemblages from Kanjera that date to soon after 2.0 Mya, show a similar pattern of preferential raw material selection. Using engineering measures of raw material qualities, they argue that edge durability and not flake predictability best explains the over-representation of certain raw material types.

Additionally, Braun *et al.* (2008) have used the Kanjera data set to show transport of selective raw materials at distances of over 10 km. They argue that early hominins not only appreciated differences in raw material qualities, but also how these raw materials were distributed across large landscapes. Most Oldowan assemblages are characterized by the use of immediately available resources with transport distances of less than 5 km and typically less than 1 km (see Goldman-Neuman & Hovers, 2009 and citations within). Some transport, however, is clear from the earliest occurrences. At Gona transport is attested to by the presence in the assemblages of flint, which is not available in the immediate drainages (Semaw *et al.*, 2003). The stone tool-induced butchery marks on the bones at Bouri without associated stone tools and in a context that does not provide raw material for stone tool production implies transport (de Heinzelin *et al.*, 1999). Similarly, the recently reported and much earlier cutmarked bones at Dikika (Braun, 2010; McPherron *et al.*, 2010) suggest transport as well. In the later Oldowan, after 2.0 Mya, there is evidence for increased transport distances. At Olduvai Gorge, nephelinite was transported a few kilometers, and quartzite was transported 2–3 km during Bed I and II times (Hay, 1976; Plummer, 2004). Gneiss is a less frequent element found in larger assemblages and would have been transported 8–10 km (Hay, 1976), and some lava artifacts in the western Bed I lake margin may have been transported 15–20 km (Blumenschine *et al.*, 2003). In the Turkana Basin raw materials may have been transported over short distances in earlier times (Toth, 1982), and over distances as great as 25 km in sites as young as 1.5 Mya (Braun *et al.*, 2008).

Insight into transport also comes from technological studies of assemblages which show incomplete reduction sequences. In a complete reduction sequence a fully cortical nodule is brought into a site, flakes are removed from it, and both the flakes and what remains of the nodule are discarded on-site. If, alternatively, prepared nodules were consistently brought into the site or if nodules, after being worked, or flakes were transported away for future use, it can potentially be detected at the assemblage level in disproportionate numbers of particular elements. So, for instance, at Koobi Fora, Toth (1987) showed that some assemblages are biased toward flakes coming from late in the reduction process, suggesting that cores were worked prior to being transported to these sites. At Bed I Olduvai Gorge, an analysis of flake counts and flake scar counts on cores by raw material type has also shown inconsistencies suggestive of both transport of cores and core reduction prior to transport on-site (Potts, 1988; Kimura, 2002; but see Braun *et al.*, 2005), and Braun *et al.* (2008) use differential reduction patterns on different raw material types to show the movement of one type of material into the site and another from the site.

Roche *et al.* (2009) argue that taken as a whole the Oldowan shows a range of motor skill control and cognitive abilities that still seems to operate at the individual or small-group level. This inference is based on the observation that there seems to be a high degree of technological variability in the Oldowan, so much so that it is difficult to identify an Oldowan approach or style (Roche *et al.*, 2009). Following on this, to directly test the possibility of cultural transmission in the first Oldowan, Stout *et al.* (2010) use assemblages from three sites at Gona in a relatively well-constrained chronological interval and operationalize Byrne's (2007) concepts of intricate complexity and near ubiquity. The premise is that complex behaviors are less likely to be invented independently, and if they are ubiquitous in a group then the most parsimonious explanation is that these behaviors have been learned through cultural transmission. As Stout *et al.* (2010) acknowledge, the difficulty is in deciding what behaviors are sufficiently complex to preclude independent invention. They then take an attribute-based analysis to lithic reduction in the Gona assemblages, and conclude that two of the Gona sites do share lithic reduction strategies of sufficient complexity and ubiquity to suggest cultural transmission.

Stout *et al.* (2010) also conclude that, taken as a whole, there is no reason to suggest that the Oldowan contains within it an earlier, so-called pre-Oldowan, phase, and they note that the evidence currently suggests that initial flaking was a kind of "threshold" event or discovery among hominins that already had the physical and mental skills to achieve Oldowan-level knapping. At the same time, the technological and raw material management strategies of the very earliest Oldowan assemblages, particularly Lokalalei 2C and Gona, along with the variability seen in these assemblages, raise the issue of whether there could be a yet earlier and simpler stage of stone tool production (Haslam *et al.*, 2009). Many have predicted that older occurrences likely exist despite the fact that decades of intensive field work has yet to move the date beyond 2.6 Mya (Panger *et al.*, 2002; Semaw *et al.*, 2003; Carbonell *et al.*, 2009). Some difficulties are to be expected in recognizing archaeologically a more simply organized technology than the Oldowan. This view has been well presented by Panger *et al.* (2002) and is largely based on separating the origins of stone tool production from the origins of Oldowan sites. Currently the earliest Oldowan sites are characterized by relatively high densities of lithic debris which make them archaeologically visible. However, if an early phase of stone tool production consisted of nothing more than a few flakes removed from a cobble and/or took place at varied locations on the landscape, it would produce an archaeological record very difficult to identify. An isolated flake or even a core and flake found on the landscape are likely not enough to demonstrate stone tool production prior to 2.6 Mya, as natural processes are also able to create such flakes and cores. Demonstration of this idea will require a pattern of multiple well-documented occurrences in contexts that could not be expected to produce flakes through natural processes (i.e., away from high-energy deposits containing gravels). Current work on the archaeological signature of extant primates may also help establish better criteria for recognizing more simple forms of tool use (Chapter 11; Haslam *et al.*, 2009).

There are other lines of potential evidence as well. McPherron *et al.* (2010) have published two bones from Dikika (Ethiopia) that date to 3.39 Mya and that show multiple instances of stone tool-inflicted marks (both cutmarks and percussion marks). The finds are not associated with stone tools, thus they provide evidence for stone-tool use but not stone tool production. One possibility is that naturally occurring sharp-edged stones and cobbles were initially used to access animal tissues prior to hominin recognition that sharp flakes could be generated through stone knapping. Thus there may have been an initial and perhaps extended period where hominins used stone tools, followed by a period where hominins sporadically knapped stone tools. Eventually this process led to an intensification and reorganization of this behavior on a landscape scale after 2.6 Mya. It is at this point that this behavior becomes archaeologically visible. This could explain the apparent sophistication of the first stone tool technologies which, while they appear suddenly, in fact developed slowly. This is an idea, however, ahead of the data, and the Dikika find is not without controversy (Domínguez-Rodrigo *et al.*, 2010), mainly because the methodology for identifying the agent responsible for surface modifications on bones is highly debated (Domínguez-Rodrigo & Barba, 2006; Njau & Blumenschine,

2006; Blumenschine *et al.*, 2007; Domínguez-Rodrigo *et al.*, 2009). At a minimum, additional confirmation is certainly required before the finds are fully incorporated into models of early hominin behavioral evolution.

Thus, there are now quite good data on Oldowan stone tool technologies and particularly on the raw material selection, management and transport. To some extent, however, the data are currently ahead of the theoretical tools to conceptualize what this means cognitively and to place this conceptualization in a comparative or evolutionary framework. In the archaeological literature, the cognitive implications of these data are assessed in terms such as "foresight," "anticipation" and "planning" (Delagnes & Roche, 2005; Roche, 2005; Harmand, 2009), but the challenge is to operationalize these concepts so that productive comparisons can be made through time and between species. So while the suggestion by Roche *et al.* (2009: 139) that the high number of flakes removed from each core at Lokalalei 2C "show foresight and anticipation while the whole operational sequence is in progress" is perhaps accurate, it is less clear whether they exceed the cognitive abilities of, for instance, chimpanzees. Is the manufacture and use of specific sticks for fishing different types of termites (Sanz *et al.*, 2009) less convincing as an example of forethought? Still, the emphasis in Oldowan studies on process (lithic production) rather than static outcomes (flakes and cores) and, importantly, the emphasis on placing the process into the landscape undoubtedly offers an opportunity to make comparisons between species and between behaviors that are otherwise materially quite different (e.g., ant dipping, nut cracking and stone tool manufacture).

Acheulian handaxes

The situation is very nearly the opposite in the subsequent Acheulian period, where multiple theoretical approaches have been applied to a single tool type, the handaxe, without an appreciation for process and often without an appreciation for variability. There is something about Acheulian handaxes that attracts cognitive interpretations. As summarized by Nowell *et al.* (2003), handaxes have been used to infer a wide range of cognitive abilities including information processing, an aesthetic sense, mathematics, complex procedural templates, symboling and linguistic abilities, effortless reflexivity, spatial cognition and modern human intelligence. The reasons why they have attracted the attention of cognitive researchers are multiple, but likely include the following. First and perhaps most importantly, handaxes are the first stone tool that is said to be shaped, meaning that the morphology of the archaeological find is more a product of the knapper's intentions than of the constraints of raw materials, fracture mechanics or function. Second, handaxes are found quite early in the archaeological record (Asfaw *et al.*, 1992; Lepre *et al.*, 2011) and persist for some 1.5 million years or so. Additionally, they are found over vast expanses including Africa, western Europe and from southwest Asia to the Indian subcontinent and east Asia. As a result, if something can be said about what handaxes might mean cognitively, then it is saying something at a species, or even multiple species, level. Third, handaxes appear to be, sometimes anyway, remarkably well made. It is as though some considerable effort and skill were expended to produce these artifacts, particularly in contrast to the artifact types from the earlier Oldowan. Finally, though it is a rather abstract property to itemize, handaxes seem to naturally draw our attention and to intrigue us with their possibilities like few other stone tools (cf. Le Tensorer, 2006, 2009). In part this no doubt draws from the fact that, of the early Paleolithic toolkit, handaxes are the most easily recognizable as stone tools, even to the non-specialist.

Each of these points, however, can be at least qualified and perhaps directly challenged with implications for the cognitive interpretations. As discussed below, the extent to which handaxes are shaped tools (Davidson & Noble, 1993; Noble & Davidson, 1996) or at least the extent to which shape is a desired or sought after property can be questioned (McPherron, 2000, 2003, 2007a), and the causes of variability in shape are still highly debated (e.g., Ashton & White, 2003; McPherron, 2003; Lycett & von Cramon-Taubadel, 2008). While it is true that handaxes are found over vast chronological and spatial expanses, it is not clear whether this is a spatially and chronologically unified phenomenon. Handaxes in China (Schick, 1994; Yamei *et al.*, 2000) and Korea (Norton *et al.*, 2006), well beyond the eastern limit of the Movius line (Schick, 1994), may suggest independent invention. Similarly, it is possible that European handaxes, which first occur hundreds of thousands of years after the initial peopling of Europe, might represent independent invention. Regardless of independent invention, the very point that handaxes were so widespread plus the fact that separate species made similar handaxes suggests that rather fundamental, low-level factors must explain their variability. As for how well made handaxes are, there is little data to demonstrate this point beyond illustrating isolated examples of very "good" handaxes, and it is very difficult to know how "good" can be measured and assessed in light of the fact that there is still great debate over what handaxes might have been used for. Finally, on a somewhat related point, the apparent accessibility of handaxes has resulted in some interesting and innovative interpretations of their significance, but often these latter are based on highly selective examples that give a perhaps distorted view of the variability in the underlying data.

Handaxes are a kind of bifacial tool, and sometimes archaeologists use the terms biface and handaxe interchangeably for these tools. They were made from a block of stone or a large flake itself struck from large blocks. Typically, and by definition, they have flakes removed from both sides. These are the so-called shaping flakes that give a handaxe its distinctive form. There are also some unifacial handaxes made on large flakes, and the number of flakes removed from the piece can vary from only a few to completely covering both faces around the entire edge. Typically, more flakes are removed from one end of the handaxe than the other, thereby allowing a tip and a base to be defined and allowing handaxes to be oriented in a consistent manner. There are also handaxes which are so symmetrical that they cannot be said to have a tip or base. Nearly all handaxes are longer than they are wide. Circular handaxes with nearly equal lengths and widths, when they exist, tend to be quite small, and they become difficult to distinguish typologically from cores.

Handaxes range in size from small scraper-like tools of only a few centimeters in length to rare finds nearly 40 cm in length (e.g., Olorgesailie). Most handaxes, however, range from 7–8 cm on the small end to 18–20 cm on the large end (McPherron, 2000).

Handaxe shapes are also quite varied, but most of this variability falls along three axes: edge shape, elongation and refinement. The first of these describes whether the handaxe is pointed or rounded in plan view (which is how they are normally illustrated). The second of these describes the ratio of the length to width. The third of these describes the ratio of the width to the thickness, with more refined handaxes having a smaller thickness for a given width. These three aspects of shape are captured in handaxe typologies and measurement systems (Bordes, 1961; Isaac, 1977; Wymer, 1968; Roe, 1968). A number of multivariate studies, including some designed specifically to capture other aspects of shape, have shown nevertheless that these three axes account for the majority of shape variability in handaxes (Callow, 1976, 1986; Wynn & Tierson, 1990). Variability within this shape space is patterned both within an assemblage of handaxes and across all handaxe assemblages. What this means is that handaxe assemblages tend to have a modal shape and this modal shape varies between assemblages. Additionally, within the shape space defined by edge shape, elongation and refinement, some combinations of shape are more frequently represented than others and some combinations occur very rarely or hardly at all.

Perhaps one of the largest challenges to interpreting patterning in handaxes is that their function is largely unknown. Generally they are described as being "all-round tools" suitable for a wide range of activities but mostly those associated with butchery. Their heft makes them well suited for breaking open joints when butchering an animal, and at the same time their sharp edges and weight make them well suited to cutting through thick portions of hide and flesh. This has been demonstrated through experimental butchery with replicated handaxes (Jones, 1980; Schick & Toth, 1994; Mitchell, 1996; Machin *et al.*, 2007). Use-wear analysis, wherein the location and character of edge damage, polishes and striations are examined microscopically and compared with experimentally replicated pieces, initially confirmed the use of handaxes in butchery based on a small sample of artifacts from sites in England (Keeley, 1980). However, subsequently the field of use-wear analysis, and particularly the type of high-powered microscopy used in these initial applications, went through an extended period of re-evaluation after blind-tests (Bamforth, 1988; Newcomer *et al.*, 1986) led to a questioning of the method's reliability. A more recent use-wear study of handaxes from the Mousterian of Acheulian Tradition also showed evidence for butchery (Claud, 2010), but it is not clear that these handaxes are actually the same kind of tool as the early Acheulian handaxes (Iovita & McPherron, 2011) and it is not clear how this study addressed the methodological challenges raised by the failed blind-tests.

Aside from butchery, there have been many other suggested uses, and it is possible that different forms had different functions. Some have argued that pointed handaxes are shaped well for digging and that they might have been used to excavate underground storage organs like roots and tubers (Domínguez-Rodrigo *et al.*, 2001), while others have argued that their symmetry and shape make them suitable for throwing at prey (O'Brien, 1981; Calvin, 1993; but see also McCall & Whittaker, 2007). It is possible, too, that handaxes were not tools at all but rather cores from which flakes could be removed to perform various activities (Davidson & Noble, 1993; Noble & Davidson, 1996), though this possibility has yet to be either demonstrated or refuted for Acheulian handaxes. Kohn and Mithen (1999), in positing a sexual selection argument to explain several aspects of handaxe variability, even suggest that handaxes were not used in a practical sense, but rather the act of handaxe manufacture was of primary importance in the social realm of attracting sexual partners. Aside from the potential problems with Kohn and Mithen's arguments, the fact that they are able to posit that handaxes were mainly unused suggests how weakly developed our functional understanding of handaxes remains. As a result, it makes it very difficult to identify which aspects of handaxes are somehow better than they need to be functionally and therefore of potential cognitive interest (cf. Kohn & Mithen, 1999; Machin *et al.*, 2007).

Against this background, the two aspects of handaxe variability that have received the most attention with regard to cognition are symmetry and standardization. The latter relates to the former (i.e., symmetry can be standardized), but standardization also applies more generally to shape, size and even the relationship between shape and size (e.g., Crompton & Gowlett, 1993; Gowlett & Crompton, 1994; McPherron, 2000). In what follows, the meaning of handaxe symmetry will be explored first and then placed into the context of standardization.

Wynn, in particular, has developed a number of arguments linking symmetry in handaxes to cognitive abilities. Initially Wynn's (1979, 1985) work drew on Piaget's stages of mental development, observed in children, which were then applied to the archaeological record to assess stages in mental development through time. More recently, Wynn (2002) has moved away from Piaget's models and focused more on the implications of artifact symmetry for spatial cognitive development and theories of perception.

Wynn argues that the symmetry seen in handaxes is real (i.e., not simply a result of our techniques of analysis) and intentional (Wynn, 2002: 394), and that it marks the origins of hominins' ability to impose symmetry on their material culture. The support for this is based on two arguments. First, Wynn argues that handaxes with more flakes removed from them are also the more symmetrical handaxes, meaning that the additional removals were intended to increase the symmetry. In other words, more finely worked artifacts are more symmetrical. Second, Wynn argues that when only a few flakes are removed from a handaxe, they are the ones that increase the handaxe's symmetry. Based on this, Wynn (2002: 398) argues that early hominins "employed cognitive abilities in frame independence, mirroring, making simple judgments of spatial quantity, and coordination of shape recognition (symmetry) with the spatial requirements of basic stone knapping."

In Wynn's (2002) model these patterns are apparent by 1.4 Mya, but are built upon after 500 Kya. At this later time hominins are able to achieve congruency in symmetry, wherein opposite edges match each other as closely as the raw material allows. Symmetry is also extended to additional perspectives, reflectional symmetry in three dimensions, wherein handaxes appear symmetrical not only in plan view but also in side view. As a result, perpendicular sections through the long axis of the handaxe are also symmetrical. Finally, Wynn argues that these later handaxes also sometimes show broken symmetry wherein the symmetry is intentionally altered in patterned ways. Handaxes with so-called S- and Z-shaped edges are cited as examples of this (White, 1998). The cognitive importance of these properties all relate to a hominin's ability to visualize in three dimensions the effect on multiple perspectives of symmetry of removing a flake from the edge of a handaxe and suggests to Wynn (2002: 398) the acquisition at this time of a "modern Euclidian set of spatial understandings."

One difficulty in Wynn's approach and others based on symmetry is identifying the selective mechanism that would have acted to increase symmetry through time (Wynn, 2002; see also Lycett, 2008). A possibility, of course, is that the ability to impose symmetry on an object was not directly selected for and was instead the result of selection on other aspects of cognition and motor control that then manifest themselves in handaxes. Alternatively, Lycett (2008) summarizes the possibilities as functional, aesthetic and sexual. The functional arguments are difficult to evaluate given competing ideas on handaxe function; nevertheless, the relationship between symmetry and butchery was recently tested with the conclusion that symmetry does not affect the efficacy of the tool (Machin *et al.*, 2007). The aesthetic argument, too, is difficult to assess, but basically rests on the notion that handaxes had symbolic or aesthetic properties, expressed through standardization and symmetry, that handaxe makers consciously attempted to maximize (e.g., Le Tensorer, 2006, 2009; Pope *et al.*, 2006).

The sexual selection argument, one that Wynn (2002) considers as well, has been more fully developed by Kohn and Mithen (1999). They argue that hominins' ability to make handaxes was a measure of their knowledge of resource distributions, ability to plan, health and social awareness, all of which would impact their mate selection and reproductive success. Of course, a better-made handaxe would have a larger impact. They argue that symmetry was an especially important factor in drawing the attention of members of the opposite sex to handaxe manufacturing skills. This line of reasoning follows from what they suggest are a number of unanswered problems with handaxes. This list includes the fact that handaxes are pervasive, occur in high frequencies at some sites, do not appear to add new functionality over the Oldowan tool kit, are more symmetrical than they need to be, and are sometimes so large that it is difficult to see how they were used. Kohn and Mithen's thesis has been criticized from a number of different perspectives (e.g., Machin, 2008; Hayden & Villeneuve, 2009; Hodgson, 2009; Nowell & Chang, 2009). A fundamental question worth asking is whether the data support these kinds of cognitive inferences.

Symmetry is certainly an attribute of handaxes, but is still very poorly documented and very poorly understood with regard to how it varies through time, within and between assemblages, across raw material types and so forth. This point has been made by Saragusti *et al.* (1998, 2005,), Nowell *et al.* (2003) and Lycett (2008), all of whom have developed separate quantitative methods to assess handaxe symmetry. The results generally support the contention that variability in symmetry is still poorly understood. Initially, Saragusti *et al.* (1998) reported a trend through time of increasing symmetry and perhaps increasing standardization in their analysis of handaxes from three Levantine sites: Ubeidiya, Gesher Benot Ya'aqov and Ma'ayan Barukh. They noted, however, that this pattern was based on only three assemblages with a total of 44 handaxes. A subsequent follow-up study using a different method for quantifying both symmetry and standardization applied to the same three assemblages, but with a larger sample size, confirmed the patterning seen in the first study (Sargusti *et al.*, 2005). However, two assemblages from Tabun, a nearby site with assemblages that date later than the others in their study, failed to follow the pattern of increasing symmetry through time. The two Tabun assemblages show equivalent levels of symmetry, and these assemblages show levels of symmetry equivalent to Ubeidiya, roughly one million years older. An analysis of the regularity of the edges, a facet of standardization, produced the same result. Saragusti *et al.* (2005) conclude that the time trend is still valid, but that the Tabun assemblages deserve further examination. An equally valid conclusion from their data is that temporal trends do not exist.

A similar story emerges from Sinclair and McNabb's (2005) study of large cutting tools (LCTs), including handaxes, in the Early Stone Age deposits of the Cave of Hearths in South Africa. Using a different method to quantify symmetry, they found that symmetry was actually rare and that variables such as size and reduction intensity (number of flakes removed from the pieces) did not explain variability in symmetry. They conclude that these artifacts were not shaped; rather, minimal working was done to achieve functional properties and additional removals did not regularize the piece, increase symmetry or make for a well "finished" artifact.

Lycett (2008; Lycett & von Cramon-Taubadel, 2008) has recently tested whether variability in both handaxe shape and symmetry are patterned as one would expect, given a model of Acheulian populations originating in Africa and then migrating into other parts of the Old World. His approach borrows heavily from models and techniques developed to track population movements through genetics. The basic principle is that founder effects will reduce variation in cultural behaviors as populations moved away from Africa. Based on an analysis of 255 handaxes from ten assemblages distributed across Europe, the Levant, the Indian subcontinent and Africa, Lycett and von Cramon-Traubadel (2008) argue that variation in edge shape fits with a drift model. They argue that variation in edge shape is structured by region, echoing a finding by Wynn and Tierson (1990). In a subsequent analysis, Lycett (2008) argues that symmetry does not correlate with distance from Africa and that there is less variation in symmetry than one would expect given neutral drift. Thus, Lycett (2008) concludes that symmetry was under a selective pressure, though what this pressure might have been is not clear and the small number of handaxes representing vast expanses of time and space mean the conclusions should be viewed with caution.

A complicating factor in all of these analyses is that a certain level of standardization is to be expected by the nature of the bifacial reduction process used to make handaxes (Hayden & Villeneuve, 2009). Symmetry can result from the repetitive application of fairly simple rule-based behavior and the ability to create objects with symmetry is not limited to hominins. The V-shape formation of migrating birds is an example of Wynn's (2002) reflectional symmetry, and a spider's web is an example of what Wynn calls rotational symmetry. In neither case, however, is it demonstrated that symmetry was intended. With handaxes this problem is compounded by a tendency to view these artifacts as implements fashioned in a single knapping episode, used, perhaps transported across the landscape to be used repeatedly, and then discarded in their original form. The tendency for archaeologists to associate the form of a stone tool excavated from a site with the intended form of the artifact when it was functioning in a cultural context is what

Davidson and Noble (1993) have called "the final artifact fallacy." There is an extensive literature showing that stone tools can have rather complicated use-lives that involve multiple episodes of re-sharpening and re-working that changes not just the size of the artifact, but also potentially the shape and the type (as defined by archaeologists) (e.g., Goodyear, 1974; Hoffman, 1985; Baumler, 1988; Dibble, 1995; McPherron, 1999).

In the case of handaxes, McPherron has argued that these artifacts were likely continually re-worked and that shape changed as a result in predictable and standardized ways. The implication is that final shape was not an intended or sought after property of handaxes, but rather a property resulting from the manufacture and maintenance process. Moreover, in the deeply stratified site of Tabun, McPherron (2003) was able to demonstrate not only that variability in shape was largely a function of size, but that size and shape varied in predictable ways with the emphasis placed on handaxes through the sequence. In levels with more handaxes (relative to scrapers) the handaxes tended to be smaller. This makes sense if handaxes are viewed not as stable forms, made once, used and then discarded, but rather as dynamic objects, reused and re-worked in response to changing and variable needs.

McPherron did not consider symmetry directly, but symmetry is built into the model. The idea is that maintenance of symmetry increases the use-life of the artifact (Hayden, 1989). In other words, by balancing removals from the left of a biface with removals from the right and removals from the front with removals from the back face, the potential to continue to extract removals is best ensured. If one edge or face is given preference over another the object loses its symmetry, but also the geometry changes in ways that make it more and more difficult to remove additional flakes. Maintaining balance or symmetry is essential to the technology of bifacial knapping and it becomes particularly important when intensive bifacial reduction is anticipated. Complicating this patterning, however, is the fact that handaxes are likely to be discarded when their symmetry is lost and their future utility declines.

These complicating factors are generally acknowledged when the cognitive importance of shaping and symmetry are discussed (e.g., Wynn, 2002; Porr, 2005), but they are generally considered insufficient to account for the observed symmetry. Arguments usually involve citing examples of individual handaxes that are incredibly well made (as judged by the preconceived standards of archaeologists), and, therefore, are highly symmetric. The studies by Sarugsti *et al.*, Sinclair and McNabb and Lycett suggest that a better understanding of the nature and cause of the underlying variability are warranted before claims of its behavioral and especially cognitive significance are advanced.

Finally, with regard to standardization, the assumption is that handaxe makers shaped their handaxes according to a mental template of what that handaxe should look like. One way of showing standardization in shape is through typology. The problem is that while there is considerable shape variation, it is widely recognized that this variation does not fall into neat categories amenable to typology. As a result, most descriptions of shapes have relied on linear measurements which are converted to shape ratios which are then divided arbitrarily into shape types (e.g., Bordes, 1961; Roe, 1968). In the European handaxe data set, this approach has focused mainly around the distinction between pointed and rounded handaxes (Roe, 1968). Handaxe assemblages can be organized according to whether they are on average more pointed or more rounded, but two points are worth noting. In any given assemblage of handaxes, from a metrical point of view, there is no clear distinction between pointed and rounded handaxes, and when handaxe assemblage averages are considered there is likewise no clear distinction between the two (McPherron, 2007a).

It is also clear that standardization varies across individual handaxes in ways that are likely related to their production and use. Multiple studies have shown that the highest levels of standardization, as measured by coefficients of variation, can be found in the base of the handaxe (McPherron, 2007a and citations within). At Tabun, for instance, throughout the sequence, handaxe size and shape varied significantly, and yet handaxes entered the archaeological record with the same width across all levels (McPherron, 2003). Handaxe makers re-worked the tips of handaxes more than the bases and as a result the overall shape varied. In other words, looking at standardization of the overall shape likely conflates areas of high standardization with areas of low standardization and focuses on a property which hominins may not have considered important at all. Studies of whole edge shape (Wynn & Tierson, 1990; Lycett, 2008, 2009) have not taken these factors into account.

In summary, it seems clear that in contrast with the Oldowan, where it is difficult to see a consistent, shared technological approach to stone production, the Acheulian handaxe represents at least a concept or mental template of how lithic production should be done. At a minimum this template consisted of a simple set of rules involving removing flakes in a balanced manner from opposite edges and opposite faces of a block of raw material. This basic concept remained incredibly stable over the time period of the Acheulian and shows very little patterned variation beyond that which can be explained in basic technological and raw material terms. Evidence for standardization and imposed symmetry beyond that which can be expected within this basic technological framework are still not yet clearly established quantitatively. Better data are needed to establish the baseline variability in these attributes before we can recognize instances that are said to exceed what would normally be expected.

Final comment

One of the important issues that arises from cognition studies based on the archaeological record of stone tools is the relevance of unusual and even unique finds. Oldowan studies tend to focus on assemblage- and even landscape-level patterning. Acheulian studies, however, tend to focus on one artifact, the handaxe, and the issue is even more acute in Neandertal and early modern human studies, where some behaviors argued to be representative of modern human behavior are known from single instances or at best a few instances sometimes widely scattered in space and time. What insights does, for instance, a handaxe too large to be easily used or carried across the landscape, even by a much stronger early hominin, give us about how these hominins might have thought about handaxes, especially in light of the fact that nearly all other handaxes were not made this way? Was someone in the past showing off, perhaps asserting their fitness in front of others, and can this instance then be generalized to other less spectacular examples? Do the extremes inform us about the mean, or are they simply examples of idiosyncratic behavior by a rather intelligent primate that should from time to time exhibit surprisingly sophisticated behaviors that, however, speak more to capacity than to actual shared behaviors essential to the adaptation? The problem, of course, is that we know which behaviors eventually became important and so the risk is that once we see evidence of these in the past we project an importance on them that is only justified by their current manifestations. From this projection archaeologists then assume that from that point forward this behavior would have at least retained that importance.

A good example of this issue is the handaxe from Sima del los Huesos (Spain). This site, still under excavation, is at the base of a deep and difficult to access karstic feature that trapped the remains of tens of individuals and a fauna dominated by carnivores (Carbonell *et al.*, 2003; Carbonell & Mosquera, 2006). How these hominins came to rest there is unclear, though the presumption is that they were deposited after death down a shaft. With these individuals, up to now, only one artifact has been discovered, and it is a handaxe. Moreover, by our standards today, it is a "well-made" handaxe. If Sima del los Huesos represents intentional burial or disposal of the dead, then a reasonable argument can be made that this handaxe was a grave offering (Carbonell *et al.*, 2003; Carbonell & Mosquera, 2006), which would make it by far the oldest example of such and, in fact, arguably the only example not associated with our species (Gargett, 1989). It would seem to support the idea that at least these hominins, at this place and time, appreciated that a handaxe could carry meaning beyond its functional significance. They apparently viewed this handaxe as special; after all, they could have tossed a well-made scraper or core in with the body instead. There is no indication, however, that this behavior spread beyond Sima de los Huesos or even that it was ever repeated there.

To a large extent this problem is dealt with by separating evidence for the capacity for certain behaviors from evidence that these behaviors formed an essential and ubiquitous (Byrne, 2007) component of the adaptation. By analogy with extant primates, bonobos can be trained to make stone tools in captivity (Schick *et al.*, 1999), but bonobos do not use tools in the wild. The problem with the archaeological record, particularly for the time periods of concern here, is that it is so fragmentary in terms of the number of data points in both space and time that it is hard to know when a behavior is truly rare (i.e., non-representative). Stone tools, because of their ubiquity and durability, remain our best opportunity to assess cognition through time, but what we need are better analytical tools that allow us to identify, for instance, the correlates of the same kinds of behavior that may have existed in Sima de los Huesos in the numerous and rich handaxe assemblages that exist in the more ordinary archaeological contexts of living sites.

Acknowledgments

I would like to thank the organizers of this workshop and volume for their invitation to participate. I would like to acknowledge the support of the Max Planck Society. A special thanks to David Braun and three anonymous reviewers for helpful discussions and suggestions on this manuscript. All faults remain my own.

References

Asfaw, B., Beyene, Y., Suwa, G., *et al.* (1992). The earliest Acheulean from Konso-Gardula. *Nature*, **360**, 732–735.

Ashton, N. & White, M. J. (2003). Bifaces and raw materials: flexible flaking in the British Early Paleolithic. In M. Soressi & H. Dibble (eds.) *From Prehistoric Bifaces to Human Behavior: Multiple Approaches to the Study of Bifacial Technology* (pp. 109–124). Philadelphia, PA: University of Pennsylvania Museum Press.

Bamforth, D. B. (1988). Investigating microwear polishes with blind tests: the institute results in context. *Journal of Archaeological Science*, **15**(1), 11–23.

Bar-Yosef, O. & van Peer, P. (2009). The chaîne opératoire approach in Middle Paleolithic archaeology. *Current Anthropology*, **50**(1), 103–131.

Baumler, M. F. (1988). Core reduction, flake production, and the Middle Paleolithic industry of Zobiste (Yugoslavia). In H. L Dibble & A. Montet-White (eds.) *Upper Pleistocene Prehistory of Western Eurasia* (pp. 255–274). Philadelphia, PA: University of Pennsylvania Museum Press.

Blumenschine, R. J., Peters, C. R., Masao, F. T., *et al.* (2003). Late Pliocene *Homo* and hominid land use from western Olduvai Gorge, Tanzania. *Science*, **299**(5610), 1217.

Blumenschine, R. J., Prassack, K. A., Kreger, C. D. & Pante, M. C. (2007). Carnivore tooth-marks, microbial bioerosion, and the invalidation of test of Oldowan hominin scavenging behavior. *Journal of Human Evolution*, **53**(4), 420–426.

Bordes, F. (1961). *Typologie du Paléolithique ancien et moyen*. Bordeaux: Imprimeries Delmas.

Brantingham, P. J. & Kuhn, S. L. (2001). Constraints on Levallois core technology: a mathematical model. *Journal of Archaeological Science*, **28**(7), 747–761.

Braun, D. R. (2010). Palaeoanthropology: Australopithecine butchers. *Nature*, **466**(7308), 828.

Braun, D. R. & Harris, J. W. (2003). Technological developments in the Oldowan of Koobi Fora: innovative techniques of artifact analysis. In J. Martinez Moreno, R. Mora & I. de la Torre Sainz (eds.) *Oldowan: Rather More Than Smashing Stones* (pp. 117–144). Bellaterra: Universitat Autònoma de Barcelona.

Braun, D. R. & Hovers, E. (2009). Introduction: current issues in Oldowan research. In E. Hovers & D. R. Braun (eds.) *Interdisciplinary Approaches to the Oldowan* (pp. 1–14). Dordrecht: Springer.

Braun, D. R., Tactikos, J. C., Ferraro, J. V. & Harris, J. W. K. (2005). Flake recovery rates and inferences of Oldowan hominin behavior: a response to Kimura, 1999 and Kimura, 2002. *Journal of Human Evolution*, **48**(5), 525–531.

Braun, D. R., Plummer, T., Ditchfield, P., *et al.* (2008). Oldowan behavior and raw material transport: perspectives from the Kanjera Formation. *Journal of Archaeological Science*, **35**(8), 2329–2345.

Braun, D. R., Harris, J. W. K. & Maina, D. N. (2009a). Oldowan raw material procurement and use: evidence from the Koobi Fora Formation. *Archaeometry*, **51**(1), 26–42.

Braun, D. R., Plummer, T., Ferraro, J. V., Ditchfield, P. & Bishop, L. C. (2009b). Raw material quality and Oldowan hominin toolstone preferences: evidence from Kanjera South, Kenya. *Journal of Archaeological Science*, **36**(7), 1605–1614.

Byrne, R. W. (2007). Culture in great apes: using intricate complexity in feeding skills to trace the evolutionary origin of human technical prowess. *Philosophical Transactions of the Royal Society B: Biological Sciences*, **362**(1480), 577–585.

Callow, P. (1976). *The Lower and Middle Palaeolithic of Britain and Adjacent Areas of Europe*. Cambridge: Cambridge University Press.

Callow, P. (1986). A comparison of British and French Acheulian bifaces. In S. N. Collcutt (ed.) *The Palaeolithic of Britain and its Nearest Neighbours: Recent Trends* (pp. 3–7). Sheffield: J.R. Collis Publications.

Calvin, W. (1993). The unitary hypothesis: a common neural circuitry for novel manipulations, language, plan-ahead, and throwing? In K. R. Gibson & T. Ingold (eds.) *Tools, Language and Cognition in Human Evolution* (pp. 230–250). Cambridge: Cambridge University Press.

Carbonell, E. & Mosquera, M. (2006). The emergence of a symbolic behaviour: the sepulchral pit of Sima de los Huesos, Sierra de Atapuerca, Burgos, Spain. *Comptes Rendus Palevol*, **5**(1–2), 155–160.

Carbonell, E., Mosquera, M., Ollé, A., *et al.* (2003). Les premiers comportements funéraires auraient-ils pris place à Atapuerca, il y a 350 000 ans? *L'Anthropologie*, **107**(1), 1–14.

Carbonell, E., Sala, R., Barsky, D. & Celiberti, V. (2009). From homogeneity to multiplicity: a new approach to the study of archaic stone tools. In E. Hovers & D. R. Braun (eds.) *Interdisciplinary Approaches to the Oldowan* (pp. 25–38). Dordrecht: Springer.

Claud, E. (2010). La tracéologie appliquée à l'étude du site de Chez Pinaud (Charente-Maritime). In J. Buisson-Catil & J. Primault (eds.) *Préhistoire entre Vienne et Charente: Hommes et societies du Paléolithique* (pp. 122–126). Chauvigny: Association des Publications Chauvinoises.

Crompton, R. H. & Gowlett, J. A. J. (1993). Allometry and multidimensional form in Acheulean bifaces from Kilombe, Kenya. *Journal of Human Evolution*, **25**(3), 175–199.

Davidson, I. & Noble, W. (1993). Tools and language in human evolution. In K. R. Gibson & T. Ingold (eds.) *Tools, Language and Cognition in Human Evolution* (pp. 363–388). Cambridge: Cambridge University Press.

de Heinzelin, J., Clark, J. D., White, T., *et al.* (1999). Environment and behavior of 2.5-million-year-old Bouri hominids. *Science*, **284**(5414), 625–629.

de la Torre, I. & Mora, R. (2009). Remarks on the current theoretical and methodological approaches to the study of technological strategies of early humans in Eastern Africa. In E. Hovers & D. R. Braun (eds.) *Interdisciplinary Approaches to the Oldowan* (pp. 15–24). Dordrecht: Springer.

Delagnes, A. & Roche, H. (2005). Late Pliocene hominid knapping skills: the case of Lokalalei 2C, West Turkana, Kenya. *Journal of Human Evolution*, **48**(5), 435–472.

Dibble, H. L. (1995). Middle Paleolithic scraper reduction: background, clarification, and review of the evidence to date. *Journal of Archaeological Method and Theory*, **2**(4), 299–368.

Dibble, H. L & Rezek, Z. (2009). Introducing a new experimental design for controlled studies of flake formation: results for exterior platform angle, platform depth, angle of blow, velocity, and force. *Journal of Archaeological Science*, **36**(9), 1945–1954.

Domínguez-Rodrigo, M. & Barba, R. (2006). New estimates of tooth mark and percussion mark frequencies at the FLK Zinj site: the carnivore-hominid-carnivore hypothesis falsified. *Journal of Human Evolution*, **50**(2), 170–194.

Domínguez-Rodrigo, M., Serrallonga, J., Juan-Tresserras, J., Alcala, L. & Luque, L. (2001). Woodworking activities by early humans: a plant residue analysis on Acheulian stone tools from Peninj (Tanzania). *Journal of Human Evolution*, **40**(4), 289–299.

Domínguez-Rodrigo, M., de Juana, S., Galán, A. B. & Rodríguez, M. (2009). A new protocol to differentiate trampling marks from butchery cut marks. *Journal of Archaeological Science*, **36**(12), 2643–2654.

Domínguez-Rodrigo, M., Pickering, T. R. & Bunn, H. T. (2010). Configurational approach to identifying the earliest hominin butchers. *Proceedings of the National Academy of Sciences USA*, **107**(49), 20929–20934.

Gargett, R. H. (1989). Grave shortcomings: the evidence for neandertal burial (and comments and reply). *Current Anthropology*, **30**(2), 157–190.

Goldman-Neuman, T. & Hovers, E. (2009). Methodological issues in the study of Oldowan raw material selectivity: insights from A.L. 894 (Hadar, Ethiopia). In E. Hovers & D. R. Braun (eds.) *Interdisciplinary Approaches to the Oldowan* (pp. 71–84). Dordrecht: Springer.

Goodyear, A. C. (1974). *The Brand Site: A Techno-Functional Study of a Dalton Site in Northeast Arkansas*. Fayetteville, NC: Arkansas Archeological Survey.

Gowlett, J. A. J. (2009). Artefacts of apes, humans, and others: towards comparative assessment and analysis. *Journal of Human Evolution*, **57**(4), 401–410.

Gowlett, J. A. J. & Crompton, R. H. (1994). Kariandusi: Acheulean morphology and the question of allometry. *African Archaeological Review*, **12**(1), 3–42.

Harmand, S. (2009). Variability in raw material selectivity at the Late Pliocene sites of Lokalalei, West Turkana, Kenya. In E. Hovers & D. R. Braun (eds.) *Interdisciplinary Approaches to the Oldowan* (pp. 85–98). Dordrecht: Springer.

Haslam, M., Hernandez-Aguilar, A., Ling, V., *et al.* (2009). Primate archaeology. *Nature*, **460**(7253), 339–344.

Hay, R. L. (1976). *Geology of the Olduvai Gorge*. Berkeley, CA: University of California Press.

Hayden, B. (1989). From chopper to celt: the evolution of resharpening techniques. In R. Torrence (ed.) *Time, Energy and Stone Tools* (pp. 7–16). Cambridge: Cambridge University Press.

Hayden, B. & Villeneuve, S. (2009). Sex, symmetry and silliness in the bifacial world. *Antiquity*, **83**(322), 1163–1170.

Hodgson, D. (2009). Symmetry and humans: reply to Mithen's "Sexy Handaxe Theory." *Antiquity*, **83**(319), 195–198.

Hoffman, C. M. (1985). Projectile point maintenance and typology: assessment with factor analysis and canonical correlation. In C. Carr (ed.) *For Concordance in Archaeological Analysis* (pp. 566–612). Kansas City, MI: Westport Publishers.

Hovers, E. & Braun, D. R. (eds.) (2009). *Interdisciplinary Approaches to the Oldowan*. Dordrecht: Springer.

Iovita, R. & McPherron, S. P. (2011). The handaxe reloaded: a morphometric reassessment of Acheulian and Middle Paleolithic handaxes. *Journal of Human Evolution*, **61**(1), 61–74.

Isaac, G. L. (1977). *Olorgesailie: Archeological Studies of a Middle Pleistocene Lake Basin in Kenya*. Chicago, IL: University of Chicago Press.

Jones, P. R. (1980). Experimental butchery with modern stone tools and its relevance for Palaeolithic archaeology. *World Archaeology*, **12**(2), 153–165.

Keeley, L. H. (1980). *Experimental Determination of Stone Tool Uses: A Microwear Analysis*. Chicago, IL: University of Chicago Press.

Kimura, Y. (2002). Examining time trends in the Oldowan technology at Beds I and II, Olduvai Gorge. *Journal of Human Evolution*, **43**(3), 291–321.

Kohn, M. & Mithen, S. (1999). Handaxes: products of sexual selection? *Antiquity*, **73**, 518–526.

Le Tensorer, J. M. (2006). Les cultures acheuléenes et la question de l'émergence de la pensée symbolique chez Homo erectus à partir des données relatives à la forme symétrique et harmonique des bifaces. *Academie des Sciences: Comptes Rendus. Palevol*, **5**(1–2), 127–135.

Le Tensorer, J. M. (2009). L'image avant l'image: réflexions sur le colloque. *L'Anthropologie*, **113**(5), 1005–1017.

Lepre, C. J., Roche, H., Kent, D. V., *et al.* (2011). An earlier origin for the Acheulian. *Nature*, **477**(7362), 82–85.

Leroi-Gourhan, A. (1993). *Gesture and Speech*. Cambridge, MA: MIT Press.

Lycett, S. J. (2008). Acheulean variation and selection: does handaxe symmetry fit neutral expectations? *Journal of Archaeological Science*, **35**(9), 2640–2648.

Lycett, S. J. (2009). Understanding ancient hominin dispersals using artefactual data: a phylogeographic analysis of Acheulean handaxes. *PloS ONE*, **4**(10), e7404.

Lycett, S. J. & von Cramon-Taubadel, N. (2008). Acheulean variability and hominin dispersals: a model-bound approach. *Journal of Archaeological Science*, **35**(3), 553–562.

Machin, A. J. (2008). Why handaxes just aren't that sexy: a response to Kohn & Mithen (1999). *Antiquity*, **82**(317), 761–766.

Machin, A. J., Hosfield, R. T. & Mithen, S. J. (2007). Why are some handaxes symmetrical? Testing the influence of handaxe morphology on butchery effectiveness. *Journal of Archaeological Science*, **34**(6), 883–893.

McCall, G. S. & Whittaker, J. (2007). Handaxes still don't fly. *Lithic Technology*, **32**(2), 195–203.

McPherron, S. P. (1999). Ovate and pointed handaxe assemblages: two points make a line. *Préhistoire Européenne*, **14**, 9–32.

McPherron, S. P. (2000). Handaxes as a measure of the mental capabilities of early hominids. *Journal of Archaeological Science*, **27**, 655–663.

McPherron, S. P. (2003). Typological and technological variability in the bifaces from Tabun Cave, Israel. In M. Soressi & H. Dibble (eds.) *From Prehistoric Bifaces to Human Behavior: Multiple Approaches to the Study of Bifacial Technology* (pp. 55–76). Philadelphia, PA: University of Pennsylvania Museum Press.

McPherron, S. P. (2007a). What typology can tell us about Acheulian handaxe production. In N. Goren-Inbar & G. Sharon (eds.) *Axe Age: Acheulian Toolmaking, from Quarry to Discard* (pp. 267–286). Indonesia: Equinox Publishing.

McPherron, S. P. (ed.). (2007b). *Tool v. Core: New Approaches in the Analysis of Stone Tool Assemblages*. Cambridge: Cambridge Scholars Publications.

McPherron, S. P., Alemseged, Z., Marean, C. W., *et al.* (2010). Evidence for stone-tool-assisted consumption of animal tissues before 3.39 million years ago at Dikika, Ethiopia. *Nature*, **466**(7308), 857–860.

Mitchell, J. C. (1996). Studying biface utilization at Boxgrove: roe deer butchery with replica handaxes. *Lithics*, **16**, 64–69.

Newcomer, M., Grace, R. & Unger-Hamilton, R. (1986). Investigating microwear polishes with blind tests. *Journal of Archaeological Science*, **13**(3), 203–217.

Njau, J. K. & Blumenschine, R. J. (2006). A diagnosis of crocodile feeding traces on larger mammal bone, with fossil examples from the Plio-Pleistocene Olduvai Basin, Tanzania. *Journal of Human Evolution*, **50**(2), 142–162.

Noble, W. & Davidson, I. (1996). *Human Evolution, Language and Mind*. Cambridge: Cambridge University Press.

Norton, C. J., Bae, K., Harris, J. W. K. & Lee, H. (2006). Middle Pleistocene handaxes from the Korean Peninsula. *Journal of Human Evolution*, **51**(5), 527–536.

Nowell, A. & Chang, M. L. (2009). The case against sexual selection as an explanation of handaxe morphology. *PaleoAnthropology*, **2009**, 77–88.

Nowell, A., Park, K., Metaxas, D. & Park, J. (2003). Deformation modeling: a methodology for the analysis of handaxe morphology and variability. In M. Soressi & H. L. Dibble (eds.) *Multiple*

Approaches to the Study of Bifacial Technologies (pp. 193–208). Philadelphia, PA: University of Pennsylvania Museum of Archaeology and Anthropology.

O'Brien, E. M. (1981). The projectile capabilities of an Acheulian handaxe from Olorgesailie. *Current Anthropology*, **22**(1), 76–79.

Panger, M. A., Brooks, A. S., Richmond, B. G. & Wood, B. (2002). Older than the Oldowan? Rethinking the emergence of hominin tool use. *Evolutionary Anthropology: Issues, News, and Reviews*, **11**(6), 235–245.

Plummer, T. (2004). Flaked stones and old bones: biological and cultural evolution at the dawn of technology. *Yearbook of Physical Anthropology*, **47**, 118–164.

Pope, M. & Roberts, M. (2005). Observations on the relationship between Palaeolithic individuals and artefact scatters at the Middle Pleistocene site of Boxgrove, UK. In C. Gamble & M. Porr (eds.) *The Hominid Individual in Context: Archaeological Investigations of Lower and Middle Palaeolithic Landscapes, Locales, and Artefacts* (pp. 81–97). London: Routledge.

Pope, M., Russel, K. & Watson, K. (2006). Biface form and structured behaviour in the Acheulian. *Lithics*, **27**, 44–57.

Porr, M. (2005). The making of the biface and the making of the individual. In C. Gamble & M. Porr (eds.) *The Hominid Individual in Context: Archaeological Investigations of Lower and Middle Palaeolithic Landscapes, Locales and Artefacts* (pp. 68–80). London: Routledge.

Potts, R. (1988). *Early Hominid Activities at Olduvai*. New York: Aldine de Gruyter.

Rezek, Z., Lin, S., Iovita, R. & Dibble, H. L. (2011). The relative effects of core surface morphology on flake shape and other attributes. *Journal of Archaeological Science*, **38**(6), 1346–1359.

Roche, H. (2005). From simple flaking to shaping: stone-knapping evolution among early hominins. In B. Brill & V. Roux (eds.) *Stone Knapping: The Necessary Conditions for a Uniquely Hominid Behaviour* (pp. 35–52). Cambridge: McDonald Institute of Archaeological Research.

Roche, H., Blumenschine, R. J. & Shea, J. S. (2009). Origins and adaptations of early *Homo*: what archaeology tells us. In F. E. Grine, J. G. Fleagle & R. E. Leakey (eds.) *The First Humans: Origin and Early Evolution of the Genus* Homo (pp. 135–147). Dordrecht: Springer.

Roe, D. A. (1968). British Lower and Middle Palaeolithic handaxe groups. *Proceedings of the Prehistoric Society*, **34**, 1–82.

Sanz, C., Call, J. & Morgan, D. (2009). Design complexity in termite-fishing tools of chimpanzees (*Pan troglodytes*). *Biology Letters*, **5**(3), 293–296.

Saragusti, I., Sharon, I., Katzenelson, O. & Avnir, D. (1998). Quantitative analysis of the symmetry of artefacts: Lower Paleolithic handaxes. *Journal of Archaeological Science*, **25**(8), 817–825.

Saragusti, I., Karasik, A., Sharon, I. & Smilansky, U. (2005). Quantitative analysis of shape attributes based on contours and section profiles in artifact analysis. *Journal of Archaeological Science*, **32**(6), 841–853.

Schick, K. D. (1994). The Movius Line reconsidered: perspectives on the earlier Paleolithic of Eastern Asia. In R. L. Ciochon & R. Corruccini (eds.) *Integrative Paths to the Past* (pp. 569–596). Englewood Cliffs, NJ: Prentice-Hall.

Schick, K. D & Toth, N. P. (1994). *Making Silent Stones Speak: Human Evolution and the Dawn of Technology*. New York: Simon & Schuster.

Schick, K. D. & Toth, N. (2006). An overview of the Oldowan industrial complex: the sites and the nature of their evidence. In N. Toth & K. Schick (eds.) *The Oldowan: Case Studies into the Earliest Stone Age* (pp. 3–42). Gosport, IN: Stone Age Institute Press.

Schick, K. D., Toth, N., Garufi, G., *et al.* (1999). Continuing investigations into the stone tool-making and tool-using capabilities of a bonobo (*Pan paniscus*). *Journal of Archaeological Science*, **26**(7), 821–832.

Semaw, S., Renne, P., Harris, J. W. K., *et al.* (1997). 2.5-million-year-old stone tools from Gona, Ethiopia. *Nature*, **385**(6614), 333–336.

Semaw, S., Rogers, M. J., Quade, J., *et al.* (2003). 2.6-million-year-old stone tools and associated bones from OGS-6 and OGS-7, Gona, Afar, Ethiopia. *Journal of Human Evolution*, **45**(2), 169–177.

Shott, M. J. (2003). Chaîne opératoire and reduction sequence. *Lithic Technology*, **28**, 95–105.

Sinclair, A. & McNabb, J. (2005). All in a day's work: Middle Pleistocene individuals, materiality and the lifespace at Makapansgat, South Africa. In C. Gamble & M. Porr (eds.) *The Hominid Individual in Context: Archaeological Investigations of Lower and Middle Palaeolithic Landscapes, Locales, and Artefacts* (pp. 176–196). London: Routledge.

Soressi, M. & Geneste, J.-M. (2011). The history and efficacy of the chaine opératoire approach to lithic analysis: studying techniques to reveal past societies in an evolutionary perspective. *PaleoAnthropology*, **2011**, 334–350.

Stout, D., Quade, J., Semaw, S., Rogers, M. J. & Levin, N. E. (2005). Raw material selectivity of the earliest stone toolmakers at Gona, Afar, Ethiopia. *Journal of Human Evolution*, **48**(4), 365–380.

Stout, D., Semaw, S., Rogers, M. J. & Cauche, D. (2010). Technological variation in the earliest Oldowan from Gona, Afar, Ethiopia. *Journal of Human Evolution*, **58**(6), 474–491.

Tostevin, G. B. (2003). A quest for antecedents: a comparison of the Terminal Middle Palaeolithic and Early Upper Palaeolithic of the Levant. In A. N. Goring-Morris & A. Belfer-Cohen (eds.) *More Than Meets the Eye: Studies on Upper Palaeolithic Diversity in the Near East* (pp. 54–67). Oxford: Oxbow.

Tostevin, G. B. (2011). Special issue: reduction sequence, chaîne opératoire, and other methods – the epistemologies of different approaches to lithic analysis. Levels of theory and social practice in the reduction sequence and chaîne opératoire methods of lithic analysis. *PaleoAnthropology*, **2011**, 351–375.

Toth, N. (1982). *The Stone Age Technology of Early Hominids at Koobi Fora, Kenya: An Experimental Approach*. Berkeley, CA: University of California Press.

Toth, N. (1987). Behavioral inferences from Early Stone artifact assemblages: an experimental model. *Journal of Human Evolution*, **16**(7–8), 763–787.

Toth, N. & Schick, K. D. (2006). *The Oldowan: Case Studies into the Earliest Stone Age*. Gosport, IN: Stone Age Institute Press.

White, M. J. (1998). Twisted ovate bifaces in the British Lower Palaeolithic: some observations and implications. In N. Ashton, F. Healy & P. Pettit (eds.) *Stone Age Archaeology: Essays in Honour of John Wymer* (pp. 98–104). Oxford: Oxbow Monograph.

Whiten, A., Schick, K. & Toth, N. (2009). The evolution and cultural transmission of percussive technology: integrating evidence from palaeoanthropology and primatology. *Journal of Human Evolution*, **57**(4), 420–435.

Wymer, J. (1968). *Lower Palaeolithic Archaeology in Britain as Represented by the Thames Valley*. London: John Baker Publishers Ltd.

Wynn, T. (1979). The intelligence of later Acheulean hominids. *Man*, New Series, **14**(3), 371–391.

Wynn, T. (1985). Piaget, stone tools and the evolution of human intelligence. *World Archaeology*, **17**(1), 32–43.

Wynn, T. (2002). Archaeology and cognitive evolution. *Behavioral and Brain Sciences*, **25**, 389–438.

Wynn, T. (2009). Whither evolutionary cognitive archaeology? Afterword. In S. A. de Beaume, F. L. Coolidge & T. Wynn (eds.) *Cognitive Archaeology and Human Evolution* (pp. 145–149). Cambridge: Cambridge University Press.

Wynn, T. & McGrew, W. C. (1989). An ape's view of the Oldowan. *Man*, **24**(3), 383–398.

Wynn, T. & Tierson, F. (1990). Regional comparison of the shapes of later Acheulean handaxes. *American Anthropologist*, **92**(1), 73–84.

Yamei, H., Potts, R., Baoyin, Y., *et al.* (2000). Mid-Pleistocene Acheulean-like stone technology of the Bose Basin, South China. *Science*, **287**(5458), 1622–1626.

Index